T0073069

# Advances in 21st Century Human Settlements

Indexed by SCOPUS

This Series focuses on the entire spectrum of human settlements—from rural to urban, in different regions of the world, with questions such as: What factors cause and guide the process of change in human settlements from rural to urban in character, from hamlets and villages to towns, cities and megacities? Is this process different across time and space, how and why? Is there a future for rural life? Is it possible or not to have industrial development in rural settlements, and how? Why does 'urban shrinkage' occur? Are the rural areas urbanizing or is that urban areas are undergoing 'ruralisation' (in form of underserviced slums)? What are the challenges faced by 'mega urban regions', and how they can be/are being addressed? What drives economic dynamism in human settlements? Is the urban-based economic growth paradigm the only answer to the quest for sustainable development, or is there an urgent need to balance between economic growth on one hand and ecosystem restoration and conservation on the other—for the future sustainability of human habitats? How and what new technology is helping to achieve sustainable development in human settlements? What sort of changes in the current planning, management and governance of human settlements are needed to face the changing environment including the climate and increasing disaster risks? What is the uniqueness of the new 'socio-cultural spaces' that emerge in human settlements, and how they change over time? As rural settlements become urban, are the new 'urban spaces' resulting in the loss of rural life and 'socio-cultural spaces'? What is leading the preservation of rural 'socio-cultural spaces' within the urbanizing world, and how? What is the emerging nature of the rural-urban interface, and what factors influence it? What are the emerging perspectives that help understand the human-environment-culture complex through the study of human settlements and the related ecosystems, and how do they transform our understanding of cultural landscapes and 'waterscapes' in the 21st Century? What else is and/or likely to be new vis-à-vis human settlements—now and in the future? The Series, therefore, welcomes contributions with fresh cognitive perspectives to understand the new and emerging realities of the 21st Century human settlements. Such perspectives will include a multidisciplinary analysis, constituting of the demographic, spatio-economic, environmental, technological, and planning, management and governance lenses.

If you are interested in submitting a proposal for this series, please contact the Series Editor, or the Publishing Editor:

Bharat Dahiya (bharatdahiya@gmail.com) or
Loyola DSilva (loyola.dsilva@springer.com)

More information about this series at http://www.springer.com/series/13196

Shabbir Cheema
Editor

# Governance for Urban Services

## Access, Participation, Accountability, and Transparency

*Editor*
Shabbir Cheema
Harvard Kennedy School
Harvard University
Cambridge, MA, USA

ISSN 2198-2546 ISSN 2198-2554 (electronic)
Advances in 21st Century Human Settlements
ISBN 978-981-15-2972-6 ISBN 978-981-15-2973-3 (eBook)
https://doi.org/10.1007/978-981-15-2973-3



This Springer imprint is published by the registered company Springer Nature Singapore Pte Ltd.
The registered company address is: 152 Beach Road, #21-01/04 Gateway East, Singapore 189721, Singapore

# Foreword

We live in a time of undeniable global challenges such as climate change, rapid urbanization and great migration flows as a result of conflicts, a growing recentralization and autocratization, in addition to decreasing access for minorities and vulnerable groups to decision-making, which is why functional local, democratic institutions are more important than ever before.

Urbanization provides opportunities but brings also major repercussions, and many mega-cities and medium-sized cities must cope with the negative impact that is related to rapid population growth. It is a local dynamic that can contribute to economic development that can bring people out of poverty, but development must be brought onto sustainable paths and create equal opportunities for people; failure to do so leads to what is described in this book as an increasing incidence of urban poverty and inequity, deteriorating quality of the urban environment and unequal access to basic urban services. Decisions relating to infrastructure, water, schools and other issues where responsibility lies at the local level must be made through people's own participation and influence. Therefore, it is a challenge to create forms of influence and responsibility in a way that ensures broad confidence in civically anchored local and regional governments.

In recent decades, decentralization has been a key element in the transformation of the state. This has meant that local and regional governments have been given responsibility for new issues, and that governance has been reformed to support the new responsibility that has been placed locally. In some respects, decentralization has been successful; in other cases, it has been problematic. Increased responsibility for vital issues in the lives of citizens requires that the local level be able to further develop and deepen democratic forms of sustainable governance and participation. Problems in meeting this responsibility are not infrequently manifested in deficiencies in the provision of equal treatment of citizens and in citizens' influence over decision-making.

Today, there is extensive research on this field, and this knowledge is important and vital in the discussion about governance in the future. Among other things, it is about finding forms of governance in times of urbanization and economic change. For globalization's negative and positive effects to be managed in a socially sustainable way, it is necessary that local and regional governments implement a rights-based approach by including weaker groups that have previously been far removed from decision-making. This volume examines three of the vital issues in urbanization and democratization: the institutional structures and processes of urban local governance to improve access to urban services; their outcomes vis-a-vis access of low-income groups to services, citizen participation in local governance, accountability of local leaders and officials, and transparency in local governance; and the factors that influence access to urban service, especially for the poor and marginalized groups, and the quality of democratic local governance. Elected representatives at the local and regional levels have a key role to play but need support when it comes to implementing programs and policies, so that urbanization brings with it sustainable and inclusive development for everyone in the cities and in communities close to the growing cities.

The work of the Swedish International Centre for Local Democracy aims to contribute to the positive development of democracy at local and regional levels. For us, this is characterized by participation, equity, transparency and accountability. In this way, the poverty that manifests itself in the form of low influence and limited rights can be broken, and living conditions are improved through people's own actions. For this reason, we are proud to be associated with the content the authors bring forward in this book. In a time when it is tempting to favor simple answers, the need is even greater for those who dare to answer the complex questions about how to create a sustainable world together and how we can achieve institutionalized local democracy.

Johan Lilja
General Secretary
Swedish International Centre for
Local Democracy
Visby, Sweden

# Preface

It is widely recognized by scholars, policymakers and development practitioners that governance—defined as the process of interaction between the state, the private sector and civil society to manage public affairs—is both an end in itself and a means to an end. As an end in itself, good governance should promote participation of citizens at all levels, accountability of public officials and access to information and transparency of decision-making. As a means to an end, governance should ensure that benefits of development are shared equitably among different segments of society. Access to services—including water, sanitation, shelter, health and education—reflects the extent to which development benefits are available to all citizens and the degree of their inclusion. Specifically, urban governance in developing countries is a vital factor in ensuring that cities and towns are sustainable, participatory and resilient and can promote economic development and increasing opportunities for all to improve human conditions.

With the rapid urban population growth over the past 50 years, cities have become centers of economic opportunities, innovations, experimentation, trade and productivity gains. Yet, urbanization has led to an increasing incidence of urban poverty and inequity, deteriorating quality of the urban environment and deficiencies in access to basic urban services, including water supply and sanitation, urban shelter, waste management, energy, transport and health. To cope with these and related challenges of unplanned urbanization, national governments and cities have initiated sets of policies, programs and institutional reforms. These include strengthening capacity and resources of urban local governments, mechanisms for participation and accountability, and targeted programs for slums and squatter settlements. They are using ICT and smart city solutions to urban challenges, supporting innovations and good practices, and promoting integrated urban planning with focus on peri-urbanization. Urban management practices in cities have yielded many lessons about pathways to promote inclusion of migrants, women, minorities and the urban poor in the process of urban governance and access to services.

Challenges and opportunities of urbanization and urban governance practices and innovations were recognized and reinforced at the global level by two landmark events culminating in global consensus. The first is the 2030 Agenda for Sustainable Development consisting of 17 Sustainable Development Goals (SDGs) endorsed by the Heads of State and Government at the United Nations. The Agenda is a combination of economic development, environmental sustainability and social inclusion. It includes SDG11 to promote inclusive, participatory and resilient cities, SDG 16 on accountable institutions and justice and SDG 5 on gender equality. The second is the New Urban Agenda adopted by the United Nations Conference on Housing and Sustainable Urban Development (HABITAT III) which endorsed the central role of governance in achieving inclusive and sustainable cities.

The idea of this book emanated from the interest of a group of scholars affiliated with research institutions in developing and developed countries to bridge gaps between theory and urban development practice by examining institutional mechanisms for participation, accountability and transparency in cities, the extent to which low-income urban residents have access to basic urban services and outcomes and impact of variety of local governance systems on improving access of all segments of society in low-income settlements. The group included (1) five scholars from China, India, Indonesia, Pakistan and Vietnam, who agreed to undertake case studies of specific cities in their countries along with a survey of households in selected slums and squatter settlements based on a survey questionnaire; (2) five leading scholars on concepts of access, participatory budgeting, public–private partnerships, gender equality and capacity development who prepared global reviews; and (3) a research team from Mo Ibrahim Foundation on Governance in Africa that undertook a survey on access to urban services and the role of local governments in Africa. The Swedish International Center for Local Democracy (ICLD) provided funding for case studies in Asia. East-West Center and ICLD convened a regional workshop on Governance for Access of Urban Services in Asia which was held in December 2017 in Yogyakarta, Indonesia.

I am grateful to a number of individuals and the partner institutions which enabled me to lead the collaborative research and prepare the manuscript: the contributors of this volume for providing their unique perspectives on different aspects of urban local governance and access to services or preparing case studies in Asia and Africa; Ana Maria Vargas and Bjorn Moller from the Swedish Center for Local Democracy (ICLD) for funding three of the Asian case studies and supporting the organization of the regional workshop for the writers of Asian case studies; Camilla Rocca and Diego Fernandez from Mo Ibrahim Foundation for analyzing their regional survey data based on the study format and preparing the regional review; and Cameron Lowry and Lillian Shimoda from East-West Center for providing research assistance and secretarial support for organizing the regional workshop. Professor Gerard Finin, Cornell University, provided very valuable comments and suggestions on an earlier draft of the manuscript. Bharat Dahiya, the Series Editor, provided useful inputs to the conceptual framework of the book. Loyola D'Silva, Publishing Editor, and Sanjievkumar Mathiyazhagan, Production Editor, provided continuous professional support through the book's copy editing and production process. The Ash

Center for Democratic Governance and Innovation, Harvard Kennedy School, provided me a fellowship to enable me to prepare the book manuscript. I am thankful to Tim Burke, Executive Director of the Ash Center, for his encouragement and support. Finally, I am grateful to my family for their patience and support.

We hope that this book will contribute to enhancing our understanding of the dynamics of relationships between institutional reforms for participatory and transparent urban local governance and access to services and bridging gaps between theory and practice of inclusive urban development. The book will be of interest to urban governance scholars, students, policymakers, practitioners, international development agencies, civil society and the private sector. The views expressed in the book are of the respective authors and not necessarily the institutions that have supported the research.

Shabbir Cheema
Senior Fellow
Harvard Kennedy School
Cambridge, USA

# Praise for *Governance for Urban Services: Access, Participation, Accountability and Transparency*

"As urbanization continues apace across the world in the twenty-first century, policy makers, development practitioners, and citizens face deep dilemmas about providing services and promoting equity in new and old environments. This edited volume provides insightful guidance about how to assess and respond to these urban dilemmas. It is a book for our times."

—Merilee S. Grindle, *Edward S. Mason Professor of International Development, Emerita Harvard Kennedy School*

"Shabbir Cheema has assembled a remarkable collection of assessments of how governments cope with the thorny problems of urban services in both developed and developing countries. Unlike the conventional urban studies that simply focus on urban politics or economics writ large, these chapters encompass both the theory of urban governance and the practical challenges of providing urban services efficiently and equitably. This comprehensive approach, vividly conveyed through case studies, provides an incisive guide for policymakers and activists dedicated to urban improvements."

—William Ascher, *Donald C. McKenna Professor of the Government and Economics, Claremont McKenna College, California, USA*

"The insightful and well documented studies in this volume provide readers with highly realistic perspectives of the challenges and promise of urban planning in the global south. As cities and their surrounding regions continue to grow, it will serve as a valuable and lasting contribution to the literature."

—Professor Gerard A. Finin, *Department of City and Regional Planning, Cornell University, USA*

"Shabbir Cheema has given us a remarkable book on peri-urban development focusing upon the disparity of access to water, sanitation, shelter, waste management, health care and education. Given hyper-urbanization resulting from domestic and international migration, these impacted case studies of India, Pakistan, Indonesia,

China and Vietnam will stimulate scholars and practitioners to unearth viable solutions for those marginalized by 'elitist capture'. This roadmap ranges from participatory budgeting, ICT and the transparent assessment of democratic decentralizaiton to barriers which need to be overcome due to residence based social policy."

—Robert Isaak, *Visiting Professor of International Management & Entrepreneurship, University of Mannheim, Germany*

"Shabbir Cheema has been a top academic and policy analyst for over 40 years, dissecting, summarizing and clarifying urban policy and practice in a variety of subtopics. Together with the group of distinguished authors, this volume brings fresh empirical data and ground-level perspective on how city administrations, often overloaded, manage to get services to expanding numbers of residents. The volume is valuable for the insights it offers in the role of governance and innovations in service delivery and access to the urban poor in Asia and Africa."

—Tim Campbell, *Creative Writer, and former Principal Urban Sector Specialist of the World Bank and Global Fellow of the Woodrow Wilson International Center for Scholars*

"Shabbir Cheema is a thought leader who has put together the state-of-the-art on governance for urban services. He rightly focuses on how urban local governance can increase access of low-income groups to basic urban services through participatory mechanisms, accountability of public officials and transparency in decision-making. The book makes a vital contribution to bridging gaps between theory and practice by offering both conceptual overviews and detailed case studies primarily from Asia and Africa. Eight sets of factors that improve access to urban services and policy recommendations discussed in the book would be of great interest to students, scholars and urban planners and practitioners engaged in democratic local governance and innovations to promote inclusive urban development."

—Robertson Work, *Author and climate/justice activist; former UNDP deputy-director of democratic governance and principal policy adviser; and Wagner Graduate School of Public Policy professor of innovative leadership for sustainable development, New York University*

# Contents

**Governance for Urban Services: Towards Political and Social Inclusion in Cities** . . . . . 1
Shabbir Cheema

**The State of Access in Cities: Theory and Practice** . . . . . 31
Jorrit de Jong and Fernando Fernandez-Monge

**Accountability Through Participatory Budgeting in India: Only in Kerala?** . . . . . 57
Harry Blair

**Public-Private Partnerships to Improve Urban Environmental Services** . . . . . 77
Bharat Dahiya and Bradford Gentry

**Gender Equality and Local Governance: Global Norms and Local Practices** . . . . . 107
Annika Björkdahl and Lejla Somun-Krupalija

**Developing Capacities for Inclusive and Innovative Urban Governance** . . . . . 127
Adriana Alberti and Mariastefania Senese

**Local Governance and Access to Urban Services: Political and Social Inclusion in Indonesia** . . . . . 153
Wilmar Salim and Martin Drenth

**Political and Social Inclusion and Local Democracy in Indian Cities: Case Studies of Delhi and Bengaluru** . . . . . 185
Debolina Kundu

**Access of Low-Income Residents to Urban Services for Inclusive Development: The Case of Chengdu, China** . . . . . 209
Bo Qin and Jian Yang

**Access to Urban Services for Political and Social Inclusion in Pakistan** . . . . . 237
Nasir Javed and Kiran Farhan

**Governance for Urban Services in Vietnam** . . . . . 255
Nguyen Duc Thanh, Pham Van Long and Nguyen Khac Giang

**Serving Africa's Citizens: Governance and Urban Service Delivery** . . . . 281
Camilla Rocca and Diego Fernández Fernández

**Local Governance and Access to Urban Services: Conclusions and Policy Implications** . . . . . 311
Shabbir Cheema

# Editor and Contributors

## About the Editor

**Shabbir Cheema** is Senior Fellow at Harvard Kennedy School's Ash Center for Democratic Governance and Innovation. Previously, he was Director of Democratic Governance Division of United Nations Development Programme (UNDP) in New York and Director of Asia-Pacific Governance and Democracy Initiative of East-West Center in Hawaii. Cheema prepared the UNDP policy papers on democratic governance, human rights, urbanization and anti-corruption and provided leadership in crafting UN-assisted governance training and advisory services programs in over 25 countries in Asia, Africa, Latin America and the Arab region. He was Program Director of the Global Forum on Reinventing Government and the Convener of the Harvard Kennedy School's Study Team of Eminent Scholars on Decentralization. He has taught at Universiti Sains Malaysia, University of Hawaii and New York University. As the UN team leader, he supported the International Conference on New and Restored Democracies, the Community of Democracies and UN HABITAT II. He has undertaken consultancy assignments for Asian Development Bank, the World Bank, U.S. Agency for International Development, Swedish International Development Agency, Dubai School of Government and United Nations. He holds a Ph.D. in political science from the University of Hawaii. He is the co-author of *The Evolution of Development*

*Thinking: Governance, Economics, Assistance and Security* (Palgrave Macmillan 2016) and the author of *Building Democratic Institutions: Governance Reform in Developing Countries* (Kumarian Press, 2005) and *Urban Shelter and Services* (Praeger 1987). He is the contributor and co-editor of the four-volume Series on *Trends and Innovations in Governance* (United Nations University Press, 2010); *Decentralizing Governance: Emerging Concepts and Practices* (Brookings Institution Press in cooperation with Harvard University, 2007); *Reinventing Government for the Twenty First Century: State Capacity in a Globalizing Society* (Kumarian Press, 2003) and *Decentralization and Development* (Sage Publications, 1984). Cheema has been a member of the advisory committees of the Swedish International Center for Local Democracy, UNHABITAT III, and the Pacific Basin Research Center and editorial boards of Urbanization and Environment and Third World Planning Review. A featured speaker at global and regional forums, he served as an advisor to the Dubai School of Government, Pakistan Institute for Economic Development, the Malaysian Academy for Leadership in Higher Education, and the UN Governance Center in Seoul, Korea.

## Contributors

**Adriana Alberti** Division for Public Institutions and Digital Government, UN Department of Economic and Social Affairs, New York, NY, USA

**Annika Björkdahl** Department of Political Science, Lund University, Lund, Sweden

**Harry Blair** South Asian Studies Council, Yale University, New Haven, CT, USA

**Shabbir Cheema** Harvard Kennedy School, Ash Center for Democratic Governance and Innovation, Harvard University, Cambridge, MA, USA

**Bharat Dahiya** Research Center for Integrated Sustainable Development, College of Interdisciplinary Studies, Thammasat University, Bangkok, Thailand

**Jorrit de Jong** Ash Center for Democratic Governance and Innovation, Harvard Kennedy School, Harvard University, Cambridge, MA, USA

**Martin Drenth** Research Center for Infrastructure and Regional Development, Institut Teknologi Bandung, Tangerang Selatan, Indonesia

**Kiran Farhan** The Urban Unit, Lahore, Punjab, Pakistan

**Diego Fernández Fernández** Mo Ibrahim Foundation, London, UK

**Fernando Fernandez-Monge** Ash Center for Democratic Governance and Innovation, Harvard Kennedy School, Bloomberg Harvard City Leadership Initiative, Cambridge, MA, USA

**Bradford Gentry** Yale University School of Forestry & Environmental Studies and Management, New Haven, CT, USA

**Nguyen Khac Giang** Mount Victoria, Wellington, New Zealand

**Nasir Javed** The Urban Unit, Lahore, Punjab, Pakistan

**Debolina Kundu** National Institute of Urban Affairs, New Delhi, India

**Bo Qin** School of Public Administration and Policy, Renmin University of China, Beijing, P.R. China

**Camilla Rocca** Mo Ibrahim Foundation, London, UK

**Wilmar Salim** School of Architecture, Planning and Policy Development, Institut Teknologi Bandung, Bandung, Indonesia

**Mariastefania Senese** Division for Public Institutions and Digital Government, UN Department of Economic and Social Affairs, New York, NY, USA

**Lejla Somun-Krupalija** Department of Political Science, Lund University, Lund, Sweden

**Nguyen Duc Thanh** Vietnam Institute for Economic and Policy Research (VEPR), University of Economics and Business at Vietnam National University, Cau Giay, Ha Noi, Vietnam

**Pham Van Long** Vietnam Institute for Economic and Policy Research (VEPR), University of Economics and Business at Vietnam National University, Cau Giay, Ha Noi, Vietnam

**Jian Yang** School of Public Affairs and Law, Southwest Jiaotong University, Chengdu, Sichuan, P.R. China

# Governance for Urban Services: Towards Political and Social Inclusion in Cities

**Shabbir Cheema**

**Abstract** Cities provide opportunities for economies of scale, products, income, services and social experimentation. However, urbanization has led to an increasing incidence of urban poverty and inequity, deteriorating quality of the urban environment and deficiencies in access to basic urban services, including water supply and sanitation, urban shelter, waste management, energy, transport and health. This chapter provides a framework to examine the role of urban local governance in promoting access to services, participation in local decision making and accountability and transparency. It presents challenges and opportunities of urbanization (Sect. 1), the paradigm shift in our thinking about development from economic growth focus to poverty alleviation and human development and about governance from traditional public administration to democratic governance in developing countries (Sect. 2), and forms, driving forces and impacts of decentralizing governance (Sect. 3). After defining access, participation, accountability and transparency, the chapter identifies four sets of intermediate-term outcomes of local governance for urban services (Sect. 4) and key features of "good" urban governance and policy challenges identified in the New Urban Agenda endorsed by HABITAT III (Sect. 5). Finally, it describes key themes and findings of each of the book chapters that follow (Sect. 6).

**Keywords** Urban local governance · Access to services · Participation · Accountability and transparency · Decentralization · Innovations · Capacity · Gender · Partnerships

Urbanization, democratization, globalization and information and communication technologies (ICT) have transformed the world over the past few decades in terms of how we live, interact and communicate with each other, exchange goods and services and participate in political systems to manage public affairs.

The urban population has been growing rapidly over the past 50 years, a trend which is estimated to continue unabated (United Nations Population Division 2012).

S. Cheema (✉)
Harvard Kennedy School, Ash Center for Democratic Governance and Innovation, Harvard University, 79 John F. Kennedy Street, Cambridge, MA 02138, USA
e-mail: shabbir_cheema@hks.harvard.edu



S. Cheema (ed.), *Governance for Urban Services*,
Advances in 21st Century Human Settlements,
https://doi.org/10.1007/978-981-15-2973-3_1

Cities provide opportunities for economies of scale, products, income, services and social experimentation. They play a vital role in productivity gains through innovation, trade, globalization of capital and the growth of service industries. Mega urban regions, urban corridors and city-regions reflect the emerging links between city growth and new patterns of economic activity. While cities occupy only two percent of the total land on the planet, they account for 70% of the global gross domestic product (GDP). However, urbanization has led to an increasing incidence of urban poverty and inequity, deteriorating quality of the urban environment and deficiencies in access to basic urban services, including water supply and sanitation, urban shelter, waste management, energy, transport and health. Urban poverty is mired in crime and violence, congestion, exposure to pollution and other similar health issues and infectious, and a lack of familiar social and community networks. In many cities of developing countries, people living in slums and squatter settlements account for 15–45% of the population, with the number of slum dwellers accounting for 60% of the total urban population in many Sub-Saharan African countries. This shows the "urban divide" between the cities of the rich and the poor. Furthermore, cities account for over 60% of global energy consumption and 70% of greenhouse gas emissions, and produce 70% of global waste (Habitat III 2017).

Over the past few decades, there has been an upsurge in the process of democratization, with steady increase in human liberty. Since 1980, over 50 military regimes have been replaced by civilian ones mostly in Africa, Asia and Latin America. In 2002, 65% of the world population lived under free or partly free states as defined by civil and political rights and 121 out of 192 countries were "electoral democracies" (Freedom House 2001–2002). The 2018 report of Freedom House showed that 75% of countries were free or partly free and that 63% of the world's population is free or partly free (Freedom House 2017). Recently, however, space for democratization has been shrinking throughout the world. Democracy is facing crisis dealing with such core norms and principles as guarantees of free and fair elections, rights of minorities, freedom of press, civilian oversight of military, and the rule of law (Freedom House 2017). Declines in political and civil liberties are reported from 71 countries, with 35 countries registering gains. The Varieties of Democracy 2018 Report shows that autocratic tendencies are increasing and even in democracies, women, minorities and migrants are disadvantaged from access to political power (Varieties of Democracy 2018).

Globalization, the third transformation, is shaping a new era of interaction and interdependence among nations, economies and people with new tools such as internet and media networks, through new actors such as the World Trade Organization (WTO) and networks of global NGOs and new rules such as multilateral agreements on trade and intellectual property (Kaul et al. 1999). It is linking foreign exchange and capital markets globally. Globalization is providing new opportunities by promoting economic growth and improvements in living conditions, among others, through trading of goods and services, increased foreign direct investment and textile exports from poor developing countries that are creating new jobs. But global wealth has not translated into equitable benefits for the vast majority of the people (UNDP 1999; Rodrik 1997; Stiglitz 2003; Rajan 2019). Services and activities vital for human wellbeing ensured through competitive markets inevitably reward those with access to skills and assets. Financial pressures are forcing governments to cut back on basic social services. There are insecurities in job and incomes due to illicit drug trade, inability of countries to train the work force to meet demands of globalization, money laundering and environmental degradation.

Finally, information and communication technology (ICT) is increasing access to knowledge and shrinking time and space (UNDP 2001; Sisk 2005) *and* fostering new patterns of interactions among the three pillars of governance i.e. state, the private sector and civil society (Rajan 2019). *ICT is* affecting the emergence of information societies that are demanding new patterns of government organization and processes. It is improving transparency and accountability, providing new communication channels between government and citizens and facilitating e-procurement to streamline purchasing and reduce administrative cost. It is serving as a tool for citizens to assess the performance of local government agencies. ICT can also equip local governments to facilitate information access such as websites, databases and consultation platforms and reach people in remote areas to engage them in governance and improve their access to services. However, access to ICTs is unequal leading to digital divide within and among countries. Therefore, in the United Nations Millennium Declaration, the General Assembly resolved to promote ICT as an important instrument for broad-based economic growth, poverty alleviation and sustainable development and to bridge the digital divide. Recognizing the significance of ICT, governments in developing countries are focusing on long-term capacity development within the public sector, creating an enabling environment for the private sector and civil society to benefit from ICT, and reengineering government's own internal operations to promote effectiveness of government programs.

Urbanization, democratization, globalization and ICT are inter-related processes that are affecting governance and development at global, national and local levels. They require an effective role of the state to provide a macroeconomic environment for creating incentives for efficient economic activities, and establishing and enforcing institutional arrangements for law and order, property rights for domestic and international investments, public-private partnerships to provide jobs and social services, and accountable, transparent and participatory institutions.

This book discusses institutions, processes and outcomes of urban local governance in developing countries. It examines participation of citizens in local decision-making, accountability and transparency in local governance, and access of low-income groups to urban services, including water and sanitation, shelter, transportation and health. This chapter discusses challenges and opportunities of urbanization (Sect. 1), the evolution of development and governance concepts (Sect. 2), decentralizing governance (Sect. 3); local governance outcomes and indicators of access to services, participation, accountability and transparency (Sect. 4) and key features of "good" urban governance (Sect. 5). Finally, it describes the focus of each of the book chapters that follow (Sect. 6).

## 1 Urbanization: Challenges and Opportunities

Many countries are being transformed from predominantly rural to urban societies. By 2030, about 56% of the population in less developed regions will be urban. As Table 1 shows, by 2050, the percentage of population living in urban areas will be about 57 in Africa, 64 in Asia, 82 in Europe, 86 in Latin America and the Caribbean and 88 in Northern America with the Asia and African regions accounting for 86% of the urban population growth. It is estimated that half of the population of Asia will live in urban areas by 2020, while Africa is likely to reach 50% urbanization in 2030 (United Nations Population Division 2012). Most of the urban population growth is likely to be concentrated in cities and towns of less developed regions, with a projected urban population increase of 1.4 billion in Asia, 0.9 billion in Africa and 0.2 billion in Latin America and the Caribbean. Most new megacities (with a population of at least 10 million inhabitants) have arisen in developing countries with 13 in Asia, four in Latin America and two in Africa. The second tier of urban agglomerations is large cities with populations ranging from 5 million to just under 10 million. The number of this category of cities is expected to increase to 59 by 2025. It is estimated that over three quarters of these cities are located in developing countries.

The most recent data from the UN Population Division (Fig. 1) shows that in 2017, 4.1 billion people or 54.9% of world's population resided in urban settlements. By 2050, urban areas will house 6.3 billion or 66.4% of world's population. The factors influencing urban population growth are internal migration due to inequalities between rural and urban areas, internal conflicts, and natural population growth. In many countries, there is an increasing concentration of urban population in large cities, and in some cases mega regions and urban corridors with global linkages. Secondary cities, with about half of the urban population in many countries, are likely to share significantly in this growth.

The contribution of cities to economic and social development is widely recognized. In the Asia-Pacific region, for example, it is estimated that cities account for

**Table 1** Percentage urban by major area, selected periods, 1950–2050

| Percentage urban | | | | | |
|---|---|---|---|---|---|
| Major area | 1950 | 1970 | 2011 | 2030 | 2050 |
| Africa | 14.4 | 23.5 | 39.6 | 47.7 | 57.7 |
| Asia | 17.5 | 23.7 | 45.0 | 55.5 | 64.4 |
| Europe | 51.3 | 62.8 | 72.9 | 77.4 | 82.2 |
| Latin America and the Caribbean | 41.4 | 57.1 | 79.1 | 83.4 | 86.6 |
| Northern America | 63.9 | 73.8 | 82.2 | 85.8 | 88.6 |
| Oceania | 62.4 | 71.2 | 70.7 | 71.4 | 73.0 |

*Source* United Nations Department of Economic and Social Affairs/Population Division. World Urbanization Prospects: The 2011 Revision

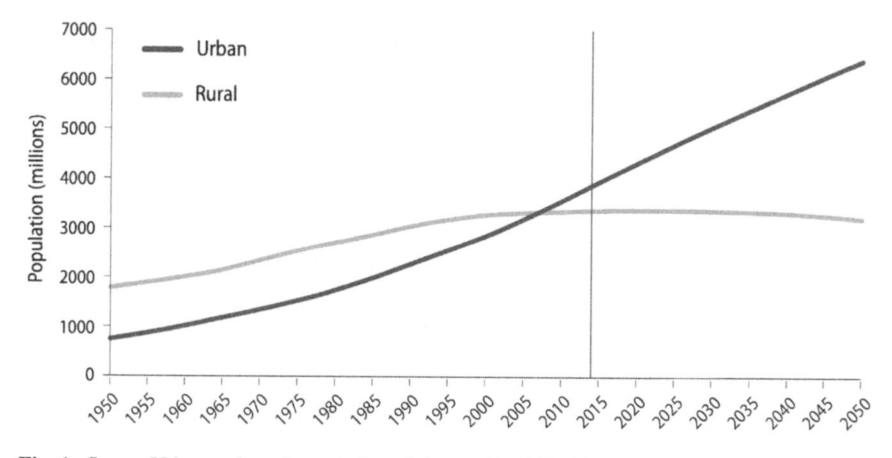


**Fig. 1** *Source* Urban and rural population of the world, 1950–2050. *Source* UNPD (2014b: 7)

about 80% of the region's economic output (Asian Development Bank 2011: p. 3). Throughout the World, cities contribute to problem solving at the local level and can "thrive in the age of populism" (Katz and Nowak 2017). Yet, sustainable urbanization faces many challenges: resources and environment (demand for land, water, energy, public finance), population growth and migration, and rising inequalities between rural and urban and within urban areas. Other challenges are inadequate access to basic urban services, weak rural urban linkages and the proliferation of peri-urban areas in the periphery of cities. Underlying each of these challenges are inadequate capacities at individual, entity and systemic levels to promote effective and inclusive urban institutions, through which actions are undertaken by governance actors from the state, civil society and the private sector to produce outcomes that can lead to the achievement of such universally accepted goals as participation, accountability, transparency and the rule of law.

Despite urban institutional reforms in many countries, urban governance deficits continue to constrain the goals of sustainable urbanization. (Hildebrand et al. 2013, pp. 24–28). National urban policies and strategies about urban growth are ambiguous. There is a lack of effective linkages between urban planning and budgeting. Decentralization policies are often not aligned with the resource base of urban local governments. Cities have inadequate financial and management skills, especially in secondary cities and towns. Participatory mechanisms to promote dialogue and partnerships between local government, civil society and the private sector are weak. Access of urban land for the poor is severely constrained which has led to increases in the number of slums and squatter settlement dwellers.

International organizations and national governments have prepared many comprehensive reports over the past 15 years on urbanization processes, urban management challenges and the impact of urbanization on development (UN-HABITAT/ESCAP 2011; UN Population Division 2012; UN-HABITAT 2013; World Bank 2013). The reports make a strong case for utilizing urbanization to promote economic development and improve living conditions of the people, especially in slums and squatter settlements. Three interrelated aspects of sustainability in urban areas are often highlighted—(1) economic sustainability, including the sustainability of cities to grow as centers of production to promote investments, create jobs, facilitate communications and improve quality of life; (2) environmental sustainability, including the provision of shelter and urban services, energy and climate change adaptation and mitigation; and (3) social sustainability, including community participation in local decision-making, equitable economic growth, rule of law and governance which is inclusive of women and ethnic and other vulnerable groups.

Inclusive and sustainable urban development policies and programs need to take into consideration increasing democratization in Africa, Asia, Latin America, and Eastern Europe. Urban governance is now defined more broadly to include the paradigm shift from urban public administration to urban democratic governance, which is characterized by the principles of participation, transparency and accountability, rule of law, subsidiarity and equity. This is discussed in the next section.

## 2 Evolution of Development and Governance Concepts

Over the past few decades, concepts of development and governance have evolved in the context of on-going and interrelated processes of urbanization, democratization, globalization and information and communication technologies.

### *2.1 From Economic Growth to Inclusive Development*

The paradigm shift in our thinking about the development and governance has been more pronounced in developing countries of Asia, Africa, Latin America and the Arab region than in the developed world. In the case of development, scholars and practitioners have focused on different dimensions over the years due to national priorities and international debates and agenda. Five phases in our thinking about development theories and practices can be identified.

First, the economic growth focus in the 1960s and 1970s under both structuralism and the socialist models of development aimed at promoting economic growth through active role of the state in the development process and interventions in economic sphere through formulation of five year and ten year plans to stimulate investments (Ascher et al. 2016). This approach was supported by the World Bank and regional development banks.

Second, pairing social and economic development emphasized broad-based economic growth to promote social development. Main elements of this phase included increased economic opportunities, participation in a democratic and pluralistic system, improving social services and narrowing gaps between developing and developed countries (Ascher et al. 2016). In 1981, the General Assembly of the United Nations adopted the third UN Development Decade where governments pledged to fulfill their commitments to a new international economic order to reduce gaps between developing and developed countries. Third, the concept of sustainable development was a major shift in the evolution in our thinking about development. The Brundtland Commission defined sustainable development as "development that meets the needs of the present without compromising the ability of future generations to meet their own needs" (UN 1987, 154). The World Summit on Environment and Development emphasized the need to promote equality, improve the quality of life and well-being, and sustain natural resources along with sustainable jobs, communities and industries. It also emphasized the need to protect human and ecosystem health, and meet international obligations (Khator 1998).

Fourth, the focus on poverty eradication and human development has been reflected in three global initiatives (1) the Millennium Development Goals (MDGs)—which included poverty eradication, universal primary education, gender equality, infant mortality, maternal health, HIV/AIDS and other diseases, environmental sustainability and the establishment of a global partnership for development; (2) the UN annual Human Development Report (HDR) first published in 1990 which advocated the concept of human development and introduced the Human Development Index (HDI)—a composite index comprised of three elements, namely, longevity, educational attainment and an income element, measured in terms of purchasing power parity (PPP) (UNDP 1990); and (3) the 2030 Agenda for Development consisting of 17 Sustainable Development Goals (SDGs)—a landmark event in the evolution of development practice in the world endorsed by the Heads of State and Government at the United Nations. The Agenda is a combination of economic development, environmental sustainability, and social inclusion. SDGs are "pathways" to inclusive and sustainable development at the local, national and global levels (UN 2016).

## 2.2 *From Traditional Public Administration to Democratic Governance*

Along with the paradigm shift in our thinking about what constitutes development and changing development agendas as discussed above, there has been a corresponding shift in the roles and approaches of three sets of actors—those from the State, civil society and the private sector—in managing public affairs. This is shown from the focus on traditional public administration in the 1950s and 1960s to development administration and New Public Management in the 1970s and 1980s, from government to governance in the 1990s, and currently democratic governance.

**Traditional public administration**—in the 1950s and 1960s, governments in developing countries and scholars of national and international development focused on the maintenance of law and order, the provision of basic infrastructure and services, and the promotion of national integration—the tasks that were considered as primary concerns of the newly independent countries after the World War II. Traditional public administration entailed a strong role for the state and centralized, legal-rational and hierarchical bureaucratic systems. It advocated the separation of politics and administration, the need to enhance efficiency, and improve public services through training civil servants and merit-based assessments of performance (Ascher et al. 2016). It was also considered to be suitable to achieve economic growth and pairing social and economic development which were two of the core development objectives during that time.

**Development administration and New Public Management**—Development administration blurred the administrative and the political as well as the public and private as governments took joint ownership and management of certain public services. To promote nation-building objectives, developing countries required a civil service with skills to undertake core development functions, including policy analysis and formulation, program planning, implementation and monitoring and service delivery. Therefore, Institutes of Public Administration training were established with technical and financial support of bilateral donors and multilateral development organizations. Within the public sector, people-centered development processes in organizations were emphasized including organizational structures and designs, leadership, training and development, team-building, and motivation. It also addressed macro-organizational issues such as organizational structures and designs to promote coherence and coordination in program planning and implementation (UNDP 1995; Blunt and Jones 1992). The objectives of "New Public Management" (NPM) were to reorganize the public sector organizations to bring management and accounting approaches closer to business methods and making them more efficient and (Dunleavy and Hood 1994); accountability based on performance management towards known benchmarks of success (Rondinelli 1990); and joint ownership and management of certain public services and smaller, decentralized units with direct links to citizens (Falconer 1997). The six core characteristics of NPM were enumerated as: productivity; marketization; service orientation; decentralization; policy orientation; and accountability for results (Kettle 2000).

**From government to governance**—It entailed a recognition that governance is more than government: governance is co-produced by three sets of actors, which are the state, the private sector and civil society. In late 1980s and early 1990s, global consensus emerged among scholars and practitioners of international development that core responsibility of the state is to provide an enabling political and legal environment, maintain public security and safety and safeguard public norms and standards. The private sector generates jobs, income, goods and services. Civil society facilitates political and social interaction and holds governance responsive to citizens. Governance comprises complex mechanisms, processes, relationships and institutions through which citizens and groups articulate their interests, exercise their rights and obligations and mediate their differences to manage public affairs (UNDP 1997). Increased role of civil society was promoted because of domestic and external factors. Governments did not have sufficient capacity to provide basic services to all citizens. Furthermore, there was an increased demand within countries for greater participation in political and developmental process and transparency and accountability of public officials. The domestic pressures were reinforced by advocacy from the global community, including bilateral donors and multilateral agencies such as UN Development Programme and the World Bank, to engage all segments of the society in making decisions that affect people. The UN Human Development Reports, the rights-based approach to development, and the Millennium Development Goals (MDGs) led to the recognition of the significance of civil society engagement in promoting participatory and inclusive development (Cohen and Arato 1994; Carothers 1999; White 1994; Alagappa 2004; World Bank 2010; Cheema and Popovski 2010; Hyden and Samuel 2011). Internal pressures due to democratization led to increased decentralization to make development outcomes more efficient and effective in terms of enhancing citizen participation and accountability and transparency of public institutions (Cheema and Rondinelli 2007).

Scholars presented different perspectives dealing with the concept of governance. These included: three forms i.e. systemic, political, and administrative (Leftwich 1995); relationships between the state, civil society and the private sector (UNDP 1997); increasing civil society engagement (Cheema 2013; Alagappa 2004); social capital (Mayer 2003; Rydin and Pennington 2000); impact of capabilities, inclusion and justice in human development (UNDP 1990; Sen 1999); inclusive urban development to cope with challenges of urban poverty and productivity (Alkire et al. 2015; Fainstein 2001; Moser 1998; Storper 2014); the need to improve service delivery and access (Mitlin 2008; Rondinelli and Cheema 2003); and "good enough" governance to recognize what is realistic (Grindle 2004; Andrews 2008). Despite different emphases, scholars and practitioners agree on guiding principles of governance—accountability, efficiency and effectiveness, openness and transparency, participation, and rule of Law (Doeveren 2011).

**Democratic governance**—Democratic governance brings two dimensions of managing public affairs together: process of interaction among actors (governance) and universally recognized core values (democracy). It is a process which is infused with such principles of democracy as participation, rule of law, transparency and accountability, representation, access, inclusion and a system of checks and balances (Cheema 2005; Ascher et al. 2016). In practice, democratic governance means the translation of these principles into certain tangible things—such as free, fair and frequent elections; a representative legislature that makes laws and provides oversight; and an independent judiciary that interprets laws, the guarantee of human rights and the rule of law, and transparent and accountable institutions. Other component or entry points for democratic governance are the decentralization of authority and resources to local governments and increased engagement of civil society in socio-economic and political development.

The evolution of democratic governance has been influenced by liberal democratic theory (Dryzek and Dunleavy 2009; Wintrop 1983) and practice and the emerging global consensus beginning the 1990s that governance should not only be democratic but also effective and efficient in engaging all groups within the society and improving their living conditions through dialogue and exchanges among the actors from the State, the private sector and civil society. A paradigm shift took place in the 1990s and early 2000 within the global community to bring the concepts of governance and democracy together in the process of designing and managing institutional reform programs. Bilateral donors and multilateral development organizations began to engage in politically sensitive issues as the protection of civil and political rights, legal framework for civil society, and free and fair elections on multi-party basis. The external development partners, thus, shifted the focus of their assistance from traditional public administration to democratic governance. At the global level an early leader in this regard was United Nations Development Programme (UNDP), the central development organization of the UN system.

The shift from traditional public administration to democratic governance at the national level has led to changing patterns of interactions at the local level in developing countries between actors from the state, the private sector and civil society. The next two sections discuss issues and challenges in designing and implementing decentralization policies and programs and the four pillars of democratic local governance outcomes i.e. access of citizens to local decision-making structures and urban services, degree of participation of different segments of society, accountability of public officials and transparency in local governance.

## 3 Decentralizing Governance

Four forms of decentralization have been identified: Deconcentration sought to shift administrative responsibilities from central ministries and departments to subnational and local administrative levels. Delegation entailed shifting by national governments of management authority for specific functions to semi-autonomous or parastatal organizations and state enterprises, regional planning and area development agencies and multi- and single-purpose public authorities. Devolution aimed to strengthen local governments by granting them the authority, responsibility and resources to provide services and infrastructure, protect public health and safety, and formulate and implement local policies. Political decentralization included elected local government/councils, organizations and procedures for civil society engagement to promote citizen participation, and central role of democratically elected local government in designing, monitoring and coordinating the implementation of local development programs, and fiscal autonomy for local governments (Cheema and Rondinelli 2007). Others have discussed the design and implementation of decentralization from three perspectives: political, administrative and fiscal (Turner and Hume 1997; Agrawal and Ribot 1999; Manor 1999; Smith 1985).

Urbanization, democratization, globalization and information and communication technologies have influenced decentralization of governance in number of ways. To effectively manage cities as engines of growth, governments needed to decentralize powers and resources to urban local governments and empower them to raise their revenues, coordinate central and provincial government agencies providing services in the cities, strengthen capacities of local governments to plan and manage projects, and make metropolitan-wide institutional arrangement to benefit from economies of scale to provide urban infrastructure and services (Dahiya and Das 2019; Cheema 2007). Democratization influenced the process of political decentralization due to pressures from below for greater involvement of sub-national and local governments (World Bank 1997; Chambers 1995). While during the 1950s and 1960s, the focus was on deconcentration and delegation, democratization shifted the focus to political and economic devolution and expanding role of civil society in the process of local development. Globalization required enhancing administrative and fiscal capacity of sub-national governments to promote investments and economic activities outside the capital cities and metropolitan areas, and facilitate the participation of individuals and enterprises to participate in the global market place (Cheema and Rondinelli 2007). ICT helped to overcome the obstacles to decentralization by improving transparency and accountability by providing new communication channels, facilitating e-procurement to streamline purchasing and reduce administrative cost, equipping local governments to facilitate information access such as city websites, databases, consultation platforms; and reaching people in remote areas to promote their engagement in local governance and to improve their access to services.

Studies show that decentralization can provide a framework for poverty alleviation through citizen engagement in planning and managing local projects, promote checks and balances between sub-national and local governments and ensure accountability of public officials and transparency at the local level (Tendler 1997; Crook and Manor 1998; Cheema and Rondinelli 2007; Work 2003; Manor 1999; Grindle 2007a, b; Shah 2006; Blair 2000). If not properly designed with political and financial support from the Center, however, decentralization can lead to "elite capture" of government programs and facilities, low capacities of local governments vis-à-vis their responsibilities, increased economic disparities among areas and regions with unequal resource base, and discrimination against minorities and migrants especially where local power-structures are inegalitarian and monopolized by authoritarian figures (Grindle 2007a, b; Ojendal and Dellnas 2013). Therefore, fair representation and downward accountability are necessary conditions for the success of decentralization (Agrawal and Ribot 1999). To sum up, decentralization is a continuous process of changing the center-local balance of power within the national context (Turner and Hume 1997; Cheema and Rondinelli 2007).

## *3.1 Driving Forces*

Many driving forces at the global, national and local levels have influenced recent trends toward greater political devolution and transfer of financial authority from the center to regions and local areas—the advent of multi-party political systems in Asia and Africa, deepening of democracy in Latin America after the latest wave of democratization and the end of military regimes, and transition from a command to market economy and democratization in Eastern Europe and Central Asian Republics after the demise of the former Soviet Union. Other driving forces were increase in ethnic conflicts and demand for greater recognition of cultural, religious and regional traditions; focus of bilateral donors, multilateral agencies and civil society on decentralized governance; and demands by groups and individuals within countries for greater control over local political processes, greater transparency, better access to services and more openness in political decision-making processes.

## 4 Democratic Local Governance Outcomes

Over the past few decades, international development organizations such as the World Bank and Regional Development Banks, the United Nations, and bilateral donors have dramatically increased their investments to strengthen national and local governance capacity of developing countries. Consequently, there is an emerging consensus among policymakers and scholars of local governance and bilateral and multilateral development organizations concerning the need to assess the outcomes and results of investments in building governance capacity. With increasing democratization, urbanization and access to information, citizens hold governments to higher standards and have higher expectations from governments, including the provision of sustainable livelihoods, economic opportunities, access to basic services, and ability to participate in decisions that affect them. Therefore, developing country governments are under pressures from their own constituents to improve the quality of governance.

The rise to prominence of governance as a key development concern has been marked by an increasing interest in the production of a huge range of governance outcomes and indicators. The early UNDP approach was to invest in governance "entry points": electoral and parliamentary process; civil service reform and innovations; legal and judicial reform; governance in crisis and post-conflict situations; transparency, accountability and anti-corruption; civil society participation and engagement; and decentralization and local governance. (UNDP 1997). The World Bank, similarly, identified six dimensions of governance process, each with a set of indicators: voice and accountability (the extent to which citizens can participate in selecting their government); political stability and absence of violence (perception of likelihood the government will be destabilized or overthrown); government effectiveness (quality of public and civil services, and credibility of government policy); regulatory quality (ability of the government to formulate and implement sound policies that permit and promote private sector development); rule of law (quality of contract enforcement, police, and courts that support the functioning of society); and control of corruption (extent to which public power is exercised for public good and not personal gain) (World Bank 1997).

Scholars and practitioners recognize challenges of showing results of investments in democratic governance at the national and local levels. First, investment in democratic governance is a long-term process, which is difficult to quantify in the short-term. Second, the transition to democratic governance is a political process. It is about political change. Explaining what, why and how of political change is a highly complex task. Third, the outcomes of governance institutions depend upon contextual factors. Similar institutional arrangements can produce different outcomes in different contexts. Ethnic heterogeneity, for example, constrains political consensus-building on public policies. Regional disparities and the size of the middle class can affect the ability of government and civil society to design and implement inclusive development policies. The legacy of civil-military relations in a country can impede strengthening of civilian political institutions and their oversight of the military. Finally, the global criteria to evaluate results keeps on changing due to rapidly evolving global norms and agendas.

Participation, accountability, transparency, access and inclusion provide a roadmap to assess the quality of local governance which is both inclusive and participatory but also effective in terms of efficiency and delivering concrete outcomes to improve human conditions. The global consensus on these guiding and fundamental principles led to the shift from a political and technocratic approaches to solve development problems to the integration of politics in development. As Carothers succinctly argues, this took place through the "three rivers of politics in development" i.e. governance, democracy, and human rights. Specialists and practitioners of each of these focused on different dimensions. Governance scholars emphasized it as a necessary intervening condition to improve human development. Democracy promotion scholars emphasized democratic institutions, the rule of law, the independence of judiciary and a system of checks and balance. Human rights specialist focused on the protection of civil and political rights and a rights-based approach to development (Carothers and Brechenmacher 2014).

The concepts of access, participation, accountability and transparency are sometimes not complementary or inter-dependent. Yet, they have sufficient convergence that has led to a unified global agenda with focus on these four principles of local governance. These concepts, therefore, are emphasized in the policy documents of developing countries and multilateral organizations and bilateral agencies. They are also the focus of some of the major global initiatives such as the Extractive Industries Transparency Initiative (EITI), the Open Government Partnership (OGP), "Global Partnership for Social Accountability" and the "Making All Voices Count "grand challenge for development". These initiatives are funded by the World Banks, United Kingdom's Department for International Development, USAID, the Swedish government, the Open Society Foundations, and the Omidyar Network, and other organizations. United Nations Development Programme (UNDP) has been providing grant funding for accountable and participatory institutions, access to urban services, and transparency and anti-corruption strategies.

### 4.1 Access

Access and inclusion are fundamental principles of democratic governance. They are included as "pathways" to inclusive development in the 2030 Agenda for Development. Seventeen Sustainable Development Goals (SDGs) are identified in the Agenda with targets and indicators. Participatory, accountable and transparent local governance and the need to improve access to urban shelter and services are emphasized in SDG 5 (gender equality), SDG 11 (resilient, participatory and sustainable cities) and SDG 16 (accountable institutions and justice).

Ribot and Peluso define access as "the *right* to benefit from things" and means, relations, and processes that enable various actors to derive benefits from resources (Ribot and Peluso 2003). Access analysis, they argue, involves (1) identifying and mapping the flow of the particular benefit of interest; (2) identifying the mechanisms by which different actors involved gain, control, and maintain the benefit flow and its distribution; and (3) an analysis of the power relations underlying the mechanisms of access involved. The mechanisms that shape access processes and relations include technology, capital, markets, knowledge, authority, identities, and social relations.

In their path-setting framework for analysis, De Jong and Rizvi define access as "the match between societal commitment and institutional capacity to deliver rights and services and people's capacity to benefit from those rights and services" (De Jong and Rizvi 2008: 4). In their definition, access includes access to basic services, intermediary organizations that deliver services, mechanisms to make citizen needs and priorities known to public agencies and equal opportunities in "politics, economy, social services, justice and more" (De Jong and Rizvi 2008: 8). They suggest four levels of analysis to understand mechanisms that lead to inequalities in access: bureaucratic behavior, agency performance, system dynamics and contextual factors. They propose individuals, organizations, institutional arrangements and social and economic forces as units of analysis to effectively cope with "bureaucratic dysfunction".

The concepts of inclusion and access are closely related. Inclusion implies equality of opportunities to participate fully in the society (European Commission 2004; Hayes et al. 2008; Stewart 2000; World Bank 2011). Economic and political inclusion entails a standard well-being to ensure that basic needs are met, access to markets and public services is provided, and political voice and feelings of respect and recognition are guaranteed. Social inclusion implies the ability and opportunity of individuals and groups to take part in society.

In this study, we define access as (1) equal opportunities for and ability of citizens to engage in local governance institutions and processes (e.g. local elections, mechanisms for public accountability and transparency, participation in local decision-making); and (2) ability to benefit from government, the private sector and civil society initiatives dealing with urban services (e.g. water and sanitation, health, education and shelter improvement).

### *4.2 Participation*

Participation entails engagement of individuals, groups and all segment of society in socio-economic and political processes. The primary rationale of local governance is to provide an institutional framework that can brings government closer to the citizens by providing them opportunities to articulate their interests and preferences, engage in making decisions that affect them and monitor and evaluate the outcomes of government policies. Local governance can also facilitate citizen participation in the process of local development program planning and management, including identifying local needs, location of basic services, and holding public officials accountable. As in the case of democratic governance at the national level, local participation is both a means to an end and an end in itself. As a means, effective mechanisms for local participation facilitate service delivery and access, mobilization of community resources, and implementation of local development projects. As an end, local participation promotes local democracy. Therefore, it is essential that people should have "substantial and equal opportunities to participate" in decisions that affect them (Fung 2004: 4).

Over the years, democratization and decentralization have led to significant progress in Africa, Asia, Latin America and Eastern Europe in holding regular elections to elect local councils and mayors, establishing branches of political parties at the local level, and engaging civil society in local economic development process, in many cases through multi-party systems. Most countries in Africa (such as Ghana, Niger and Uganda) and Asia (India, Indonesia and the Philippines) have mechanisms below formal local government to engage citizens in the local decision-making process through consultations. Gram Sabha in India and Barangay in the Philippines are examples.

Many developing countries have introduced innovative and inclusive urban policies and programs that accommodate marginalized groups in urban governance. In Kuala Lumpur, and Quezon City, urban authorities are building kiosks for sidewalk vendors in legally sited areas in consultation with vendors. In Bandung, Bangkok and Manila, community-upgrading programs now provide housing and basic services in situ, rather than evicting squatters. Orangi Pilot Project in Karachi is focused on urban sanitation and drainage in partnership with local communities. In India, the 73rd and 74th amendments to the Constitution specified the roles to be played by community-based organizations (CBOs), and women in local governance. In the Philippines, the 1987 revised Constitution upholds the right of community based organizations (CBOs) to become directly involved in governance.

In some countries, local governments have initiated mechanisms to facilitate the participation of citizens in economic policy making and, specifically, getting them engaged in the process of budgeting and auditing. Porte Alegre in Brazil undertook a "path-setting" experiment with participatory budgeting and has influenced similar processes in other countries such as India. In Abra, the Philippines, the "Concerned Citizens of Abra for Good Governance (CCAGG)", an NGO, investigates projects for sub-standard materials, poor construction techniques and fraud contracting procedures. It is an example of a participatory public audit.

New forms of participation and learning are taking place in cities. In addition to participation in the first person (such as local elections of mayors and councils and community organizations), participation in the virtual person is rapidly increasing (such as G2C—Government to Citizen; G2B—Government to Business; C2C—Citizen to Citizen; C2G—Citizen to Government; and G2G—Government to Government (Campbell 2013). This is being facilitated by an increasing use of ICT. In addition, cities are learning from each other through national, regional and global networks through which innovations in access to shelter and services, "smart city" solutions and designing and implementing participatory urban development programs are being shared.

## 4.3 Accountability

Public accountability implies holding individuals and organizations responsible for their decisions and actions, as measured through their performance, reports and feedback from citizens. It has three dimensions. Financial accountability is the obligation of anyone handling resources, public office or any other position of trust, to report on the intended and actual use of the resources or of the designated office. The tools of financial accountability include open and transparent public procurement, annual performance reports and declaration of assets. Political accountability consists of having branches of government (executive, legislative & judicial) watch over each other; separating institutions that raise and spend funds from those that execute spending decisions; and the independence of the judiciary. Administrative accountability implies critical systems of control internal to the government and ensuring proper functioning of checks and balances supplied by constitutional governments. It also includes civil service standards and incentives, ethics codes, criminal penalties and administrative reviews. Accountability takes place in two directions: vertical accountability between citizens and institutions of government through such instruments and tools as civil society, grassroots organizations, media, professional association and decentralization framework; and horizontal accountability between branches and institutions of government through checks and balances, subsidiarity and separation of powers.

Accountability, along with participation, can improve effectiveness at the local level. Based on her study of selected municipalities in Mexico, for example, Grindle argues that one of the explanations for higher level performance of municipalities is the extent to which citizens are "mobilized to participate and demand accountability" (Grindle 2007a, b: 67). In such situations, citizens and community groups put pressure on public officials to provide better services and demand to be actively engaged in the process of policy and program design and implementation processes. They identify local needs, organize themselves, communicate local priorities, and contribute their time, local expertise and financial resources for local development.

Over the past few decades, local governments throughout the world have tried different mechanisms, tools and instruments of local government accountability and transparency to promote and sustain devolution and strengthen trust between citizens and local governments. These include effective anti-corruption bodies, transparent and accountable system of public procurement, and participatory budgeting and auditing. Other accountability mechanisms are the engagement of civil society in local decision-making, promotion of ethics and integrity among local public officials and local leadership commitment to accountability and transparency.

One of the reasons for corruption at the local level is the lack of established mechanisms and procedures for accountability in the public procurement system. One of the examples to cope with this issue is the establishment of a new electronic system by the Government of Andhra Pradesh in India to offer the tenders online and handle the procedures electronically. The government of Seoul City adopted the Integrity Pact (IP), by which the city government and companies submitting bids agree neither to offer nor to accept bribers in the public contracts. Civil society engagement and an on-going dialogue among the local actors provide other mechanisms to ensure accountability of local officials. In Bangalore, India, for example, since 1994 Public Affairs Center has brought out "Citizens Report Cards"[i] (CRP) that rated and compared agencies on the basis of public satisfaction and responsiveness of public agencies. The Philippine Centre for Investigative Journalism (PCIJ) is an independent, non-profit media agency specializing in investigative journalism. The PCIJ provides training for investigative reporting to full-time reporters, free-lance journalists and academics. Another local accountability mechanism is to strengthen ethics and professionalism among officials in order to combat local corruption. Some of the mechanisms that are being used to promote ethics and integrity at the local government level are conflict of interest laws, disclosure of income and assets, lobbyist registration, whistle blower protection, codes of conduct and ethics training. Such mechanisms are especially important in large urban local governments and metropolitan regions that build and maintain large infrastructure projects because of potential grand corruption in such projects.

### 4.4 Transparency

Transparency implies openness of local governance processes and procedures and conduct and actions of public officials. This is achieved through the freedom of information legislation, access to governmental political and economic activities and decisions, open records or sunshine laws, and set rules on access to information and records held by government officials. There are two dimensions of transparency policies: public sector transparency to build and sustain transparency in government's internal processes and mechanisms for decision-making and use of public resources; and "targeted transparency policies" to ensure that societal standards and norms are maintained in services provided through the private sector such as drug safety, food labeling, consumer products and environmental protection (Fung 2004). Transparency within the public sector can improve quality and access to services, promote political legitimacy of the government, enhance citizen trust in government and create fairness and competition for better quality. Policies designed to ensure accountability of the private sector vis-à-vis public mandates are equally important in such areas as the protection of health and safety of citizens.

Freedom of press is necessary to promote transparency. It facilitates the identification of citizen preferences, citizens' responses to government policies and programs and the protection of civil and political rights of people. Freedom of press enables the discussion of alternative points of view about various policy options, engagement of different segments of society in issues of national interest and safeguarding the abuse of power by public officials.

Local leaders play an important role in promoting transparency and accountability at the local level by setting an example and enforcing laws. Examples of leaders who have had a transformational impact on the accountability mechanisms in their respective local government are Maclean-Abaroa in La Paz, Bolivia and Mayor Lerner in Curitiba, the capital of Parana state in South Eastern Brazil. Mayor Lerner became an international reference for integrated and holistic planning, being recognized by the international media, experts and development institutions as an award-winning example of how local elected local officials can mobilize stakeholders and communities to tackle development challenges and create an environment of accountability and transparency of local governance.

The United Nations Programme on Human Settlements (UN HABITAT) has identified four strategies with specific tools to promote transparency and accountability in local governance—(1) assessments and monitoring of programs to ensure that timely actions are taken based on local learning and experimentation, through such tools as Report Cards, and the Public Record of Operation and Finance (PROOF); (2) access to information and public participation so that citizen are engaged in local decision-making, including through information laws, management and computerization, media training, and public education; (3) the promotion of ethics, professionalism and integrity that can guide individual behavior in organizations and at systemic level, including disclosure of income and assets, codes of conduct and ethical campaign practices; and (4) institutional reform to mainstream and sustain norms and behavior patterns of transparent and accountable local governance, including Complaints and Ombudsman Office, Oversight Committees, Independent Audit Office, Anti-Corruption Agency, and Participatory Budgeting (United Nations Programme on Human Settlements 2004).

## *4.5 Indicators*

The rise to prominence of governance as a key development concern has been marked by an increasing interest in measurement and the production of a huge range of governance indicators (UNDP 2002; Freedom House 2016; Variety of Democracy 2016; Swedish International Center for Local Democracy 2015, 2017). Table 2 presents objectives and some indicators of medium-term (2–3 years) outcomes of access, participation, accountability and transparency.

**Table 2** Intermediate-term outcomes of governance for urban services

| Principles | Objectives | Medium-term outcomes |
|---|---|---|
| Access | Urban services (such as water, waste management, health, shelter and education) are accessible to all including women, minorities, migrants and residents of slums and squatter settlements | • Degree of change in household coverage of water and waste management, health, shelter, education and other urban services<br>• Percentage of the local government (LG) budget vis-a-vis population spent on improving shelter and urban services in low-income settlements<br>• Extent of engagement of community- based organizations in design and implementation of local projects on shelter improvement and services<br>• Degree of change in land title regularization in slums and squatter settlements<br>• Degree to which vulnerable groups have opportunities for sustainable livelihoods and access to credit for shelter improvement |
| Participation | Local governments actively work for citizens' increased participation in local decision making processes | • Extent to which mechanisms for citizen engagement and dialogue are integrated within planning and design stage of LG projects<br>• Percentage increase in communication and feedback including complaints from community-based organizations on LG projects<br>• Integration of mechanisms to safeguard the participation of women, minorities, migrants and vulnerable groups in decision-making<br>• Extent to which decentralized governance provides an effective institutional framework for citizen participation<br>• Legal and regulatory framework that promote civil society engagement |

(continued)

Table 2 (continued)

| Principles | Objectives | Medium-term outcomes |
|---|---|---|
| Accountability | Local governance institutions and processes provide for enabling citizens to hold public officials accountable | • Integration of mechanisms for providing citizens with info on who is responsible for which decisions<br>• Degree of LG focus on public education and awareness raising about accountability and anti-corruption<br>• Sharing of timely and accessible information by LG to all stakeholders about local development programs that affect local people directly<br>• Extent to which procedures for financial and administrative monitoring, auditing and evaluation are mainstreamed<br>• Extent of community level audit of local services provided by the government, the private sector and civil society |
| Transparency | Local governments actively work for increased transparency | • Integration of mechanisms for citizens' increased access to information<br>• Integration of mechanisms for increased transparency in decision-making<br>• Extent to which procedures for resource allocation to different areas and sectors are institutionalized and mainstreamed<br>• The extent to which the systems of LG public procurement are open and transparent |

*Sources* Adapted from Swedish International Center for Local Democracy (2015, 2017), United Nations Development Programme (2004), Cheema (2013)

## 5 Good Urban Governance

Over the past few decades, governments of developing countries, United Nations and bilateral development organizations have advocated different strategies and approaches to promote the role of cities as engines of economic growth and identify policy and program options to reduce urban poverty, improve the quality of the urban environment and implement participatory programs to engage citizens in urban development. Various approaches have included "housing by people" during the 1960s and 1970s, sites and services and slum upgrading from the 1970s, the recognition of the International Year of Shelter for the Homeless by United Nations

**Table 3** Policy challenges towards good urban governance

| What | Why/How |
|---|---|
| Increasing complexity of urban governance | The integration of different levels of government and a wide range of formal and informal actors in urban governance need to be understood and incorporated in urban policymaking and implementation |
| | There is a need to acknowledge the challenges associated with the diversity of local conditions and contexts and new urban forms |
| | A differentiated policy approach is needed to ensure the co-existence of metropolitan areas, intermediary cities, small towns, rapidly growing cities and shrinking cities |
| Absent or inadequate decentralization | Often local governments have limited powers and resources and lack professional staff and revenue raising capacities. Implementation of national decentralization policies has shown mixed results due to local context and complexity of decentralization |
| | To ensure optimal design and implementation of decentralization policies in different contexts, the need is for actions to cope with inadequate planning processes, weak inter-agency coordination, economic inefficiencies, backlogs in budget spending, weak local capacity, high transaction costs, and experimentation at the local level |
| Ineffective legal and institutional frameworks | Very few countries have developed and implemented comprehensive and coherent national urban policies |
| | The need is to connect legal, administrative and fiscal frameworks; clarify the distribution of responsibilities between different levels of government; and correct contradictory regulations |
| The metropolitan challenge | Cities have physically grown beyond their administrative boundaries, and their economies have become more globalized, attracting flows of goods, capital and migrants from rural areas and from outside the national borders |
| | The need is to improve systems and processes of coordination at the metropolitan scale to ensure cost-effective responses to urban problems, especially for coping with spill-overs and peri-urban areas |

(continued)

**Table 3** (continued)

| What | Why/How |
|---|---|
| Inequality and exclusion | Current urbanization processes are reinforcing inequality and exclusion – particularly for women, youth, the elderly, minorities and the urban poor |
| | The promotion of participation, accountability and transparency in public institutions and access to services and livelihoods is essential to ensure equality and inclusion |
| | To promote equality and inclusion in cities and towns, actions needed include: the removal of structural constraints on citizen participation; and legislation that recognizes civil society organizations, guarantees and promotes participation, and allows access to public information and data to promote informed citizenship |
| Weak frameworks for service delivery partnerships | Public partnerships with other actors including the private sector, nongovernmental organizations and community based organizations can facilitate service delivery and access and critical aspects of urban development such as effective program design, feedback from program implementation and community ownership of government initiatives |
| Insufficient monitoring and evaluation of urban policies | Local governments have inadequate access to localized data and thus are not able to take informed decisions and better prioritize local policies |
| | The need is for national governments to promote the involvement of local governments and stakeholders including civil society in the process of defining, implementing and monitoring urban and regional policies and plans |
| Rapid technological change | Technological change poses complex and interrelated urban challenges that require city institutions to adapt. The potential of "smart city" solutions and use of big data to design, monitor and evaluate urban programs should be utilized |

*Source* Abstracted from Paper 4 of the New Urban Agenda of HABITAT III; Dahiya and Das (2019), and Cheema and Rondinelli (2007) in 1987 to provide shelter to the poor and the homeless, and the decentralization of services since the 1980s (Dahiya and Das 2019). Other approaches were participatory budgeting, local agenda 21 for sustainable development and bottom up planning in the 1990s, and resilient cities since 2010 (Satterthwaite 2016; Cheema 1993). The United Nations Conference on Human Settlements (Habitat I) held in Vancouver in 1976 and HABITAT II in Istanbul in 1996 advocated multi-level, multi-stakeholder and multi-sectoral, integrated urban development. HABITAT III held in Quito in 2016 was a landmark event because it forged a global consensus on the New Urban Agenda (NUA) that identifies policy challenges that require concrete actions with an integrated view of cities as the drivers of global environmental changes (Table 3).

## 6 About the Book

This book discusses governance for urban services. It raises three questions:

(1) What are the urban local governance structures and processes and urban policies and programs in developing countries to improve access of low-income groups to urban services?
(2) What are the outcomes of institutional structures and urban programs vis-a-vis improved access of low-income groups to urban services, citizen participation in local decision-making and local development, accountability of local leaders and government officials, and transparency in local governance?
(3) What are the factors and barriers that influence urban service delivery and access and the quality of democratic local governance, especially the inclusion of women, minorities, migrants and the urban poor?

Chapters that follow discuss the scale of deficits in access of the marginalized groups to urban services including water and sanitation, health, education and other social services in cities; government decentralization policies and programs that attempt to accommodate the needs of marginalized groups; and institutional mapping to understand and assess the functional mandates and boundaries of the agencies that provide services and regulate cities. The studies identify good practices and innovations in service delivery and access and structural and institutional barriers to full engagement of women, migrants, minorities and the urban poor in processes and mechanisms of local governance.

The methodology of the country case studies consisted of institutional analysis including review of all relevant legislation, policy documents and published and unpublished materials; interviews with government officials, local leaders, and other key informants; and primary surveys of selected slum and squatter settlements in 9 cities in China, India, Indonesia, Pakistan and Vietnam to assess views of residents of these settlements and identify barriers to political and social inclusion of the marginalized groups. The survey questionnaire was collaboratively designed by the group of collaborating research institutions in Asia. Findings of the regional survey on access to urban services in Africa by Ibrahim Foundation on Governance in Africa are also included in the book.

This chapter has presented the conceptual framework of the study focused on governance for urban services to promote political and social inclusion. Chapters 2–6 present conceptual and global perspectives on access, participation, multi-stakeholder partnerships, gender equality, and innovations in inclusive urban local governance. Chapters 7–12 include case studies in China, Indonesia, India, Pakistan and Vietnam and a regional survey of access to urban services in African countries. The last chapter presents main conclusions and policy implications of local governance reforms in democratic and one party systems that affect access to urban services in cities, especially for marginalized groups.

Jorrit de Jong and Fernando Monge in Chap. 2 identify four dimensions of access for a comprehensive understanding of how public institutions can influence it: access to policymaking, access to the economy, access to public services, and access to accountable government. They argue that "bureaucratic dysfunction" affects access to benefits especially for citizens from disadvantages groups and that understanding mechanisms though which it leads to inequitable practices requires analysis at the four levels of any given political-administrative system: context, system dynamics, agency performance, and bureaucrat behavior. They present innovations in access to urban benefits from around the world and argue that cities and city leaders have both responsibility and potential to effectively cope with the bureaucratic dysfunction to promote access to urban benefits.

In Chap. 3, Harry Blair examines participatory budgeting (PB) as an engine for accountability. He uses as a lens the World Bank's principal-agent model of state accountability for public service delivery. He discusses the relationship between participatory budgeting and social accountability, the Brazilian origin of participatory budgeting and its implementation India. Based on experience in Kerala and other Indian cities, he argues that PB has been successful in Kerala but its implementation has not been sustainable in other Indian cities. He points out that essential ingredients of success of participatory budgeting are strong state support, CSO engagement, competitive politics and educated populace.

Bharat Dahiya and Bradford Gentry, in Chap. 4, examine public-private partnerships to improve urban environmental services. They discuss why PPPs are needed, how they improve the urban environmental services, key features of successful partnerships and how to use PPPs to improve the urban environment. They argue that despite an increase in the number of projects, the private capital flows to urban water sector declined from US $37 billion during 1990–2000 to US $25 billion between 2001 and 2010. They attribute this to functioning of public sector and political systems; private sector and commercial realities; and opposition to private sector involvement.

In Chap. 5, Annika Björkdahl and Lejla Somun-Krupalija present a theoretical framework to translate global ideas into local practices, with focus on Sustainable Development Goal 5 (gender equality). They present the case study of Bosnia and Herzegovina to analyze institutional mechanisms and tools to change perceptions and behavior, mainstream gender equality processes, and cooperation for gender equality. They argue that obstacles to the implementation of SDG #5 include inadequate political will, lack of funding to implement relevant activities, inadequate awareness of SDGs, patriarchal structures and instruments and ineffective strategies for gender mainstreaming.

Adriana Alberti and Mariastefania Senese, in Chap. 6, examine capacities needed for inclusive and innovative urban governance. They describe two dimensions of the global context: global consensus on the 2030 Agenda for Sustainable Development and rapid pace of urbanization that have led to new challenges and opportunities. Citing specific innovations from around the world, they argue that there is a need and strong potential for a holistic approach to develop urban governance capacities through transformational leadership and changing mindsets of public servants, institutional and organizational innovations, partnership building and citizen engagement, knowledge sharing and management, and promoting "smart" cities.

In Chap. 7, Salim and Drenth describe Indonesia's urbanization and decentralization policies, the slum development programs and agencies, and key features of the two selected cities i.e. Bandung and Surakata. They also present main findings of the survey of slum settlements focused on community organization and participation, access to urban services and overall satisfactions of households concerning organizational performance and institutional features of local governance. They argue that in Indonesia the decentralization policies have facilitated participatory planning, improved access to urban services, and led to innovations and good practices of participation, accountability and transparency. However, there are still barriers to social and political inclusion such as lack of trust between government and communities, diverse political and economic interests, mindset of communities, allocation of about 40% of the city budget on personnel expenditure, and chaotic data management system.

Debolina Kundu, in Chap. 8, examines political and social inclusion and local democracy in India. She presents an overview of urbanization and government policies and programs to improve the access to urban services, the scale of macro level deficits in urban services and housing and micro-level results based on surveys in select slums and squatter settlements in Delhi and Bangalore. She argues that urbanization has been exclusionary in nature leading to inegalitarian distribution of benefits to the urban poor and marginalized groups. The factors that have influenced access to urban services are democratic local governance and NGO participation, the lack of coordinated governance, low capacity and resources of local governments, poor implementation and targeting of government programs and elite capture of services.

In Chap. 9, Bo Qin and Jian Yang examine access of low income residents to urban services in Chengdu, China. They describe patterns of urbanization, and institutional system of urban service delivery, including the *Hukou* System and Residence Permit Reform. They attribute the success of Changdu in providing services to continued reforms to improve public services for both local residents and migrants, rapid economic growth that allowed more investments by the local government, and continued public service assessments. They briefly describe good practices and innovations to improve access, participation and accountability. Qin and Yang argue that improving public services is the focus of governments in transitional China and the quality of basic public services in Chengdu is rapidly improving. However, low-income groups still lack channels for influencing public policy-making. They emphasize value of public leadership and fiscal capacity of local governments as key factors that influence access to urban services.

Nasir Javed and Kiran Farhan, in Chap. 10, describe national and city context of access to services including decentralization policy, deficits in urban services, functional overlaps and coordination, and institutional mapping in Pakistan. They also present findings of household surveys in selected locations. They argue that there are macro-level political, social and institutional barriers to inclusion and access to services and that the role of urban local governments continues to be marginal due to unwillingness of Provincial Governments to decentralize powers and resources to local governments. Yet, progress has been made over the past ten years to promote democratic institutions and that there are a number of good practices and innovations that aim to enhance participation, accountability and transparency at the local level.

In Chap. 11, Nguyen Thanh, Pham Long, and Nguyen Giang discuss the case of Vietnam. They describe national and city context including urbanization and the household registration system, or *hộ khâu*, government decentralization and urban development policies and programs, and institutional mapping and functional mandates and boundaries of the agencies. The authors argue that despite the economic boom after adopting the market economy reforms, Vietnam has remained a highly centralized state, with strict control over the social, economic, and social life of its citizens affecting participation and accountability; state agencies are still the primary public services provider; and access to basic services like electricity, water, health and education is now much better compared to the pre-*Doi Moi (Reform)* era both at the national level and in the selected low-income settlements in Hanoi and Ho Chi Min City. The factors that have contributed to effective performance are human resource capacity, integrated policy framework, political support and coordination of political system.

Camilla Rocca and Diego Fernández, in Chap. 12, examine governance and urban service delivery to serve Africa's citizens. Based on their survey in African countries, they point out that, despite variations among different countries in the region, there are alarming deficits in access to services such as 76% access to electricity, and 42% to basic sanitation. Poor waste management is linked to public health hazards and environmental damage. In 2014, more than half (55.3%) of sub-Saharan Africa's urban population lived in slums. They argue that the picture regarding decentralization remains fragmented; and the level of satisfaction of the public with basic services is declining. They also argue that local governments and urban authorities face several obstacles related to their relations with central governments, their autonomy and capacity, and their ability to bring government closer to citizens to improve transparency and accountability.

In the last chapter, Shabbir Cheema presents conclusions and policy implications concerning local governance and access to urban services. He discusses the extent of urban service deficits in Asia and Africa and citizen's perceptions about government programs based on household surveys. He identifies eight sets of factors and institutional reforms to fully utilize the potential role of cities in promoting political and social inclusion and access of residents of low income urban settlements to urban services: local government resources and capacity, agency overlaps and coordination, local participatory mechanisms, accountability and transparency, engagement of migrants, women and minorities, innovations, information and communication technologies, and peri-urbanization.

## References

Agrawal A, Ribot J (1999) Accountability in decentralization. J Dev Areas 33:473–503
Alagappa M (2004) Civil society and political change in Asia: expanding and contracting democratic space. Stanford University Press, Palo Alto, CA
Alkire S, Roche JM, Seth S, Sumner A (2015) Identifying the poorest people and groups: strategies using the global multidimensional poverty index. J Int Dev 27(3):362–387. https://doi.org/10.1002/jid.3083
Andrews M (2008) The good governance agenda: beyond indicators without theory. Oxf Dev Stud 36(4):379–407. https://doi.org/10.1080/13600810802455120
Ascher W, Brewer G, Cheema GS, Heffron J (2016) The evolution of development thinking: governance, economics, assistance and security. Palgrave Macmillan, London
Asian Development Bank (2011) Asia 2050: realizing the Asian century. Asian Development Bank, Manila
Blair H (2000) Participation and accountability at the periphery: democratic local governance in six countries. World Dev 28(1):21–39
Blunt P, Jones ML (1992) Managing organisations in Africa. Walter de Gruyter, Palo Alto, CA
Campbell T (2013) First person, virtual and collective: the new participation in rising cities. In: Cheema GS (ed) Democratic local governance. United Nations University Press, Tokyo
Carothers T (1999) Civil society. Carnegie Endowment for International Peace, Washington DC
Carothers T, Brechenmacher S (2014) Accountability, transparency, participation and inclusion: a new development agenda?. Carnegie Endowment for International Peace, Washington DC
Chambers R (1995) Rural development: putting the last first. Prentice Hall, London
Cheema GS (ed) (1993) Urban management: policies and innovations in developing countries. Praeger Publishers, New York
Cheema GS (2005) Building democratic institutions: governance reform in developing countries. Kumarian Press, Bloomfield, CT
Cheema GS (2013) Democratic local governance: reforms and innovations in Asia. United Nations University Press, Tokyo
Cheema GS, Popovski (2010) Engaging civil society. UN University Press, Tokyo
Cheema GS, Rondinelli DA (eds) (2007) Decentralizing governance: emerging concepts and practices. Brookings Institution Press, Washington DC
Cohen J, Arato A (1994) Civil society and political theory. MIT Press, Cambridge, MA
Crook R, Manor J (1998) Democracy and decentralization in south asia and west Africa: participation, accountability and performance. Dev Dialogue 26(2):114–120
Dahiya B, Das A (2019) The new urban agenda in Asia. Springer, Singapore
De Jong J, Rizvi G (2008) The state of access. Brookings Institution Press, Washington DC
Doeveren V (2011) Rethinking good governance. Public Integrity 13(4):301–318. https://doi.org/10.2753/pin1099-9922130401
Dryzek J, Dunleavy P (2009) Theories of democratic state. Palgrave Macmillan
Dunleavy P, Hood C (1994) From old public administration to new public management. Public Money Manage 14(3):9–16
European Commission (2004) Joint report on social inclusion. Report 7101/04. European Commission, Brussels
Fainstein SS (2001) Competitiveness, cohesion, and governance: their implications for social justice. Int J Urban Reg Res 25(4):884–888. https://doi.org/10.1111/1468-2427.00349
Falconer P (1997) Public administration and the new public management: lessons from the UK experience. Linea. http://www.Vus.Uni-Lj.si/Anglescina/FALPOR97.Doc
Freedom House (2002) Freedom in the world. Freedom House, Washington DC
Freedom House (2016) Freedom in the World. Washington DC
Freedom House (2017) Freedom in the world. Freedom House, Washington DC
Fung A (2004) Empowered participation: reinventing urban democracy. Princeton University Press, Princeton, NJ

Grindle M (2004) Good enough governance: poverty reduction and reform in developing countries. Governance 17(4):525–554
Grindle M (2007a) Going local: decentralization, democratization, and the promise of good governance. Princeton University Press, Princeton, NJ
Grindle M (2007b) Local governments that perform well. In: Cheema GS, Rondinelli DA (eds) Decentralizing governance: emerging concepts and practices. Brookings Institution Press, Washington DC
Habitat III (2017) HABITAT III website: http://habitat3.org/the-new-urban-agenda
Hayes A, Gray M, Edwards B (2008) Social inclusion: origins, concepts and key themes. Australian Institute of Family Studies, prepared for the Social Inclusion Unit, Department of the Prime Minister and Cabinet, Canberra
Hildebrand M, Kanaley T, Roberts B (2013) Sustainable and inclusive urbanization in Asia-Pacific: strategy paper. UNDP policy paper, New York, pp 24–28
Hyden G, Samuel J (eds) (2011) Making the state responsive: experience with democratic governance assessments. UN Development Programme, New York
Katz B, Nowak J (2017) The new localism: how cities can thrive in the age of populism. Brookings Institution Press, Washington DC
Kaul I, Grunberg I, Stern M (eds) (1999) Global public goods: international cooperation in the 21st century. Oxford University Press, New York
Khator R (1998) The new paradigm: from development administration to sustainable development administration. Int J Public Adm 21(12):1777–1801
Leftwich A (1995) Bringing politics back in: towards a model of the development state. J Dev Stud 31(3):400–427
Manor J (1999) The political economy of democratic decentralization. World Bank, Washington DC
Mayer M (2003) The onward sweep of social capital: causes and consequences for understanding cities, communities and urban movements. Int J Urban Reg Res 27(1):110–132. https://doi.org/10.1111/1468-2427.00435
Mitlin D (2008) With and beyond the state—co-production as a route to political influence, power and transformation for grassroots organizations. Environ Urbanization 20(2):339–360. https://doi.org/10.1177/0956247808096117
Moser C (1998) The asset vulnerability framework: reassessing urban poverty reduction strategies. World Dev 26(1):1–19. https://doi.org/10.1016/S0305-750X(97)10015-8
Ojendal J, Dellnas A (eds) (2013) The Imperative of good local governance. United Nations University Press, Tokyo
Rajan R (2019) The third pillar: how markets and state leave the community behind. Penguin Books, New York
Ribot J, Peluso N (2003) A theory of access. Rural Sociol 68(2):153–181
Rodrik D (1997) Has globalization gone too far?. Institute for International Economics, Washington DC
Rondinelli D (1990) Financing the decentralization of urban services in developing countries: administrative requirements for fiscal improvements. Stud Comp Int Dev 25(2):43–59
Rondinelli DA, Cheema GS (eds) (2003) Reinventing government for the twenty first century: state capacity in a globalizing society. Kumarian Press, Bloomfield, CT
Rydin Y, Pennington M (2000) Public participation and local environmental planning: the collective action problem and the potential of social capital. Local Environ 5(2):153–169. https://doi.org/10.1080/13549830050009328
Satterthwaite D (2016) A new urban agenda? Environ Urbanization 28(1):3–12. https://doi.org/10.1177/0956247816637501
Sen A (1999) Development as freedom. Oxford University Press, London
Shah A (ed) (2006) Local governance in developing countries: public sector governance and accountability series. World Bank, Washington DC

Sisk J (2005) Creating and applying knowledge, innovation and technology. In: Rondinelli GS, Cheema DA (eds) Reinventing government for the twenty-first century. Kumarian Press, Bloomfield, CT
Smith BC (1985) Decentralization: the territorial dimension of the state. Allen and Unwin, London
Stewart A (2000) Social inclusion: an introduction. In: Askonas P, Stewart A (eds) Social inclusion: possibilities and tensions. Macmillan, London
Stiglitz J (2003) Globalization and its discontents. W.W. Norton, New York
Storper M (2014) Governing the large metropolis. Territory Polit Gov 2(2):115–134. https://doi.org/10.1080/21622671.2014.919874
Swedish International Center for Local Democracy (2015) Results-based management. Swedish International Center for Local Democracy, Stockholm
Swedish International Center for Local Democracy (2017) Democratic local governance indicators: how do we measure results of ICLD support?. Swedish International Center for Local Democracy, Stockholm
Tendler J (1997) Good governance in the tropics. Johns Hopkins University Press, Baltimore, MD and London
Turner M, Hume D (1997) Government, administration and development: make the state work. Macmillan Press, London
UNDP (2004) Human development report. UNDP, New York
UN-HABITAT (2013) The state of world's cities 2012/2013: prosperity of cities. UN-Habitat, Bangkok
UN-HABITAT/ESCAP (2011) The state of Asian cities. UNESCAP, Bangkok
United Nations (1987) The world summit on environment and development. United Nations, New York
United Nations (2016) The Sustainable development goals report. United Nations, New York
United Nations Development Programme (1990) Human development report 1990. Oxford University Press, New York, New York
United Nations Development Programme (1995) Human development report 1995. Oxford University Press, New York
United Nations Development Programme (1997) Governance for sustainable human development: a policy paper. Management Development and Governance Division, Bureau for Development Policy, New York
United Nations Development Programme (1999) Human development report 1999. Oxford University Press, New York
United Nations Development Programme (2001) Making new technologies work for human development. http://www.undp.org/hdr2001/
United Nations Development Programme (2002) Deepening democracy in a fragmented world. Oxford University Press, New York
United Nations Human Settlements Programme (2004) Tools to support transparency in governance. UN-HABITAT, Nairobi
United Nations Population Division (2012) World Urbanization Prospects: The 2011 Revision
Varieties of Democracy (2016) Global report 2016. Stockholm
Varieties of Democracy (V-Democracy) (2018) Annual democracy report: democracy for All? Varieties of Democracy, Stockholm
White G (1994) Civil society, democratization and development: clearing the analytical ground. Democratization 1(2):375–390
Wintrop N (1983) Liberal democratic theory and its critics. Croom Helm, London
Work R (2003) Decentralizing governance: participation and partnership in service delivery to the poor. In: Rondinelli DA, Cheema GS (eds) Reinventing government for the twenty-first century: state capacity in a globalizing society. Kumarian Press, Bloomfield, CT
World Bank (1997) The State in a changing world. World Bank Annual Report, Washington DC
World Bank (2010) Development and climate change: World Bank annual report. World Bank, Washington DC

World Bank (2011) Vietnam urbanization review. World Bank, Hanoi, Vietnam
World Bank (2013) Planning, connecting and financing cities. World Bank Annual Report 2013, World Bank, Washington DC

**Shabbir Cheema** is Senior Fellow at Harvard Kennedy School's Ash Center for Democratic Governance and Innovation. Previously, he was Director of Democratic Governance Division of United Nations Development Programme (UNDP) in New York and Director of Asia-Pacific Governance and Democracy Initiative of East-West Center in Hawaii. Cheema prepared the UNDP policy papers on democratic governance, human rights, urbanization and anti-corruption and provided leadership in crafting UN-assisted governance training and advisory services programs in over 25 countries in Asia, Africa, Latin America and the Arab region. He was Program Director of the Global Forum on Reinventing Government and the Convener of the Harvard Kennedy School's Study Team of Eminent Scholars on Decentralization. He has taught at Universiti Sains Malaysia, University of Hawaii and New York University. As the UN team leader, he supported the International Conference on New and Restored Democracies, the Community of Democracies and UN HABITAT II. He has undertaken consultancy assignments for Asian Development Bank, the World Bank, U.S. Agency for International Development, Swedish International Development Agency, Dubai School of Government and United Nations. He holds a Ph.D. in political science from the University of Hawaii. He is the co-author of *The Evolution of Development Thinking: Governance, Economics, Assistance and Security* (Palgrave Macmillan 2016) and the author of *Building Democratic Institutions: Governance Reform in Developing Countries* (Kumarian Press, 2005) and *Urban Shelter and Services* (Praeger 1987). He is the contributor and co-editor of the four-volume Series on *Trends and Innovations in Governance* (United Nations University Press, 2010); *Decentralizing Governance: Emerging Concepts and Practices* (Brookings Institution Press in cooperation with Harvard University, 2007); *Reinventing Government for the Twenty First Century: State Capacity in a Globalizing Society* (Kumarian Press, 2003) and *Decentralization and Development* (Sage Publications, 1984). Cheema has been a member of the advisory committees of the Swedish International Center for Local Democracy, UNHABITAT III, and the Pacific Basin Research Center and editorial boards of Urbanization and Environment and Third World Planning Review. A featured speaker at global and regional forums, he served as an advisor to the Dubai School of Government, Pakistan Institute for Economic Development, the Malaysian Academy for Leadership in Higher Education, and the UN Governance Center in Seoul, Korea.

# The State of Access in Cities: Theory and Practice

**Jorrit de Jong and Fernando Fernandez-Monge**

**Abstract** This chapter presents a framework to analyze access to rights and services in urban settings. Following De Jong and Rizvi's (2008) definition of access as *the match between societal commitment and institutional capacity to deliver rights and services and people's capacity to benefit from those rights and services*, the chapter examines the different dimensions that underpin access in urban settings. It argues that efforts to deal with the bureaucratic dysfunction that impedes access should be grounded in an approach that looks at context, system, agency and individual levels of analysis. Such conceptual approach highlights the adaptive nature of dealing with bureaucratic dysfunction to enlarge access to urban benefits, putting an emphasis on the role of leadership in innovating to make it possible. The chapter tests these propositions by examining examples of recent innovations to manage bureaucratic dysfunction and associated lack of access from cities across the world. Some lessons are drawn from the analysis: (i) leaders who can articulate the public value proposition, can enable the necessary legitimacy and can build operational capacity are a fundamental pillar of any effort, (ii) focusing in an agency or a narrow set of agencies may leave key stakeholders out, rendering efforts to increase access unsustainable, and (iii) engaging frontline workers has to be a central part of any effort, but it cannot fail to act at the context and societal level, so that the deeper forces inhibiting access to urban benefits are deactivated in the long term.

---

This chapter is an adapted and expanded version of previous work by Jorrit de Jong, in particular his books *Dealing with Dysfunction: innovative problem-solving in the public sector* (2016), *The State of Access* (with G. Rizvi, 2008) and *Agents of Change: Strategy and Tactics for Social Innovation* (with S. Cels and F. Nauta, 2012).

---

J. de Jong (✉)
Ash Center for Democratic Governance and Innovation, Harvard Kennedy School, Harvard University, 79 John F. Kennedy Street, Mailbox 74, Cambridge, MA 02138, USA
e-mail: jorrit_dejong@hks.harvard.edu

F. Fernandez-Monge
Ash Center for Democratic Governance and Innovation, Harvard Kennedy School, Bloomberg Harvard City Leadership Initiative, 79 John F. Kennedy Street, Mailbox 74, Cambridge, MA 02138, USA
e-mail: fernando_monge@hks.harvard.edu



S. Cheema (ed.), *Governance for Urban Services*,
Advances in 21st Century Human Settlements,
https://doi.org/10.1007/978-981-15-2973-3_2

**Keywords** Access · Urban rights and services · Bureaucratic dysfunction · Public institutions · Innovation · City leadership

## 1 A Theory of Access in Cities

The year 2000 not only inaugurated a new millennium; it also marked a shift to a completely different world. For the first time in the history of humankind, people living in urban areas outnumbered their rural neighbors. If current trends continue, by 2050, for every person living in the countryside there will be three living in cities. It is fair to say that the future of humankind will likely be urban.

Urbanization has many champions touting the benefits and opportunities associated with proximity and cross-cultural exchange (Glaeser 2012; Katz and Nowak 2018). They observe that, throughout history, people have clustered in cities to exchange goods and ideas and reaped the benefits of interactions that more spread-out human settlements did not offer. Through such closeness, the story goes, cities reduce costs, foster collective and commercial enterprise, and provide access to high-quality jobs and services.

While there is truth in this story, many who live in or move to cities to reap the benefits of urban life are excluded from economic opportunity, public services, and opportunities to participate in policymaking or hold their governments accountable. Instead they experience the darker side of urbanization—the pollution, crime, congestion, social exclusion, and poverty that have also characterized cities around the world throughout history. Economic inequality is often more pronounced in cities than in the countries in which they are located in CAF and UN-Habitat (2014), with about one in three urban dwellers worldwide living in slum-like conditions (UN-Habitat 2010). This has led some commentators to talk about the "urban divide" affecting many cities, or even warn of a "new urban crisis" (Florida 2017).

If cities are to realize their promise, their benefits need to accrue more equitably. This was a theme of the New Urban Agenda, the primary international roadmap for sustainable urban development (Habitat III 2017). If we, as scholars and practitioners of urban policymaking and leadership, want to make sure cities are zones of opportunity rather than exclusion, we need to invest in access.

We define access as *the match between societal commitment and institutional capacity to deliver rights and services and people's capacity to benefit from those rights and services* (De Jong and Rizvi 2008). Like Ribot and Peluso (2003), who conceptualize access as "the ability to benefit from things," we acknowledge that having the right to something does not guarantee access, and that factors like knowledge and social group affiliations affect people's level of access. Our definition of access adds to this the idea that the "supply side" of the equation is equally important. Ensuring access to the benefits of city life requires not only the demand from city dwellers, but also the societal commitment and institutional capacity to deliver rights and services.

The purpose of this chapter is to introduce two frameworks that will enable scholars and practitioners to better diagnose the state of access in cities. First, we provide an overview of the different dimensions of access in cities. This first exercise shows that public institutions are key to enabling or providing access to city benefits in all its underpinning dimensions, and that bureaucratic dysfunction is at the core of the lack of access in cities. Second, we analyze how public institutions impede access by looking at four different layers: context, system dynamics, agency performance, and bureaucrat behavior. This framework helps us dissect the relationship between bureaucratic dysfunction and access in cities.

Despite this somewhat grim analysis, we remain true to our stubborn optimism and conclude with examples of innovations that are deactivating bureaucratic dysfunction and increasing access in cities across the world.

## 2 Accessing Urban Benefits

Our concept of urban benefits goes well beyond public services such as water and sanitation. We understand urban benefits as all the perks that come with living in cities. Some of these are material, such as better-paying jobs and amenities, but others, such as the joy of social connection or the fulfillment derived from being engaged in meaningful political endeavors, are not.

At their best, cities are the cultural centers where people can meet to enjoy conversation over the latest in fancy cafés, or take in a play in a theater, a concert in the street, or paintings in a museum. They are economic engines, bustling markets not only of goods and services, but also of jobs. They are breeding grounds for public political participation, home to political groups of every description, open town halls, protests and rallies aplenty, and myriad opportunities for citizens to interact with policymakers. They allow for a diversity of ideas and identities, and their residents are generally more tolerant than their suburban and rural peers. It is not by coincidence that almost all political revolutions have started in city squares.

In this sense, many of the benefits cities provide or enable are genuinely urban or are made more accessible—at least in theory—by the density of city populations. At the same time, physical proximity is only one dimension, and arguably not the most important one, of accessibility to these benefits. Societal commitment and institutional capacity to deliver or enable these benefits may vary broadly. Some communities may agree that providing free education is a shared goal but lack institutional capacity to realize it. In other contexts, there might not be agreement in the community about whether everyone should receive certain services that institutions do have the ability to deliver. Understanding these tensions is important for identifying the forces that may be limiting access in cities (De Jong and Rizvi 2008).

It is intermediary institutions that provide many of the benefits of city life. These institutions may be public (e.g., health services in a public hospital), private (e.g., jobs in a company) or nonprofit (e.g., a free concert organized by a philanthropic institution). As many scholars have noted (Kettl 2002; Stoker 2006), the shift from government to governance has engaged a constellation of actors in work formerly monopolized by governments. This transition has been particularly salient in cities, with urban public institutions operating as platforms or nodes in a network of benefit providers (Goldsmith and Eggers 2005). Notwithstanding this general trend, all these intermediaries operate within a public-sector framework. Whether as direct deliverer or regulator, the public sector sets the rules that determine how, where, and to what extent these benefits show up in a city.

Decentralized city policies and decisions, combined with chronic dysfunctions in the old institutions of city government can create or exacerbate access problems. Practitioners and observers in different policy domains have observed this phenomenon. For example, urban mobility specialists have identified the increase in automobile-centered policies and infrastructure in Africa and Asia as a problem that may not only harm the poor, but also create path dependence in urban infrastructure and built environments with long-lasting consequences. The problems of car-based policies as well as the alternative approaches to improved urban mobility are well-known, but when they turn to providing solutions for policymakers, experts recognize that the problem does not lie in "*what* to do but on *how* to do it" (Sclar et al. 2014).

Considering all this, lack of access looks like an adaptive problem rather than a technical one (Heifetz 1994). It has more to do with tackling the barriers of bureaucratic dysfunction (De Jong 2016) than with designing technically impeccable policies. That work will require a more comprehensive understanding of access and the dimensions that underpin it: access to policymaking, access to the economy, access to public services, and access to accountable government (De Jong and Rizvi 2008).

## *2.1 Access to Policymaking*

People's ability to access the benefits of city life are a function of their ability to shape the policymaking process and its outcomes. As noted by Acemoglu and Robinson (2012), inclusive public institutions are fundamental to economic development, while those controlled by self-interested elites pose a barrier to communities' wellbeing. If the diverse groups populating the city, with their various goals, interests, and perspectives, do not have a say in decisions over budget allocation, investment in public services, or the regulation of economic activities, this will undermine their access to city benefits. When one identity group dominates city government, whether that identity is economic, ethnic, religious, racial, or something else, the outcome is generally a divided and unequal city. The history of segregation in many US cities are a good proof of this perverse link (Trounstine 2018).

Democratic processes (e.g. elections) and institutions (e.g. city councils) are cities' central mechanism for increasing the base of those who have a say in how benefits are created and shared. Through these, individuals express their preferences and values. Democracy, however, is not perfect nor it will be, and its promise can never be realized if people do not exercise their most basic right: voting. A study looking at 144 large US cities and 340 separate mayoral elections calculated a mean participation of 25.8%

of eligible voters (Holbrook and Weinschenk 2014), and other research finds even lower turnout rates (Portland State University 2016). Participation is not only low, but also unevenly distributed. Minority, low-income, and young citizens vote less on average, and their lack of electoral participation disadvantages them even further. Uneven turnout by these groups results in fewer redistributive policies in favor of "non-voting" groups (Hajnal 2009). The flaws of the formal electoral system also highlight the importance of the role of civil society and the media in improving access to policy-making structures.

Access to policymaking is not limited to democratic institutions, however, and other mechanisms for expressing opinions and exercising influence have appeared in cities around the world. Participatory budgeting, for example, which started in Porto Alegre, Brazil, in 1989, is now a relatively common practice. Many other cities across the world have experimented with interesting models of direct and deliberative democracy, often using digital tools (World Bank 2016).

## 2.2 Access to the Economy

Cities are places of economic opportunity. Ed Glaeser has written widely about the advantages cities provide for companies and workers. He has documented, for instance, how workers in metropolitan areas with populations above 1 million earn on average 30% more than workers in rural areas in the US (Glaeser 2012). These benefits, however, have not been equally distributed across cities (Moretti 2012) nor within cities (Florida 2017).

Improving access to the urban economy starts with guaranteeing the right to start a business, find a job, or raise capital.[1] To open a business, entrepreneurs need to acquire licenses and permits from several city government departments. Complicated regulations, corruption, or costs can make obtaining those licenses extremely burdensome. Most attempts at regulatory reform, however, have focused on the macro-level and neglected the micro-level economy operating in urban neighborhoods.

City governments' approach to labor regulations, local taxation, and land management also have implications for economic access. The ability to raise capital is key to launching a business. Few people have enough savings to make the investments required to set up a business. Typically, access to financing is secured with assets, especially immovable assets like land or buildings. Banks and financial institutions will accept those guarantees only if there is legal security of ownership. This, in turn, depends on government's ability to keep an orderly and clear registry of land ownership, uses, and transactions. Such systems are seldom the norm in emerging countries, compromising citizens' ability to raise capital as well as the secure and safe tenure of housing.

[1]In the case of the urban poor, it starts with access to even more basic assets such as housing and shelter.

These are just a few examples of how the policies and practices of city governments affect access to the economy. If bureaucratic dysfunction cripples the government's ability to regulate activities, act as register, or tax property and economic activity, citizens' access to the economy suffers.

## *2.3 Access to Public Services*

"The faces of limited access are perhaps most visible in the delivery of public services" (De Jong and Rizvi 2008). For many people, public forms of transport, health, and education are the only way to cover these basic needs, and without them, cities can quickly turn from places of opportunity to cages of marginalization and poverty. Failures in the delivery of these services are among the main drivers of inequality in many cities across the world.

Corruption, poor performance, silo mindsets, and unresponsiveness are familiar to many citizens, and also cited as key problems by city officials around the world (Cruz et al. 2019). Common wisdom presumes these flaws in public-service bureaucracies, but private and non-profit service providers exhibit them too. Indeed, as we will later explore, the surge of public-sector innovations over the past decade suggests that notions of the public sector as a uniquely bureaucratic and dysfunctional realm are, at the very least, exaggerated (Altshuler et al. 1997; Borins 2001; De Vries et al. 2016).

Real or perceived pitfalls in the institutional capacity of governments are not the only reason some people do not have access to public services. A fuller picture of the problem includes the consequential fact that different communities and groups have different ideas about the right level and mix of public services to be delivered. At the same time, citizens' misconceptions, fear, or lack of knowledge may prevent them from using some services for which they are eligible and from which they could benefit. This problem is more common among disadvantaged groups—precisely those who need public services most. And this, of course, further aggravates inequality and entrenched poverty in cities.

## *2.4 Access to an Accountable Government*

The final issue underlying access to city services and benefits reminds us that a narrow focus on the institutional or managerial aspects of public institutions will produce only a partial account of reality and, therefore, unsatisfactory solutions to access problems. Because institutions and their managers have systemic biases and limited knowledge, they need feedback about their performance from citizens (De Jong and Rizvi 2008).

Creating accountability mechanisms is closely tied to the question of citizens' access to policymaking beyond elections. If citizens can only hold politicians accountable once every election cycle, the feedback loop is very limited. Not only are the intervals too long, but voting is also an inadequate way to express which specific aspects of a government (departments, policies, etc.) pleased them and which disappointed them. Responsive city governments need other mechanisms of accountability to learn about their performance and ensure the legitimacy of their activities.

As the closest level of government, cities have implemented instruments to widen the availability of accountability mechanisms. Easy-to-remember telephone hotlines (such as 311 in the US) citizens can use to ask questions and register complaints became a symbol of accessible and responsive local governments at the turn of the millennium. In the digital era, these hotlines are evolving into platforms where citizens can interact with the government and among themselves using voice, text, social media, and other apps (Goldsmith and Crawford 2014). New digital instruments are enhancing accountability by increasing the number of channels through which citizens can co-produce public outcomes, but also generating large amounts of data that, if publicly available, can allow researchers and activists to analyze the workings of city institutions.

## 3 Determinants of Access

Now that we know all the dimensions of access to city benefits that public institutions can influence, we can consider the nature of their failures to make progress on the issue. When city governments fail to notice and address structural barriers to access, they enter the domain of bureaucratic dysfunction. In this section, we offer an analytical framework to further understand how the performance of public institutions determines the state of access in cities.

Bureaucratic dysfunction affects access by creating unnecessary burdens, wasting citizens' time and money, and producing arbitrary decisions about who gets benefits and who does not. In the process, it erodes public trust in government (Peters 1995). As we have seen, the burdens of dysfunction are unequally distributed. Citizens from disadvantaged groups often lack the knowledge, time, and resources to navigate "red tape," further exacerbating inequality in outcomes and uptake of benefits and services.

Understanding the mechanisms through which bureaucratic dysfunction leads to these inequitable practices and restricts access requires analysis at the four levels of any given political-administrative system: context, system dynamics, agency performance, and bureaucrat behavior (De Jong and Rizvi 2008).

### *3.1 Context*

Any analysis of the determinants of access that focuses exclusively on public institutions' capacity to deliver services will likely miss important factors that help explain why some cities have higher levels of access than others. Because this view also therefore overlooks potential solutions to access problems, we start with a look at broad contextual factors that affect societal commitment to access and people's ability to derive benefits.

A city's history, culture, social norms, local power structures or religious traditions establish the canvas on which the goals communities define for themselves are drawn. In a city where tribal affiliations determine political affiliation but also property rights, ideas underlying access to resources (e.g., a well) will be radically different from the individualistic and rights-based ideas of public life and the commons seen in a European city, for example.

Contextual factors also shape citizens' ability to benefit from services and entitlements (Ribot and Peluso 2003). Even when proscribed by law, ethnicity, caste, and race are still *defacto* bases for discrimination in employment, bank services, and public service delivery. Socio-economic background and neighborhood residency also factor into the likelihood individuals will be recruited for political participation (Strömblad and Myberg 2012).

## 3.2 System Dynamics

The next level of analysis, between contextual factors and agency performance, is system dynamics. Most theorists and practitioners now recognize that access to benefits is no longer (if it ever was) solely the responsibility of governments (Goldsmith and Eggers 2005; Kamarck 2007; Kettl 2002). Provision of even the most basic public services involves a constellation of actors from public, private, and voluntary sectors connected through new governance mechanisms (Torfing et al. 2012). These systems add interorganizational networks, regulated markets, and voluntary initiatives to the traditional command-and- control systems of decision making and service delivery.

Understanding the determinants of access at this level of analysis requires consideration of at least two things. The first is the usual disconnect between collective responsibility and organizational autonomy. While network partners may agree to deliver a service or enable a benefit collectively, coordination of that effort is hardly guaranteed. Whether poor management, information asymmetries, or tensions and uncertainties around turf or decision-making authority (or all of the above) are behind the lack of coordination, it tends to be poor and underprivileged people who fall between the cracks. We have seen social welfare programs in large cities leave entire disadvantaged groups out of the system because all partners believed someone else was taking care of them (De Jong and Rizvi 2008).

The second, related point to consider is that even when partners in a public service delivery network establish strong accountability regimes among themselves, accountability to *the public* is diffused and dispersed across the network. It is hard to say what entity, ultimately, is responsible for an access failure. The weak visibility and selective composition of the networks makes them less accountable to citizens (Papadopoulos 2007). Studies at the local level have also found that networks' decision-making processes can shield them from public control through political mechanisms, or reduce the influence of elected or politically-appointed public officials with responsibilities in relevant policy fields (Aars and Fimreite 2005). This latter issue can lead to a lack of interest or engagement in certain network activities and their impact on the part of public agencies (Bogdanor 2005), leaving citizens with no recourse to influence those decisions.

## 3.3 Agency Performance

Structures, strategies, tactics, and culture at the agency level are another major factor determining whether certain groups will enjoy access to services and entitlements or not (De Jong and Rizvi 2008). Traditionally organizations, whether public, private, or nonprofit, operate under certain design principles: formalization (no informal transactions); standardization (equal treatment regardless of individuals' status); specialization (division of labor); hierarchy (an accountable chain of command); expert officialdom (hiring based on merit); and, particularly in the public sector, distinction between public and private spheres (protecting the public interest and preventing the use of public powers for private gains) (Albrow 1970; Etzioni-Halevy 2012). These bureaucratic principles served as a basis of legitimacy and effectiveness for public agencies, but some of them came under increasing criticism in the early 1990s, giving way to new market arrangements in public service delivery (Howard 1995; Osborne and Gaebler 1992). Some of these reforms also tried to tackle the juridical and budgetary constraints, scrutiny by the media and high expectations from politicians and the public to which these organizations were subject to (Wilson 1989), by generating semi-autonomous agencies at arms-length from politicians (Pollitt et al. 2004).

At the core of these criticisms is the inability of bureaucratic forms to deal with new challenges in innovative ways, and each of these bureaucratic design principles can indeed pose a challenge to innovation (Cels et al. 2012):

- *Formalization*. While bureaucracies seek to formalize all aspects of process, innovation requires thinking outside of familiar forms, and the creative process is hard to document or fit within pre-existing forms.
- *Standardization*. Standardized procedures across bureaucratic organizations make disruptions of the status quo—the essence of innovation (Sørensen and Torfing 2011)—undesirable, and risky for workers.
- *Specialization*. Dividing tasks and resources among departments may increase efficiency, but when solving problems requires working across organizational boundaries, compartmentalized agencies struggle to design new solutions that bring together different capabilities and perspectives.
- *Hierarchy*. Clearly defined authority lines and structures that emphasize compliance make it difficult for frontline or middle managers to put innovative ideas about how to tackle challenges into practice.

- *Expert officialdom.* Bureaucratic officials are experts in the technical and legal application of rules, but this can lead them to prioritize following rules over solving problems and can reduce their organizations' creativity in responding to changing contexts and problems.
- *Distinction between public and private spheres.* A strict separation between the private domain of officials and their public duties may discourage them from bringing valuable opinions or relevant interests to their work, or from designing solutions that could possibly be interpreted as advancing a private interest.

Despite this awkward relationship between bureaucratic structures and innovation, one problem with the criticisms of bureaucracy is that they give more attention to the design characteristics of organizations than to the values in which they are rooted (De Jong 2016). The values associated with these design characteristics are rational decision making, integrity, effectiveness, efficiency, transparency, accountability, and fairness. This confusion between design principles and underlying values has also resulted in inadequate evaluations of whether the new forms of public service administration have resulted in greater innovation or increased citizens' access to benefits (De Jong and Rizvi 2008).

## *3.4 Bureaucrat Behavior*

Public agencies are essential determinants of access to benefits, but they cannot operate without the individuals working within them. Frontline public workers are the ultimate level of interaction between public organizations and citizens—both as r*ecipients* of services and as *obligatees* under regulations (Moore 1995; Sparrow 1994). These street-level bureaucrats develop strategies to cope with limited resources, ambiguous policies, demands for accountability, and the complex lives and needs of clients on a daily basis (Lipsky 1980a, b). These coping strategies by frontline bureaucrats can hamper access in a variety of ways (De Jong 2016; Lipsky 1980a, b):

- *Imposing costs of services on clients.* The costs (not only in monetary terms but also in terms of time, etc.) that clients have to bear to obtain certain services can decrease demand and limit access, in particular for those without time and money to spare.
- *Withholding information.* Individuals working in agencies can decide in which way, to what extent, and to whom they make information available. Because the availability of information determines the knowledge of potential beneficiaries, it also determines their access to goods and services.
- *Psychological strategies.* Psychological nudges can be built into intake interviews, control mechanisms, or general attitudes of frontline workers. Clients may withdraw if these strategies affect their perceptions or feelings.

- *Queuing*. For clients in great need, or who have nowhere else to turn for services, first come, first serve policies compound waiting costs associated with accessing services.
- *Categorization*. Determining a set of characteristics for use in processing clients, and mapping clients in terms of some qualifying or disqualifying characteristics often leads to prioritizing clients who are most likely to succeed in terms of bureaucratic success indicators (and therefore least likely to experience access barriers).
- *Worker bias*. For all the emphasis on standardization, prejudice or personal values and beliefs may still influence the ways in which frontline workers use their discretionary powers. This can lead to bias and unequal access for certain groups.

Sometimes, the reluctance to use the leeway afforded by discretion (Kruiter and De Jong 2008) can also become an important barrier to access. Frontline workers may not want to deal with the uncertainty, anxiety, and vulnerability they associate with change, and therefore consciously or unconsciously interpret rules more strictly than written or intended. This is where leadership comes into the picture. Public leaders can reframe the issues and bring together the resources that are required to resolve bureaucratic dysfunction and remove or respond to barriers to access at the level of context, system, organization, and individual. The power to expand access to the benefits of city life ultimately lies in skillful local leadership.

## 4 Bureaucratic Dysfunction and Local Leaders as Change Agents

By now, it should be clear that, in our view, one cannot understand access in cities without understanding bureaucratic dysfunction in city government. Bureaucratic flaws deprive people of their rights and benefits, sometimes on purpose and sometimes by omission, in what Michael Lipsky called "bureaucratic disentitlement" (1980b). These are the people whose perspectives we adopt as we consider the role of leadership in overcoming access barriers produced by bureaucratic dysfunction. Mirroring our definition of access, we characterize bureaucratic dysfunction as *the mismatch between state capacity to deliver services and enforce regulation and people's capacity to benefit from public services and the enforcement of rules and regulations as individual clients or citizens in general or both* (De Jong 2016).

Bureaucratic dysfunction can have different impacts depending on the citizen's position vis a vis the government. As recipients, they can have trouble accessing services or obtaining benefits. As *obligatees*, they may find it difficult or costly to comply with certain laws or regulations. The real-life manifestations of these flaws can be found in long waiting times, cumbersome procedures or incomprehensible paperwork, among others. Value losses for both individuals and the public associated with dysfunction often go unacknowledged. And even if recognized, it may also be hard to connect the root of the problem (within the system or the agency, for example) to the incurred cost. Identifying problems and understanding the cost of either solving or not solving them is especially difficult in a city context, where many agencies and institutions coproduce services and entitlements. For example, one cannot think about tackling homelessness—one of the most extreme forms of lack of access—without looking at all of the potential dimensions of the issue and all of the agencies and entities involved (health providers, shelters, police, etc.).

In these multi-sector, multi-department, multi-agency contexts, each organization typically works on the problem from its own perspective, and progress stalls because nobody steps up to redefine the problem and rally everyone around a new, shared understanding. To overcome silo mentalities, someone has to authorize those working on the problem to take a step back, see the bigger picture, and come back and redesign approaches. That requires leadership.

Dealing with bureaucratic dysfunction in cities calls for leaders—mayors and senior members of local governments—who can secure legitimacy and support, build or organize sufficient operational capacity, and make a compelling public value proposition (Moore 1995). Those leading organizations in this kind of strategic action have to build and provide enough political cover for their frontline workers to feel authorized to exercise their discretion to pull up barriers to access. Leaders also have to convince other stakeholders in the system that the change is necessary and will be safe, and gather, structure and align the resources and capacities needed to tackle the challenge. This almost always means attracting some skills and resources to the organization, rearranging others to direct them towards the goal, or engaging in and strengthening partnerships. None of this will be feasible without leaders able to articulate clearly the status of the problem, why and how it should be solved, and what value will be generated for the community.

This reflection leads us to the following conclusion: To deal with bureaucratic dysfunction and increase access in cities, innovation is key. City leaders need to think of themselves as agents of change, as promoters of the transformations that will tear down the walls to citizens' access to the benefits offered by cities. This is nothing new. Change agents in cities around the world are already responding to bureaucratic dysfunction to make sure it does not result in disentitlement for their citizens.

## 5 Innovations in the Access to Urban Benefits

Throughout the chapter, we have unpacked dimensions of access and their intimate relationship to a functioning bureaucracy. We have explained how bureaucratic design principles and bureaucrats' behaviors may hinder the innovation required to increase access. We have also reminded readers that bureaucratic design principles should not be confused with the values that inform them and that leadership holds the key to fresh approaches to access problems. There are instances, in fact, when organizations and the innovators within them have been able to uphold the values underlying bureaucracy while circumventing the design principles that might have stymied their solutions. In the process, these agents of change deactivated bureaucratic dysfunction and expanded access.

In this section, we examine innovations that have dealt with bureaucratic dysfunction and improved access to urban benefits in cities around the world. We have selected cases from both developed and developing countries. Since the dimensions of access are so intertwined, categorizing them would be difficult. For example, the case of the community land trust in San Juan, Puerto Rico, shows how a community increased its access to policymaking, but also, in the process, gained a different type of access to its economic assets. Similarly, the mobile assistance centers in municipalities in Brazil not only increased citizens' ability to access public services, but also improved the accountability of local governments.

The vignettes illuminate how various determinants eased or constricted access throughout the innovation process. In some cases, contextual constraints are more evident. In most, the role of leadership, either visibly or invisibly, is what tipped the balance in favor of increased access. The cases of the transparent procurement rules in Bogotá and the improvement of business registration processes in Amsterdam clearly illustrate the role of leadership. In other cases, contextual or agency determinants dominate the story, but leaders' actions behind the scenes help achieve progress.

Overall, these vignettes summarize the kinds of effort that public leaders and their teams are putting into tearing down the walls of bureaucratic dysfunction every day, helping cities fulfill their promise of becoming places of access to opportunity.

## 5.1 Empowering People to Access Elementary Education in Mumbai

Obstacles to accessing certain urban services may be intentionally kept from the public eye. Victims of inequality may be very well aware that they are not receiving the entitlements they deserve, but lack the power or motivation to do something about the situation on their own. In these cases, it is the "people's ability" side of the equation that needs to be addressed.

The *Balsakhi* program for remedial elementary education in Mumbai, India, worked to overcome exactly this problem of entrenched, seemingly insurmountable inequalities. The Indian educational system needed greater help for children in the third and fourth grades who had yet to master reading or arithmetic. But, of course, cash-strapped school districts could not find the money to hire new teachers. The only possibility seemed to be to let some kids fail, while others succeeded. This hierarchy in the school system seemed to some an inevitable fact of life, with citizens unable to change the status quo.

Unwilling to accept inequality as unavoidable, the NGO Pratham, in collaboration with municipal governments, began to hire local women from urban communities to teach basic skills to children in upper elementary school who still needed help with their reading and math. By relying on amateur tutors, the program managed to keep costs down while involving community members in improving the education of their children.

Because the *Balsakhi* method was so easily replicable, it spread from its origin city in Mumbai to 20 cities throughout India, with remarkable results in improving children's learning. Undaunted by inequity of access to education in India's cities, Pratham's innovators were able to cost effectively level the playing field for students who needed extra help, and they did so by engaging members of the community who were unable to organize for education reform on their own.

## 5.2 Better Business Licenses in Amsterdam

Cities are sites of economic opportunity, and as such, they are a magnet for people. Yet, not all groups benefit equally from those opportunities. Immigrants are among the groups identified by the World Bank as most vulnerable to exclusion from these opportunities and most likely to join the ranks of the urban poor (World Bank 2017). In developing countries with rapid urbanization, large numbers of rural inhabitants are moving to the cities. In developed countries, cities are constantly growing with the arrival of people from neighboring countries fleeing conflict, repression, or a lack of economic opportunity.

In many places, these people are excluded from economic rights due to their migrant status or absence of official residence or working permits.[2] This can reflect the need for societal commitment to enable migrants' access to the economy. Even immigrants who are not excluded through specifically-targeted laws may experience obstacles in their access to the economic gains offered by cities. For example, a high percentage of migrant populations in cities are self-employed or small business owners. To operate their businesses, these entrepreneurs must obtain all required licenses and permits.

In Amsterdam, many immigrants work in the hotel and restaurant industry. Despite the economic value of the hotel and restaurant sector in Amsterdam, the regulation for acquiring a bar, hotel, or restaurant license in the city is prohibitively complicated. An applicant needs to obtain licenses and dispensations from more than 18 different authorities. These regulations are confusing and pose obstacles to access city services for aspiring entrepreneurs. The obstacles are high enough to make it practically impossible to open a shop in Amsterdam while faithfully following the rules. The rules and regulations are sometimes contradictory, effectively prohibiting access. The rules have an additional excluding effect: through their complexity, they pose disproportionate obstacles for those who are less familiar with navigating bureaucratic regulations, such as immigrants and people from lower socio-economic classes.

[2] According to the World Bank (2017) such policies are particularly popular in Asia and Africa, where the majority of urban growth is projected to occur. At the same time, current political developments in the US and Europe show a trend towards more restrictive immigration policies.

Erik Gerritsen, city manager of Amsterdam, tried to remedy this issue with the HoReCa1 project (HoReCa is the Dutch acronym for the hotel, restaurant and café industry). Through this "one stop shop," applicants could fill out a 20-question application to determine which of the 40 government documents the applicant needed to open a new business. It provided a form to apply for the seven local licenses necessary for the city of Amsterdam. This project combined digital application procedures with back-office coordination mechanisms using civil servants' knowledge. While reviewing applications, civil servants could identify overlaps and inconsistencies in bureaucratic legislation that had gone unnoticed by individual agencies. Increasing the discretionary powers of frontline workers enabled them to rationalize their processes and make the application criteria more transparent to entrepreneurs and civil servants themselves (De Jong 2016).

Erik Gerritsen was able to articulate a public value proposition and convince key agencies to join their efforts. As a result, the barriers faced by entrepreneurs were removed, reducing the value lost by the city and business owners. It also became part of a city-wide program to improve access to all city services. At the same time, other important stakeholders, like banks or semipublic organizations, did not participate in the effort, limiting the reach and impact of the city's project. This serves to remind local leaders that constantly fine-tuning and incorporating key partners in collaborative efforts is a never- ending endeavor for public organizations, their leaders and frontline-staff to increase access in cities.

## 5.3 *Waste Recyclers in Colombia*

Urban waste recyclers live in extreme poverty. They survive by collecting garbage, which they use, if they can, to satisfy their own needs. The remainder of the trash is separated, classified and sold to industrial recyclers. Usually they earn less than a dollar a day for their extremely hard work, and they subsist on the waste of society.

In 1996, a study conducted by Colombia's Ombudsman Office of the nation's five largest cities found that 14% of waste recyclers were between 8 and 18 years old. Sixty-nine percent of these children had been surviving on this activity for more than a year, 22% were illiterate, 65% had left school to work, and only 13% were currently enrolled in school. Fifty-eight percent of them contributed some or all of this money to the household.

After a general strike of Bogotá's cleaning and waste management services, a decisive moment arrived: the Administration of Bogotá issued an urgent appeal to the recyclers to help the city manage its waste disposal. The waste recyclers organized themselves in response to the city's plea to mitigate what could have been a major sanitation issue. After demonstrating a strong ability to organize themselves, the recyclers gained enough confidence to bid in the upcoming contract for municipal cleaning and waste management services. The recyclers organized the *Asociación de Recicladores de Bogotá* (ARB), a registered and accredited association with strategic investor partnerships that gave the recyclers the financial backing necessary to put in a bid for contracts with the city.

Their entry into the business market, and subsequent competition in the bidding process, however, ran into several obstacles. First, according to Colombian Law 142/94, access to the public service market was limited to stock corporations (which are for-profit organizations) or to industrial and commercial enterprises of the State. In practice, this provision prevented the ARB—an association of non-profit and solidarity economy cooperatives—from extending its waste management activities from smaller municipalities into Bogotá or other major Colombian cities.

A second legal impediment to the recyclers' livelihoods was a National Decree signed in 2002, which determined that all waste, once outside the house or building of the citizens who produced it, was no longer municipal property but was the private property of the corporation holding waste management concessions. The waste recyclers' extremely marginal *modus vivendi* was thereby not only disregarded but was moreover criminalized. A third obstacle only emerged when the terms for Bogotá's waste management bidding process were published. The demands placed by the Administration on contractors bidding for disposal rights were so narrow that the ARB was deprived of the opportunity even to compete in the bidding process.

By taking a constitutional approach and emphasizing the right to equal treatment of for-profit and non-profit organizations under the law, ARB's lawyers paved the way for recognition of waste recycler cooperatives as important market actors in Colombia. In filing a Writ of Human Rights Protection, the ARB used legal means to challenge the city of Bogotá's unfairly exclusionary bidding process. In this case, the Constitutional Court became an important means of recourse for a disadvantaged group seeking access to their basic human rights.

In ARB's case, lawmakers in Bogotá attempted to infringe on the rights of the city's waste recyclers, and it was community organizers representing this group that agitated for improvements in their legal and financial status. The key innovation here was the insistence, through the knowledgeable use of court systems already in place, that justices in Bogotá acknowledge the fundamental rights of all organizations and all individuals to receive equal treatment under the law. The ARB's lawyers worked within the system to demand that the legislation excluding the city's waste recyclers from its formal economy be overturned. They used the institution itself to prove their case, reforming the legal system to include waste recyclers *from the inside*.

## 5.4 Improving Neighborhood Safety in North Carolina

Street-level drug dealing is toxic to a neighborhood because of the direct nexus between drug dealing and violence. Despite efforts by law-enforcement to deal with the issue, targeted neighborhoods in North Carolina consistently showed high rates of drug trafficking, leaving residents without access to basic law and order within their own neighborhoods. As a result of these problems, the High Point community (which is largely African American) felt excluded from the protection they deserved. They felt disappointed in and distrustful of the police.

Determined to make a difference, the executive staff of the High Point police department collaborated with David Kennedy from Harvard University. They devised a strategy working on multiple levels. The Overt Drug Market Strategy had 11 key elements: (1) mapping, to determine where the most serious offenses were concentrated; (2) mobilizing commitment of community through public meetings to identify and inform community stakeholders; (3) surveying by police and probation officers to identify those involved in street drug dealing; (4) formal identification of offenders and their areas of activity; (5) incident review; (6) undercover investigation of each location and offender; (7) contact with the offender's family to invite them to join law enforcement in asking offenders to quit; (8) the call-in, involving a face-to-face call between offenders, law and enforcement, and the community (to overcome anonymity); (9) issuance of a deadline of three days after the call in for offenders to quit dealing; (10) enforcement; and (11) follow-up visits about a month after the call-into ensure that former offenders were being given the help they needed to resist returning to drug dealing.

By identifying known drug dealers and encouraging them to give up dealing, the High Point police department improved access on two fronts. Offenders could access public programs and resources to assist them to find livelihoods without breaking the law, while community members had access to safer streets and a more responsive criminal justice system. Yet, the most significant obstacle to High Point's Overt Drug Market Strategy was the lack of employment opportunities for offenders. Jobs were scarce across the board, and most street-level drug offenders had minimal qualifications. These social issues were not going to be solved quickly; however, working in collaboration with area service providers, the police approached each offender's situation by offering support for whatever was requested, including mentoring, partially funded apprenticeship programs, training, or treatment needs.

While employment topped the list of needs, smaller issues came into play as well, such as transportation and education. A local African American congregation partnered with the police, providing mentors willing to walk with offenders for as long as necessary, providing help in any way possible.

The single most important achievement of High Point's drug strategy was the collapse of drug markets in the targeted neighborhoods, and the disappearance of associated violence and visible danger to residents on sidewalks and street corners. This was accomplished with very few arrests. In fact, arrests declined by 12%. The 1998 institution of the violent offender notification process, on which the Overt Drug Market Strategy was based, resulted in consistent reductions in High Point's violent crime rate as it decreased 47% from 1303 in 1997 to 682 in 2005.

## 5.5 The Community Land Trust in San Juan, Puerto Rico

There is, perhaps, no better symbol of urban exclusion than the inequality shown by informal settlements and segregation. The marginalization of these neighborhoods is not just economic, but political. Often, the inhabitants of these areas do not have influence on the decisions that are imposed on their futures. Across the world, however, citizens are finding new ways to organize themselves and devise mechanisms to defend their interests, opening new avenues to influence policymaking.

That is the case of the Caño Martin Peña Community Land Trust (CLT) in San Juan, Puerto Rico.[3] The area of Caño Martin Peña comprises eight neighborhoods formed by informal settlements in a water channel since the 1930s. Rural migrants settled in these watersides by filling the water with garbage and debris. Over the years, the land they claimed reached the estuary. By the 2000s, more than 27,000 residents lived in these neighborhoods, with an average per capita income of about half of San Juan's average. The area lacked proper urban sewage and drainage systems and was constantly being flooded by fecal waters, so the mayor and the city announced plans to dredge the canal and renew the neighborhoods.

The community of Caño Martin Peña, however, was suspicious of the plans. The neighborhoods are located very close to San Juan's luxury commercial district, and they were wary that the project would be used to displace many families. This had happened to other neighborhoods in the city before. Developers had acquired land from individual owners, building unaffordable housing and capturing all the increase in value. To avoid the same fate, the community of Caño Martin Peña created a Community Land Trust. In this ownership system, each resident owns the building and the CLT, a non-profit organization, owns the land on which the building is erected. The CLT has a governance structure to ensure the community makes decision over the land collectively, posing an obstacle to clientelist practices by politicians and to developers buying plots individually at lower prices from members of the community to capture the value increase. This way, the CLT takes power away from the government and private developers and strengthens that of the community.

CLTs like the one in Caño Martin Peña are innovative instruments to increase the access to political decision-making by communities, and they have also been implemented in other parts of the world, such as Kenya.[4] By increasing the political power of the community, defenders of these mechanisms argue that they can increase housing affordability and prevent involuntary displacement, gentrification, absentee ownership, and predatory owners.

Despite its promises, however, the Caño Martin Peña CLT had its challenges. Part of the community agreed with the public value proposition of the CLT, but other members preferred the individual titles to their parcel of land. The authorization

[3]This example is based on a Harvard Kennedy School Case produced by Quinton Mayne and Patricia García-Ríos, *Caño Martín Peña. Land Ownership and Politics Collide in Puerto Rico*. HKS number 2082.0 (2016).

[4]See, for example, http://cltnetwork.org/community-land-trusts-kenya/ and http://www.hic-gs.org/document.php?pid=2548.

environment was also shaky. When politicians opposed to the CLT were elected to the governorship and congress in 2009, individual titling resumed, and laws were passed to return the land of the CLT to municipal ownership. In 2012, the party supporting the CLT came back into power at city hall, and the CLT was restored.

The project of the CLT was also slowed down by bureaucratic dysfunction at the agency level. Despite its political backing since 2012, the project of dredging the channel still needed funding. In 2015, the Caño Martín Peña won the United Nations World Habitat Award, but after the impact of Hurricane Maria in the spring of 2018, the public health and environmental hazards of the neighborhoods made the news again—a good reminder that for improved access in policymaking to have lasting impact, it needs to be operationalized and combined with innovations at the agency or individual level.

## *5.6 Tackling Gender-Based Barriers to Urban Services in Quezon City*

Gender-based violence is a massive problem across the world, seriously impeding the freedom and safety of women. Studies have shown that women are more likely to be harassed in certain locations of urban settlements, such as toilets or public transportation. This deeply impedes women's mobility and access to public services and jobs in cities (World Bank 2011). In several countries in East Asia and Africa, efforts to increase health and safety of women have been implemented through early-warning systems and community-level engagement processes (World Bank 2017).

In Quezon City, the biggest city in Manila's metropolitan area and the largest municipality in the Philippines, a survey found that 3 in 5 women have experienced sexual harassment at least once in their lifetime.[5] In an effort to tackle the issue, in 2016 the city hall of Quezon passed a law establishing a fine of up to 200 US dollars and jail sentences of up to a year for sexual harassment in public spaces.[6]

The initiative was launched in partnership with UN Women and with the collaboration of women's groups and community groups. This systemic approach was important prior, during and after passing the law. Although approving the law was the responsibility of the city alone, other activities were necessary for the successful design and implementation of the initiative. For example, Safe Cities Manila, part of the UN's Safe Cities Initiative, conducted surveys to urban dwellers to better understand women's safety concerns and how they impacted their access to urban services such as public transport.

---

[5]https://www.sws.org.ph/downloads/media_release/pr20160311%20-%20Baseline%20Study%20Topline%20Results%20FINAL.pdf.

[6]https://apolitical.co/solution_article/philippines-largest-city-will-fine-men-200-street-harassment/.

One of the biggest hurdles faced by the initiative related to the context and culture in the Philippines, where sexist practices prevail. This directly affected the community's societal commitment to tackle harassment as a barrier to public services and women's dignity in general. Furthermore, some of the women themselves saw the law as unjustly penalizing men. To deal with these dimensions of access (societal commitment and people's ability to benefit), the city of Quezon and its partners launched media and awareness campaigns that showed how fear could affect women's mobility.[7]

The collaboration to deal with sexual harassment also extended to frontline workers; policemen could play a central role as street-level implementers to ensure women were safe. To make sure they handled the cases properly, the city of Quezon and UN Women partnered to train the local police on how to respond to harassment sensibly and on how to prosecute the cases.[8] Similar initiatives have been launched with the support of UN Women in 27 cities across the world, including Cape Town, Ho Chi Minh, Kigali, Maputo, New Delhi and Rabat, among others (UN Women 2017). Despite these initiatives, there is a long way to go in cities, north and south, east and west, to ensure women are safe and have the same access to urban benefits as their male counterparts.[9]

## 5.7 *Bringing Public Services Closer to Citizens in Brazil*

As in much of the world, public services in the state of Bahia, Brazil have traditionally been delivered by disparate government agencies at different locations and with very different service standards. In some cases, citizens seeking a single service must visit multiple agencies. Often, state residents discover the documentation needed for a given service only after visiting a string of government agencies on multiple occasions. Citizens have regularly encountered poor customer service and lack of professionalism in government offices, compounding the difficulty of access to basic services.

If bureaucracy impedes service delivery to city residents, it doubly affects the services available to citizens living in remote areas. Their physical remoteness from government offices means that many of the rural residents in the state have no documents at all, making access to services next to impossible.

To improve access to services and to bypass state bureaucracy, the Bahia State Government created Citizen Assistance Service (SAC) Centers, a pioneering initiative that became a true revolution in public service delivery. SAC Centers were

---

[7] https://apolitical.co/solution_article/philippines-largest-city-will-fine-men-200-street-harassment/.

[8] https://apolitical.co/solution_article/philippines-largest-city-will-fine-men-200-street-harassment/.

[9] https://www.theguardian.com/cities/2018/dec/13/what-would-a-city-that-is-safe-for-women-look-like.

veritable "one stop shops": full-service, multi-purpose complexes partnered with federal, state, and municipal agencies as well as private companies, offering services most in demand by citizens. These centers helped bring service delivery directly into the community.

The state government placed SAC Centers in convenient locations for the public, such as in shopping malls and major public transportation hubs, as well as crowded low-income neighborhoods. One emphasis of the SAC Centers was professionalism. Applicants for services at these centers could expect the same level of courtesy they might expect from private enterprises. By networking service delivery horizontally, SAC Centers gave their users the impression of a single, unified system.

The SAC Project brought public administration closer to its constituents, improved the relations between state authorities and citizens and established a new level of professionalism in public service delivery. The SAC Project's superior performance became a benchmark in terms of public service delivery in Brazil, which led several agencies in other states to ask for assistance from SAC's innovators in implementing the model in their own offices. SAC Centers were established in 22 out of Brazil's 26 states and internationally in Portugal and Colombia. It is also worth noting that the SAC Centers inspired some partner agencies to improve public service in their home offices and to train their civil servants in customer service.

## 5.8 More Transparent Public Procurement in Bogotá's Schools

In some countries, public procurement and contracting can be used to exclude entrepreneurs or small business owners who are not connected politically. Cumbersome laws and confusing rules are often used to keep the majority of businesses from accessing the economic benefits of government contracts. In these systems, the companies that offer better services at a more competitive value are ignored, and the local government grants the lucrative contracts to its cronies. Of course, this also means exclusion of poorer people, who often depend on quality and affordable prices to acquire benefits. This was exactly what happened in the education system of Bogotá, where school meals were costly and substandard due to a perverse public procurement system.[10]

Given the opaque contracting system, businesses and entrepreneurs did not have any recourse against the government's abuses and a group of contractors that colluded with it. In short, there was no accountability in how the government spent its funds, and there was no way for citizens to understand, monitor and complain about it.

In 2016, the city's new head of the education secretariat, Maria Victoria Angulo, committed to reverse that. Once again, a sole department's involvement was not enough to carry out the reform. She partnered with the national agency for public

[10]https://apolitical.co/solution_article/super-power-procurement-four-ways-open-contracting-is-making-cities-smart/.

procurement, *Colombia Compra Eficiente*, which, under the leadership of another woman, Maria Margarita "Paca" Zuleta, was promoting a revamp of public procurement across the country. Together, the city hall and *Colombia Compra Eficiente* introduced open data and open contracting reforms for the purchase of school meals in Bogotá's education system. The goal was to increase the transparency on how the 170 million US dollar funds were allocated to pay for the school meals of more than 800,000 students.

The main reform broke down the distribution channel, with separate bidding processes for meal production, packaging and distribution. That way, they increased transparency of the costs of each of the delivery stages. They also established framework agreements with the cap price of each product and signed them with several suppliers. Each food item had several suppliers under each new framework agreement. That way, if one provider could not fulfill the purchase order at the agreed price, another could take over. Before the reform, the supplier that had won the contract had enormous negotiating power and could impose high prices on a government under pressure to ensure kids received meals.[11]

At the start of the reform, the context in Bogotá was far from friendly towards the proposed changes. Suppliers, who controlled the system, strongly resisted. They launched media campaigns against the reform, trying to gain the opinion of parents. Other companies, potential entrants to the new system, were not completely convinced that it would work either. In a clear effort to articulate a vision about the public value of the innovation through a strategy of discrediting the status quo (Cels et al. 2012), Maria Victoria Angulo used data to show the faults of the previous system. Thanks to the partnership with *Colombia Compra Eficiente*, they could also use their big database of previous contracts to calculate fair prices and potential suppliers.[12]

Transparency and accountability were enhanced in at least two ways. First, a team of 347 people were in charge of visiting production and packing plants, as well as schools to check on the quality of the food. Second, and most importantly, they published all the data online, and encouraged data and tech-savvy activists, academics and analysts to check on it and use it to create interesting applications for parents to monitor the meals each week and have a say in the nutrition of their children.[13]

Thanks to the changes introduced in Bogotá, the number of school meal providers increased four-fold and the costs of the products were cut by an estimated 10–15%. Given the success of the program, the city education department expanded the open contracting framework to other areas such as cleaning services and school materials.[14] It also represented a way in which access to accountable governments can be increased through the use of new digital instruments, allowing citizens to co-produce public outcomes.

---

[11]https://medium.com/open-contracting-stories/the-deals-behind-the-meals-c4592e9466a2.
[12]https://medium.com/open-contracting-stories/the-deals-behind-the-meals-c4592e9466a2.
[13]https://medium.com/open-contracting-stories/the-deals-behind-the-meals-c4592e9466a2.
[14]https://medium.com/open-contracting-stories/the-deals-behind-the-meals-c4592e9466a2.

## 6 Conclusion

Cities can greatly increase the access to material and immaterial benefits for citizens, but they can also become places of exclusion and marginalization. Whether they remain as a place of opportunity for both the rich and poor depends, to a large extent, on their ability to successfully manage bureaucratic dysfunction.

This first requires understanding that access depends on societal commitment and institutional capacity, but also on peoples' capacity to benefit from rights and services. Scholars and practitioners alike cannot lose sight of the four dimensions that underpin access in cities: access to policymaking, access to the economy, access to public services and access to an accountable government. Accounts or initiatives that only focus on one narrow aspect of access will be incomplete.

Similarly, efforts to deal with bureaucratic dysfunction should be founded in sound analysis. In recent years, a plethora of one-size-fits-all remedies for dysfunctional bureaucracy—deregulation, smaller government, adopting private sector models for customer service, etc.—have been prescribed without proper diagnosis. We argue that a more grounded approach would be one that looks at the context, system, agency and individual levels of analysis.

Such an approach needs to recognize that "dysfunction" is a socially constructed problem with political implications, and that value trade-offs are made in addressing it. Acknowledging the adaptive nature of dealing with dysfunction to enlarge access to urban benefits also puts an emphasis on the role of leadership in innovating to make it possible.

The examples from cities across the world used to analyze recent innovations to manage bureaucratic dysfunction and associated lack of access confirm these hypotheses and provide important lessons. One cannot forget about the political dimension of the process, nor ignore that the leaders who can articulate the public value proposition, can enable the necessary legitimacy and can build operational capacity are a fundamental pillar of any effort. If one exists without the other, the political battle may be won but the war may be lost at the implementation stage. Focusing in an agency or a narrow set of agencies may leave key stakeholders out, leaving efforts to increase access unsustainable. Finally, engaging frontline workers has to be a central part of any effort, but it cannot fail to act at the context and societal level, so that the deeper forces inhibiting access to urban benefits are deactivated in the long term.

Cities and city leaders hold great promise and an even greater responsibility to improve the lives of millions of citizens across the world. Whether it is to help a migrant set up a business, a community to have drinking water, a school girl to eat nutritious meals or a woman to travel to work safely, city leaders will have to tackle bureaucratic dysfunction. Without innovative leaders able to mobilize resources and problem-solving capacity, it will be impossible to do so. With these leaders, the high hopes for accessible cities will be closer to becoming a reality.

## References

Aars J, Fimreite AL (2005) Local government and governance in Norway: stretched accountability in network politics. Scand Polit Stud 28(3):239–256
Acemoglu D, Robinson JA (2012) Why nations fail: the origins of power, prosperity, and poverty. Random House, New York
Albrow M (1970) Bureaucracy. Palgrave, London
Altshuler AA, Behn RD, Altshuler AA (1997) Innovation in American government: challenges, opportunities, and dilemmas. Brookings Institution Press, Washington DC
Bogdanor V (2005) Joined-up government. Oxford University Press, Oxford and New York
Borins S (2001) Leadership and innovation in the public sector. J Intellect Capital
CAF, UN-Habitat (2014) Construction of more equitable cities: public policies for inclusion in Latin America. UN and Development Bank of Latin America (CAF), New York
Cels S, De Jong J, Nauta F (2012) Agents of change: strategy and tactics for social innovation. Brookings Institution Press, Washington DC
Cruz NF, da Rode P, McQuarrie M (2019) New urban governance: a review of current themes and future priorities. J Urban Aff 41(1):1–19
De Jong J (2016) Dealing with dysfunction: innovative problem solving in the public sector. Brookings Institution Press, Washington DC
De Jong J, Rizvi G (2008) The state of access. Brookings Institution Press, Washington DC
De Vries H, Bekkers V, Tummers L (2016) Innovation in the public sector: a systematic review and future research agenda. Public Adm 94(1):146–166
Etzioni-Halevy E (2012) Bureaucracy and democracy. Routledge, New York
Florida R (2017) The new urban crisis. Basic Books, New York
Glaeser E (2012) Triumph of the city. Penguin Books, New York
Goldsmith S, Crawford S (2014) The responsive city: engaging communities through data- smart governance. Wiley, New York
Goldsmith S, Eggers WD (2005) Governing by network: the new shape of the public sector. Brookings Institution Press, Washington DC
Habitat III (2017) The new urban agenda. Retrieved 4 Feb 2019 from http://habitat3.org/the-new-urban-agenda/
Hajnal Z (2009) America's uneven democracy. Cambridge University Press, Cambridge
Heifetz R (1994) Leadership without easy answers. Harvard University Press, Cambridge
Holbrook TM, Weinschenk AC (2014) Campaigns, mobilization, and turnout in mayoral elections. Polit Res Q 67(1):42–55
Howard PK (1995) The death of common sense. Random House, New York
Kamarck E (2007) The End of government as we know it: making public policy work. Lynne Rienner, Boulder, CO
Katz B, Nowak J (2018) The new localism: how cities can thrive in the age of populism. Brookings Institution Press, Washington DC
Kettl DF (2002) The transformation of governance: public administration for twenty-first century America. Johns Hopkins University Press, Baltimore
Kruiter AJ, De Jong J (2008) Providing services to the marginalized. Anatomy of an access paradox. In: De Jong J, Rizvi G (2008) The state of access. Success and failure of democracies to create equal opportunities. Brookings, Washington D.C.
Lipsky M (1980a) Street level bureaucracy: dilemmas of the individual in public services. Russell Sage Foundation, New York
Lipsky M (1980b) Bureaucratic disentitlement in social welfare programs. Soc Sci Rev 58(1):3–27
Moore MH (1995) Creating public value: strategic management in government. Harvard University Press, Cambridge
Moretti E (2012) The new geography of jobs. Houghton Mifflin Harcourt Publishing Company, New York

Osborne D, Gaebler T (1992) Reinventing government: how the entrepreneurial spirit is transforming the public sector. Penguin, New York
Papadopoulos Y (2007) Problems of democratic accountability in network and multilevel governance. Eur Law J 13(4):469–486
Peters BG (1995) The politics of bureaucracy: an Introduction to comparative public administration. Routledge, New York
Pollitt C, Talbot C, Caulfield J, Smullen A (2004) Agencies—how governments do things through semi-autonomous organizations. Palgrave Macmillan, London
Portland State University (2016) Who votes for mayor?| A project of Portland State University and the Knight Foundation. Retrieved 23 Jan 2019 from http://www.whovotesformayor.org/
Ribot JC, Peluso NL (2003) A theory of access*. Rural Sociol 68(2):153–181
Sclar ED, Lonroth M, Wolmar C (eds) (2014) Urban Access for the 21st century: finance and governance models for transport infrastructure. Routledge, New York
Sørensen E, Torfing J (2011) Enhancing collaborative innovation in the public sector. Adm Soc 43(8):842–868
Sparrow MK (1994) Imposing duties: government's changing approach to compliance. Praeger, Westport
Stoker G (2006) Public value management: a new narrative for networked governance? Am Rev Publ Adm 36(1):41–57
Strömblad P, Myberg G (2012) Urban inequality and political recruitment networks. Urban Stud 50(5):1049–1065
Torfing J, Peters BG, Pierre J, Sørensen E (2012) Interactive governance: advancing the paradigm. Oxford University Press, Oxford
Trounstine J (2018) Segregation by design by Jessica Trounstine. Cambridge University Press, Cambridge
UN Women (2017) Safe cities and safe public spaces: global results report. UN Women, New York
UN-Habitat (2010) State of the world's cities 2010–2011. UN-HABITAT, Bangkok
Wilson J (1989) Bureaucracy. What government agencies do and what they do it. Basic Books, New York
World Bank (2011) World development report 2012. Gender equality and development. World Bank, Washington DC
World Bank (2016) Evaluating digital citizen engagement: a practical guide. World Bank, Washington DC
World Bank (2017) East Asia and Pacific cities: expanding opportunities for the urban poor. World Bank, Washington DC

**Jorrit de Jong** is Faculty Director of the Bloomberg Harvard City Leadership Initiative, a joint program of Harvard Kennedy School and Harvard Business School in collaboration with Bloomberg Philanthropies, where he is responsible for executive education programs for mayors and their senior leadership teams in 240 cities worldwide. He also oversees the research agenda, the curricular materials development portfolio and a comprehensive program of ongoing field support to cities. Dr. De Jong is Senior Lecturer in Public Policy and Management at Harvard Kennedy School (HKS) and Academic Director of the Innovations in Government Program at the school's Ash Center for Democratic Governance and Innovation. His research and teaching focus on the challenges of making the public sector more effective, efficient, equitable and responsive to social needs. A specialist in experiential learning, Jorrit has taught strategic management and public problem solving in degree and executive education programs at Harvard and around the world. Jorrit holds a Ph.D. in Public Policy and Management (VU Amsterdam), a Master in Philosophy (Leiden) and a Master in Public Administration (Leiden). He has written extensively, including the books The State of Access: Success and Failure of Democracies to Create Equal Opportunities (Brookings, 2008, co-edited); Agents of Change: Strategy and Tactics for Social Innovation (Brookings, 2012, co-authored); and Dealing with Dysfunction: Innovative Problem Solving in the Public Sector (Brookings, 2016).

**Fernando Fernandez-Monge** is a senior associate with the Bloomberg Harvard City Leadership Initiative and an adjunct professor of urban innovation at the IE School of Global and Public Affairs. He also collaborates with the World Bank, where he was a governance specialist. He is a graduate of the Harvard Kennedy School and the Autonoma University of Madrid. He has written several teaching cases and academic and policy papers on urban governance, public innovation and city leadership and management. His work has been published or featured in Apolitical, Governing, World Economic Forum, El Pais, Agenda Publica, among others.

# Accountability Through Participatory Budgeting in India: Only in Kerala?

**Harry Blair**

**Abstract** Since its beginning in Brazil in 1989, participatory budgeting (PB) has spread worldwide to several thousand local governmental units (LGUs) in all continents, celebrated for its success in combining citizen involvement and state accountability in delivering public services. While PB has been adopted in most places by individual LGUs on their own initiative, in India the state of Kerala implemented PB throughout all its governmental units from rural villages and urban wards up through district in one "big bang" move in 1996. Over the succeeding two decades and more, PB has become securely institutionalized, surviving numerous changes of ruling party at state level. Outside of Kerala, however, few LGUs of any sort have implemented PB and it has not flourished in any of the adopters. Using the World Bank's principal-agent model of state accountability for public service delivery, this paper will explore Kerala's experience at PB and more briefly look at its lack of success elsewhere in India.

**Keywords** Participatory budgeting · Accountability · Public service delivery · Decentralization · India · Kerala

## 1 Introduction

Over the past three decades, participatory budgeting (PB) has become a major institution throughout the world, renowned for its ability to combine citizen involvement and state accountability in delivering public services. From its beginning in the city of Porto Alegre, Brazil, in 1989, PB has spread to several thousand local government units in both developed and developing countries.[1] In most if not all other countries,

[1] The latest published count estimates between 1269 and 2778 PB "traceable experiments," depending on how they are counted (Sintomer et al. 2013: 11). In this listing, Kerala counts as just one case.

H. Blair (✉)
South Asian Studies Council, Yale University, PO Box 208206, New Haven, CT 06520-8206, USA
e-mail: harry.blair@yale.edu


S. Cheema (ed.), *Governance for Urban Services*,
Advances in 21st Century Human Settlements,
https://doi.org/10.1007/978-981-15-2973-3_3

local government units (LGUs)—primarily urban areas—have instituted PB on an individual basis, but in India the entire state of Kerala, with a population of some 30 million, adopted PB at all governmental levels from its more than a thousand village panchayats and urban wards to its 14 districts in one "big bang" in 1996. For this reason, Kerala offers an excellent arena for exploring the advantages and shortcomings of PB as a mechanism incorporating participation and accountability in state provision of public services. Kerala PB has been primarily undertaken in its rural areas, comprising about three-quarters of the state's population, though its towns and cities also have less well funded PB programs. Otherwise, only a few Indian cities have thus far initiated PB. Pune began PB in 2006, a decade after Kerala, and Delhi followed almost a decade later, with a pilot PB effort in 2014, later expanded. But none of these urban programs proved sustainable. Accordingly, analysis in the present paper will focus mainly on Kerala, with some attention to the Pune and Delhi cases. The paper will undertake such an exploration of the Indian experience at PB, employing as a lens the World Bank's principal-agent model of state accountability for public service delivery.

## 2 The World Bank's Routes to Accountability for Public Service Delivery

The Bank's *World Development Report* for 2004 (WDR 2004) took as its central theme ways to enhance accountability for the delivery of public services. Using a "principal-agent" approach, the Report offered two routes to link citizens (here the principals) demanding public services with the state at all levels (the agents) which supply those services. The two routes can be captured graphically, as in Fig. 1.

The **long route** accounts in a very real sense for the expansion of public services widely defined in the advanced democracies over the last couple of centuries. Competitive elections have been the key ingredient here. Competing political leaders have had to promise ever more public services over time in their attempts to get elected, and citizens have held the winners to account by re-electing them or turning them out of office (exercising "voice" in Fig. 1). Political leaders hire government officials and direct them to provide services (the "compact" of Fig. 1). Citizens thus act as principals making demands on their politician agents, who in turn act as principals making demands on the agent/bureaucracy to provide services. In this fashion citizens in the West gained education, transport, sanitation, regulation of public health and commerce, civil rights protection, etc. In India since 1947, improvements in food production, literacy and life expectancy can similarly be attributed in significant measure to politicians' need to compete for votes by responding to citizen demands for a better life.

The long route has problems, however. The first lies in its very name: the advances in public well-being achieved through it have taken a very long time to become realized. Scheduled Castes and Tribes in India have struggled since the colonial

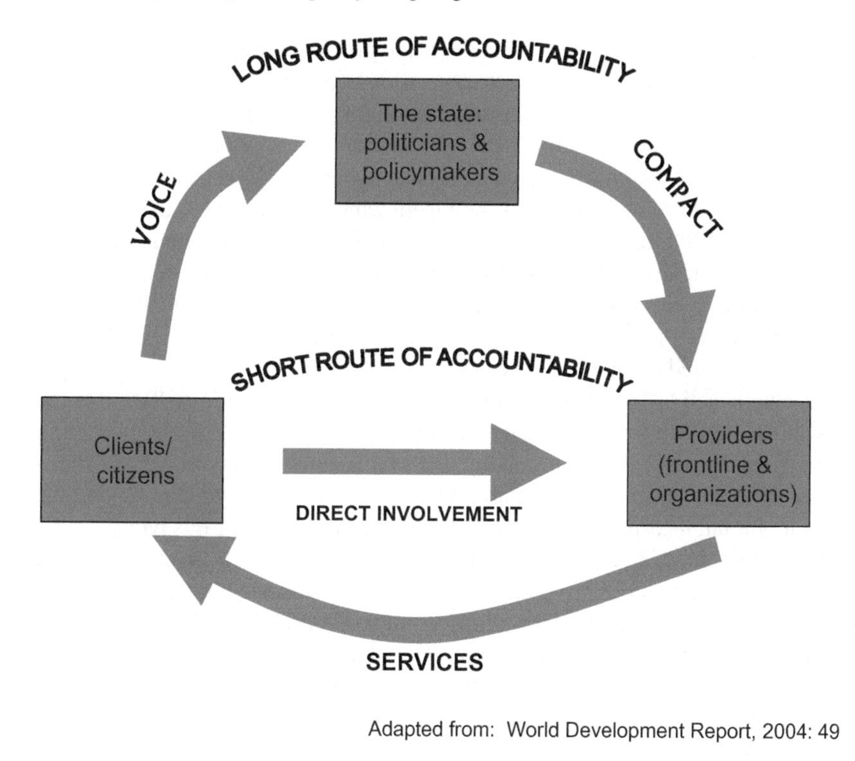


**Fig. 1** The long and short routes to accountability for public service delivery

era for equal rights and dignity, and reservations for the Other Backward Classes were achieved only after decades of effort. Secondly, elections offer a chance to hold leaders to account only every few years, and when they are held they offer the citizenry no more than a blunt policy tool (e.g., stop inflation, curb corruption). Even at the local level, single issues (absent teachers, clogged drains) tend to get buried beneath bigger ones (mayoral corruption, rising crime).

Thirdly, the long route can become infected with clientelism as political office-holders direct benefits to specific groups in return for their votes (subsidies to farmers, hiring preferences to specific ethnic groups) or financial campaign support (overlooking bank defaults) rather than pursuing programs providing universal benefits (better schools, mosquito abatement). Fourth, the principal-agent roles can become reversed, especially at lower levels: bureaucrats can turn themselves into principals and their supposed political masters into agents by manipulating policies (turning infrastructure projects into graft opportunities for themselves), while political leaders capture voters with petty patronage (mishandling relief funds). Finally, the long route utterly depends on honest, "free and fair" elections, without which accountability at the ballot box is lost.

While the long route links citizens and service providers only indirectly through elections, the **short route** offers a direct connection between the two sides, as shown in Fig. 1. In fact, the short route can take two paths: "choice" and "voice."[2] The "choice" path replicates a market in which the consumer chooses between vendors to purchase a product. Parents using vouchers can choose a school for their children, householders can choose which ration shop from which to purchase subsidized rice. On the "voice" path, citizens can contact the state directly with a demand (under a right-to-information law) or opinion (citizen report cards). The short route has some real advantages over the long route in exacting accountability from state institutions. It can deal with specific citizen demands and can function within a relatively short timeframe as opposed to an electoral term of office.

An especially active "voice" mechanism is participatory budgeting (PB), whereby citizens in essence cross the boundary between demand and supply by becoming directly involved in state decision-making. PB as it is known today traces its origins to the city of Porto Alegre in Brazil around 1990 and has since spread around the globe. In its Indian incarnation it will become the central focus of this paper.

To be sure, the short route has its own problems. For one thing, in direct market-like transactions it often assumes erroneously that consumers have essential information about the goods or services on offer (indeed a problem seriously affecting the market itself). Secondly, lack of competition among vendors may invalidate any "choice" options. A third difficulty arises with the kind of role-reversal that can occur in the long route: providers can become the principals and consumers the agents (e.g., clientelism with favored customers at a rice ration shop). Fourth, while the long route can function (or malfunction) at any level, the short route is basically a local one, at least if positive outcomes are expected: direct contact between citizen and state can only work on a small-scale.[3] Finally and most critically, short route mechanisms depend on political leaders for their creation and maintenance. Absent continued strong and continuous political support, even the most thoroughly institutionalized short route engine will stall and become ineffective.

The stage is now set to explore participatory budgeting as a short route mechanism, but first, in order to give a more complete picture of the range of institutions in widespread use to exercise democratic accountability from the state, it would be good to mention briefly three others.

**Civil society**, which can be defined as organizations that are not part of the state or the market and that further the interests of their members, can follow the long route

---

[2]The use of "voice" with both the long and short routes (see Joshi 2007; WDR 2004) but with different meanings in each context is not helpful. But there seems no better expression in either framework. In any case, the remainder of the present paper will be dealing with the short route use of the term.

[3]Mass demonstrations on a wide scale could be considered a short route mechanism, but such activity is virtually always negative, seeking to undo or reverse some perceived state malfeasance, aiming to pressure the state to desist, as with the recent "yellow vest" protests in France (e.g., Friedman 2018; Viscusi 2018). In the extreme case, the demonstrations seek to replace or overthrow the state itself (e.g., in the fall of communist rule in Eastern Europe around 1990). But short route activity toward more positive ends is essentially local.

by advocating for their wants and needs with the political leaders (often called "lobbying") or pursue the short route by doing so directly with public service providers (which can lead to corruption).[4] Such advocacy has the advantage that it can be undertaken at any time (no need to wait for an election) and can focus on topics of particular interest to the organization as distinct from the fuzzy and fluid array of policy ideas that officeholders and parties must deal with. But state accountability through civil society depends on organizational strength and resources and so is not equally open to all, unlike the short route mechanisms which, as state institutions, are when properly operating available to all citizens.

In recent decades, legal redress has become an effective instrument for demanding accountability, particularly in the form of **public interest** lawsuits, whereby an individual or group sues the state for failing to implement laws and regulations that are already on the books but have not been enforced. Perhaps the most famous such suit in India was brought by the attorney M. C. Mehta, who sued the state for not protecting the Taj Mahal in Agra from airborne pollutants, finally obtaining a judgment from the Supreme Court in 1993 directing a cleanup of the Taj (Mehta 1997). But aside from taking a long time (over a decade for Mehta), such suits are expensive—clearly not feasible for ordinary citizens.

Arguably the most important instrument for exercising state accountability is a **free media**, for so long as there is no public awareness of government misdeeds, it is easy to cover them up and continue state misbehavior. But the media must not only be free but have the interest, expertise and resources to undertake investigative journalism—a combination that generally exists only at the national level or in the very largest metropolitan centers. And media freedom is of course dependent on state tolerance of unpleasant revelations about its behavior. So as with the short route, here too strong state support is needed, only In this case it is negative support—the state must refrain from interfering with the media.

## 3 Participatory Budgeting and Social Accountability

Within the domain of democratic social accountability, PB is one of a number of short route mechanisms for promoting state accountability to citizens. As such, its success or failure in any given case can be judged against a set of three goals that Carmen Malena and her co-author have laid out for social accountability (Malena et al. 2004; Malena and McNeil 2010). And to these three goals can be added a fourth, longer term objective. Collectively, the goals can be summed up as:

(1) Better governance. Citizens can "access information, voice their needs and demand accountability."
(2) Empowered poor people. Poor citizens can begin to take charge of their own futures.

[4]It can be argued that civil society offers a third, middle route to accountability in addition to the long and short routes (Blair 2018).

(3) Improved service delivery. Citizen needs and public services can better match each other.
(4) Enhanced well-being. To the extent that the first three goals are realized, citizen well-being should over time improve with better health and education and longer lives.

These four goals will comprise a good test of how participatory budgeting has worked in India. They also encapsulate the two purposes generally posited for pursuing democracy to begin with. The first two goals assert in effect that democracy is an end in itself, that it is good for citizens to engage with their government and especially good for less privileged citizens to gain agency for themselves. The second two hold that democracy essentially constitutes the best means to further ends: superior public services from the state and enhanced lives for its citizens. The essay will return to these thoughts later on.

## *3.1 Brazilian Origins of Participatory Budgeting*

As noted at the beginning of this chapter, PB had its start in the city of Porto Alegre, Brazil, from which it has spread worldwide. In that expansionary process, all manner of state-sponsored programs have come to be labeled "participatory budgeting," ranging from those like the Porto Alegre original model featuring deep citizen involvement in state fund allocation to some that merely advertise a promise that the state will listen to citizen input. Sintomer et al. (2008, 2013), who have become the de facto record keepers of PB's global spread, have identified fully six distinct PB types, so it will be necessary to pin down what will be discussed in this paper. Accordingly, it will be worthwhile to look briefly at PB as it was introduced in Porto Alegre and continued for its first dozen years and more.[5]

The original Porto Alegre model, introduced by the city's mayor in 1990 has three layers, beginning with annual neighborhood assemblies in which citizens discuss, debate and vote on priorities for investing municipal funds on capital projects in their area. They also elect two representatives to assemblies sitting at the next higher level, that of the city's 16 regions. At these open Regional Budget Forums, the elected representatives in turn consolidate and rank the neighborhood priorities into a list for their region, and elect two delegates to the third assembly, the citywide Participatory Budget Council (PBC) which further consolidates and prioritizes the regional input in accord with a weighting formula into a city plan. The formula is a complex one, combining local preferences, the extant level of facilities in relation to need (e.g., housing units lacking sanitary water), and population size. The PBC plan then goes to the municipal council for deliberation and approval, and finally to the mayor's office. The PBC monitors implementation over the coming year, as do also the lower two

[5]For a somewhat more extended discussion of the model, see Blair (2013: 146–149). It has been extensively analyzed, e.g., in Baiocchi and Ganuza (2017).

tiers, and the process starts over again the next year.[6] This whole system is clearly a complicated one and requires significant technical support at all levels from the mayor's office (De Souza Santos 1998).

Porto Alegre's PB can claim a number of achievements. As for participation, the World Bank found that one-fifth of all the city's citizens had participated in PB and that almost one-third of what was defined as the poor population was taking an active part in PB (World Bank 2008: 23, 28).[7] If both these estimates are approximately correct, poorer citizens engaged at a higher rate than those better off. And poor participants were as active as the non-poor, both in speaking at meetings (Baiocchi 1999: 9) and getting elected to serve at regional meeting and the PBC (CIDADE 2010).

In terms of process, PB made significant progress in replacing a traditional patron-client structure with a budget system focusing on neighborhood wants and objective needs. Previous pork patronage in which municipal council members would direct budget allocations to specific individuals or groups was largely eliminated under PB (Koonings 2004: 85–91). And the sums involved were significant: PB determined around half of all municipal investment spending, or about 7% of total municipal spending (World Bank 2008: 48, 56; also Melgar 2014: 127). Of equal importance, the Porto Alegre experience showed that poor people could overcome the disincentives to cooperate in political activity (transaction costs, risks of embarrassment, etc.) when such engagement offered perceptible gains in public services like piped water and sewage treatment (see Abers 1998, 2000; also World Bank 2008; Pateman 2012). Furthermore, PB had a real impact in reducing poverty rates over its first dozen years and more (World Bank 2008). In other words, PB fostered a real degree of empowerment to poor people who had previously been excluded from meaningful political participation.

Porto Alegre also offers some cautionary lessons for PB. Introduced there in 1990, the program continued intact under successive administrations of the leftist Partido dos Trabalhadores (PT or Workers Party). Even after the PT lost power in the 2004 municipal elections to a center-right party, the new administration maintained the program, but materially changed its operation. Its principal change was to introduce a new participation model featuring much wider inclusion, specifically civil society organizations and the private sector, with the objective of attenuating the potential class-conflict inherent in the PB approach and increasing resources available for development. PB itself was retained, but a new program titled Governança Solidára Local (Local Solidarity Governance) became the main vehicle for allotting investments.[8]

---

[6]The Porto Alegre model has been explained in detail many times, for instance in Koonings (2004), also Wainwright (2003) and Avritzer (1999).

[7]The very poorest stratum was much less involved, however, in large part because of transaction and opportunity costs.

[8]The city of Belo Horizonte, which adopted the Porto Alegre PB model in 1993, underwent a similar de-emphasis on PB combined with a widening to include a middle-class constituency and a decrease in funding after a non-PT administration took over the municipality. See Montambeault (2019).

In the new model, investment planning became a top-down enterprise, directed from the municipal administration, in contrast with the bottom-up process characteristic of PB. Business interests gained a larger voice, for example securing state support for displacing poor families to build shopping malls. And social services like community kitchens that had been funded through PB were off-loaded to self-funded NGOs and philanthropic sources. The new administration saw these moves as "de-politicizing" PB, while PB's supporters perceived them as politicizing the program in a very different direction.[9] As things turned out, the same center-right party retained office in the next several elections, so its participatory model remained in place.

The new center-right administration's policy changes can be seen as an example of elite capture or perhaps better "re-capture"—in this case a return to an earlier era of clientelistic governance and suppression of poor people's participation in governance. But it can also be seen as pluralistic democracy in action. Just as when the PT gained power in 1990, it rewarded its political base by instituting a new model of PB, so too the PT's center-right successor rewarded its base by changing "the rules of the game" (in this case the municipal investment game).

There was a difference here, however. The PT pulled a thitherto marginal constituency into local governance, which diluted but did not eliminate elite influence; the PB budget mostly comprised additional municipal funding rather than redirecting old funding. In contrast, the successor administrations decreased the overall municipal investment budget and within it decreased PB's role in allocating it. One would hope that pluralistic democracy would lead to negotiation and compromise between political players, but clearly this is not invariably the case, especially when stark class interests are concerned, as in Porto Alegre.

A second lesson to be drawn from Porto Alegre is that PB as a governmental mechanism is critically dependent on political support from the top. Even though it appeared to have become well institutionalized over three successive municipal administrations, and the incoming administration found it politically necessary to retain it at least in name, the new mayor had relatively little difficulty in fundamentally reorienting it after taking office.

Widely regarded as the most rigorous PB model,[10] the Porto Alegre exemplar in its earlier years had a number of characteristics, against which the Kerala model and other Indian versions can be tested[11]:

---

[9]This and the preceding paragraph are largely drawn from Melgar (2014). See also Baiocchi and Ganuza (2017).

[10]See Sintomer et al. (2013: 14&ff). In contrast with Porto Alegre's "Participatory democracy" model, the weakest of Sintomer et al's six types is labeled "Multi-stakeholder participation" and includes private interests generally in a dominant position. Participation is largely a management tool. This type is characteristic of PB systems in Eastern Europe and Africa. The other four types have one or more elements of the Porto Alegre model.

[11]The first five points are based in Sintomer et al. (2013). The last two are my own, based on the Porto Alegre experience in early years. After the PT lost power in 2004, these two characteristics significantly weakened (though they did not completely disappear).

- Principal focus on financial/budgetary process.
- An annual replication of the PB cycle.
- Public discussion, deliberation and prioritization, amounting to de facto decision-making on state spending by an unelected body.
- Citywide coverage (not just a neighborhood).
- Monitoring during the year and public review at the beginning of next year's cycle.
- Initiation by a left-of-center government with an agenda to include an underclass or excluded communities (working class in Porto Alegre, women, Dalits and Adivasis in Kerala).
- Continued political will from the top to support the PB program.

## 4 Participatory Budgeting in Kerala

Following the 73rd and 74th Amendments to India's Constitution in 1993 (which established requirements for decentralized local governance in all rural and urban jurisdictions), the Kerala government in 1996 established the People's Campaign for Decentralized Planning.[12] Skipping any sort of pilot experimentation, the Communist Party of India (Marxist) or CPM government went for a "big bang" approach throughout the state, including all 65 urban structures and three tiers of rural organization (panchayats at the village [grama], block and district levels), collectively numbering more than 1200 elected bodies. In a word, PB was to mean that a significant portion of what had been centrally allocated development funds would henceforth be programmed at the local level (Heller et al. 2007: 628).

The Kerala PB program has operated more or less like the Porto Alegre model. In the rural areas, every gram panchayat has 10–12 wards, each with a population averaging something over 2500. The yearly cycle begins here with open meetings (grama sabha), which are facilitated by trained key resource personnel. In the ward-level grama sabhas, priorities are established and two delegates selected to the next higher panchayat level, where they meet with elected local government officeholders and bureaucrats in a series of development seminars to forge a unified panchayat budget aggregating all wards. Task forces and sectoral working groups (e.g., for education, infrastructure, poverty reduction, watershed management) are formed to plan and implement projects to be taken up. The projects selected by the working groups are then prioritized into a plan document, which is vetted at higher level by Block and District Planning Committees for technical viability. Once approved, the projects are implemented and monitored by the grama panchayats. The elected PB delegates monitor project implementation throughout the cycle and then report to the grama sabha at the beginning of the next cycle.[13]

[12]Information in this and the next paragraph is largely taken from Heller et al. (2007). For more detail, see Isaac and Franke (2002). Heller (2012) discusses similarities between Porto Alegre and Kerala.

[13]The most complete description of Kerala's PB process can be found in Isaac and Franke (2002).

Kerala's PB program was fortunate to be able to draw on assistance from the Kerala Sastra Sahitya Parishad (KSSP), a large NGO that had long been active throughout the state. The KSSP trained some 100,000 key resource persons as facilitators and technical advisors for the grama sabhas. More than 10% of the electorate participated in the grama sahbas in the program's first two years. By the second year, Dalit and Adivasi attendance had increased to more than their share of the population, and in the fourth year a survey showed Dalit participation at 14% of the total, as against their 11.5% of the population. About 40% of attendees were women. (Heller et al. 2007: 636). Another study found grama sabha attendance even higher among Dalits at 34% (PEO 2006: xiv) and further that among those below the official poverty line, participation was higher (30%) than for those above it (18%). After the first couple of years, upper and middle class participation declined further at the grama sabhas (Rajesh 2009: 12). Altogether, in contrast with so many studies finding a pattern of elite capture in decentralization programs (Blair 2000), Kerala's program appears to have remained free of this problem. As the most obvious evidence here, program benefits have not accrued to those better off (Heller et al. 2007).[14]

Relative to other PB systems, funding for Kerala's program has been quite generous. At the beginning the state devolved one-third of total development plan outlay[15] to PB, a percentage that remained constant up through 2003, even as the total state plan outlay increased over the years. After that, the PB allotment continued to rise, but as a share of the even more rapidly rising state plan outlay, it dropped in subsequent years from one-third to one-fourth or one-fifth.[16] The exact amounts depended in part on whether the CPM-led Left Democratic Front (LDF) or the Congress-led United Democratic Front (UDF) happened to be in power and exercising a greater or lesser enthusiasm for PB, but the important point here is that by the time the UDF succeeded the LDF in 2001, the program had become sufficiently institutionalized that it was kept intact. In the subsequent 2006 election, the LDF returned to office, and in the next two elections of 2011 and 2016, power changed both times. There were some minor changes, with funding allotments bobbing up and down as bit depending on which coalition held power at any given time, accompanied by cosmetic name changes, reflecting the primary party's ideology: PB began under CPM leadership as the "People's Campaign for Decentralized Planning," then changed with the Congress in charge to the "Kerala Development Plan" in 2001, and to the "People's Plan" after the CPM returned to office. In short, PB has survived four turnovers while remaining essentially intact. The contrast with Porto Alegre's experience is stark.[17]

---

[14]Heller et al. (2007: 636–637) attributes this pattern to Kerala's history of lower class mobilization beginning in the late 19th century and more recently a progressive land reform. See also Rajesh (2009). To this might also be added the high level of literacy throughout the state, which enables greater scope for local accountability.

[15]Total development plan outlay has held steady over the years at about 18% of GOK spending (Sebastian et al. 2014: 22), so one-third of that would come to around 6%.

[16]Data from GOK, Economic Review (various years).

[17]PB's track record in Belo Horizonte, Brazil's other major city taking up the program in earlier years, is very similar to Porto Alegre's experience when the PT lost power. See Montambeault (2019). Heller (2012) presents an extensive comparison between the Kerala and Brazilian histories with PB.

When PB began in 1996, Kerala was about one-quarter urban and three-quarters rural in population. PB allotments somewhat skimped on the urban side, with 14%, while the rural areas received some 86%. As the state became more urban over time, the allocations changed also, with about a quarter going to the cities and three-quarters to the countryside. Within the rural funding, about two-thirds went to the grama sabha level in the earlier period and continued at that rate in the more recent years. Block and District levels each received about one-sixth. The smaller urban share was divided between roughly three-fifths going to the 87 municipalities and two-fifths to the state's five corporations.[18]

At all levels both rural and urban, PB project spending was to be apportioned to three sectors: productive, services, and infrastructure, which in recent years have received roughly 10, 60 and 30% of the PB budget respectively. As its name indicates, "productive" sector is intended to promote development directly, and in the rural PB budgets this has meant primarily agriculture and irrigation, along with animal husbandry and dairy, collectively accounting for more than 85% of allocations to the sector. The "service" sector refers to public service provision and funds a wide mélange of activities from education and health to child welfare, sanitation and electrification. Finally, "infrastructure" funds mostly (about two-thirds) went to roads, with the rest to public building construction and electrification (GOK 2018).

## *4.1 Problems with Kerala PB*

While PB has helped improve infrastructure in rural areas, principally by spreading roads, and has contributed to reducing poverty (though it is difficult to tell by how much[19]), its investments in the productive sector have not yielded much return (PEO 2006; Harilal and Eswaran 2016). In fact, agricultural output in Kerala has decreased over the past several decades, from 1376 thousand metric tons of rice in 1972–73 to 549 thousand in 2015–16—a staggering drop (GOK 2018: 44). A national Program Evaluation Office report attributed PB's poor performance in the productive sector to "inadequate capacity of PRI [Panchayati Raj Institutions, i.e., PB] members to draw up production plans on a scientific basis" (PEO 2006: xiv). In other words, PB participants lacked the technical expertise to make effective investments in agriculture.

But should they be expected to possess this level of knowledge? Or is it rather that the state bureaucracy has redirected the principal-agent essence of the World Bank's short route so that the two roles are reversed? Whereas it is the PB members that should be identifying priorities and holding state officials accountable for delivering

---

[18]Analysis and evaluation have followed an even greater rural orientation. With the exception of George and Neunecker's (2013) study, virtually all of it has focused on the rural side, with urban attention confined to statistical data (e.g., GOK 2018). The PB process itself was essentially similar in both rural and urban areas.

[19]This point will be taken up later on in the paper.

public goods and services, here it becomes the state authorities as principals who are holding the PB units as their agents accountable for following all 14 required steps for submitting proposals. Elite capture is not the problem here but rather it is what Harilal and Eswaran (2016) have called "bureaucratic capture." In terms of the World Bank's model short route, providers and citizen roles have reversed: the providers become principals, and the citizens have become their agents, doing their bidding.

A different line of explanation would hold that the problem here concerns the identity of anticipated beneficiaries. PB members can see themselves as benefiting from investments in the service or infrastructure sectors, but allocating funds to agriculture will advance only farmers, who are in any case declining in numbers. Accordingly, they want to reduce allocations to the productive sector and increase them to the other two sectors.

The lengthy PB process has led to another and even more serious problem in that fund outlays have been excessively slow. Harilal (2013) reported that generally by November (two-thirds of the way through the fiscal year), only about one-third of the year's plan outlay had been spent, and that the remaining two-thirds was mostly spent in March toward the fiscal year's end.[20] And invariably less than the full allocation was actually expended, so that a good many projects were not finished. Finally in 2017, the state introduced a "new methodology" to speed up PB annual plan formulation that it claimed worked to complete the process by mid-June that year (GOK 2018).

Yet another problem arose from the state's line departments like health and education, whose officers found themselves subjected to a dual authority, answering not only to their departments administratively but now also operationally to panchayati raj councils for work done on PB-funded activities (Chathukulam and John 2002: 4919–4920). Dual authority has been an abiding problem for most if not all decentralization initiatives, as civil servants in line departments have resented having to answer to locally elected officials they regarded as professionally less competent (see e.g., Blair 1985, also Blair 2000), so it is not unexpected that it has surfaced in Kerala also. What is perhaps surprising is that more analysts have not noticed it.

A more profound issue stems from a failure to realize the LDF's deepest original hope for PB in Kerala: that the broad involvement of a conscientized and mobilized citizenry would lead to a higher and sustainable popular engagement with public policy at a societal level, i.e., moving beyond the individual incentives citizens have shown for involvement in PB. This ambition evidently was something of an obsession with E. M. S. Namboodiripad, the CPM chief minister who led the move to establish PB initially, but did not come to fruition (Rajesh 2009).

It might seem that Kerala PB's "big bang" start would have created problems in that testing the PB system was not feasible. Because the entire state entered the program at a single stroke, the very possibility of a randomized control trial was eliminated, for there was no place in the state that could serve as a control; every local government unit belongs to the treatment group. In the event, though, lack of any

[20]In India as in the subcontinent generally, the government's fiscal year runs from 1 April to 31 March.

kind of pilot program and testing has not seriously hampered PB's accomplishments, as will be seen in the following subsection.

## 4.2 *PB's Success in Kerala*

Despite its problems, PB has succeeded in a number of important ways, improving services and infrastructure, and bringing a significant degree of democratic empowerment to a constituency that had been systematically denied any role in deciding how public money would be spent.

Progress has been impressive in many ways, for instance poverty, both rural and urban, stood at about 25% in the mid-1990s, but 10 years later had declined to about 13 and 20% respectively (GOK 2018: Appendix 1.23). PB may well have had a significant role here, but without a treatment-and-control comparison, there is no way to tell how much or little a role played.

Another measurement issue arises in that the state is so advanced socio-economically that in most measures it has arrived at a level where further improvement becomes successively more elusive. By the early 1990s, some 96% of villages already had medical facilities, and infant mortality had declined to 17 per thousand live births (as against 14% and 80 per thousand in India overall)—levels that would be hard to improve upon (Keefer and Khemani 2005: 16). So PB cannot be expected to generate great improvement in these indices statewide. But given that with PB a central purpose is to enable each local government unit to choose what it needs, the expectation has to be that grama sabha A will choose to enlarge a dilapidated primary school, whereas grama sabha B will want to build a road connecting it with a neighboring village, and grama sabha C will decide to improve garbage disposal. So while the aggregate number of schoolrooms, miles of roads and functioning waste disposal systems may increase only slightly, these three grama sabhas will have substantially improved the services and infrastructure they most needed. And this, after all, is a central purpose of PB.

Rather than try to measure quantifiable change in particular sectors like education, electrification and sanitation, a better way would be to assess citizen satisfaction with service delivery and infrastructure where they live. Heller et al. (2007: 632) conducted just such a survey in 72 panchayats covering the first five years of PB, finding that in every category a majority of respondents reported at least "some" improvement and in a number of subsectors (e.g., roads) a "significant" improvement. Perhaps more importantly for PB's future continuation, a large majority of political leaders from the main opposition party in each panchayat felt that the situation had improved in every subsector (including primary education, roads and sanitation. Representatives from Dalit organizations did so as well (Heller et al. 2007: 634).

As an inherently subjective matter in the mind of each individual, "empowerment" is difficult to assess, but Heller et al. (2007: 642) did endeavor to do so by asking their respondents how likely had it become after five years of PB that women and the

Dalits/Adivasis would "voice their needs and demand responses from elected representatives and elected officials." Among all respondents and among only women, the answers were virtually identical: just over 55% replied "somewhat more likely," and just over 40% said "much more likely." When asked whether the PB campaign had empowered women to enter the public arena and raise developmental issues related to women, again respondents overall and women in particular gave almost identical answers: 32% said there had been a "slight change" and 66% reported "drastic change" (ibid. 642). It seems fair to conclude then, that PB had contributed materially to a sense of empowerment among citizens who (aside from voting) had not previously participated in the political arena at local level.

## 4.3 Causes of PB's Success

Surely a major reason behind Kerala's success in PB has been that the general quality of governance is so good relative to the rest of India. Beginning in 2016 the Public Affairs Centre in Bengaluru has calculated a Public Affairs Index measuring such aspects of governance as essential infrastructure, social protection and environment, and in each year Kerala has come out as the best governed of all the country's states large and small (PAC 2018). Such recent assessments do not guarantee that Kerala has always had the highest quality of governance in India, of course, but the state's past attainments in indicators of well-being should serve as strong indicators that its governance has been very good indeed over recent decades and that this quality of good governance has been instrumental in providing an institutional environment within which PB could flourish.

A second factor has been the nature of political competition in Kerala. Arguably the major reason for PB's decline in Porto Alegre and other Brazilian cities like Belo Horizonte was that the PT and opposition parties had their main political bases in quite different constituencies. The PT's base lay in the working class, while opposing parties found their support in middle- and upper-class voters, with little crossover. So whichever party or coalition of like-minded parties was in power understandably pursued policies favoring its base: a strong PB under the PT, a weak PB under the PT's opposition. In Kerala on the other hand both the CPM-centered LDF and the INC-based UDF sought to appeal to the entire spectrum of voters, including in both cases the working class, Dalits and Adivasis. Accordingly, both Fronts supported PB, though perhaps with greater intensity when the LDF was in office than when the UDF held power, and PB as an institution has fared well for more than two decades.[21]

In their comparative study of Kerala and Uttar Pradesh, Keefer and Khemani (2005) found that after a long period in which the INC dominated Uttar Pradesh against weak and fragmented opposition (and thus had little incentive to engage in serious pro-poor policies), more recently the state has been contested between three parties, each with its own base: the Bharatiya Janata Party appealing to upper

[21] See Keefer and Khemani (2005: 18–21).

caste Hindus, the Samajwadi Party to middle Hindu castes and Muslims, and the Bahujan Samaj Party to Dalits, with some but minimal crossover. Each party when in power has pursued policies favoring its base, none has tried to govern with an eye to benefiting all three constituencies—all a direct contrast with Kerala.

The powerful incentives for both Kerala parties to support PB have meant strong leadership at the top favoring PB, in contrast with the tepid backing given to PB by incoming leaders in Porto Alegre and Belo Horizonte. In both the latter cases, PB had become sufficiently institutionalized that it could not easily have been eliminated outright, but it could be and was watered down and weakened. Thus in Kerala it has been the combination of party incentive for broad spectrum appeal and political will at the top that has enabled PB to endure essentially intact in Kerala.[22]

## 5 PB in Other Urban Indian Settings

Thus far no other state in India has followed Kerala's path with PB.[23] However, several large cities have given it at least a brief tryout before abandoning or severely weakening it. Among them are Mumbai, Bengaluru, Pune, Mysore and Cochin, and in addition the Union Territory of Delhi, which includes New Delhi. The most ambitious of these experiments took place in Pune, which was also the only city still using PB well into the current decade, although in much attenuated form. Some analysis of the Pune experience has appeared, along with a brief description of the short-lived Delhi experiment. Given the absence of any serious external analysis of Kerala's urban PB history, it will be worthwhile to provide a review of these two cases.

The Pune Municipal Corporation is organized into four zones, divided into 14 administrative wards, which are subdivided into 76 prabhags (averaging around 42,000 population in the 2000s).[24] The PB program began in 2006, with newspaper advertisements soliciting citizen suggestions for public investment. Two established NGOs then publicized the new system through local organizations like neighborhood associations, Lions Clubs, senior citizens organizations, etc. Citizens submitted their suggestions on forms, which were collected, collated, classified and costed out at the administrative ward level. A committee composed of elected prabhag counsellors then approved/rejected the suggestions. In PB's first year, at this point public prioritization meetings were held at ward level, at which all those who had submitted proposals were invited. Attendees were then divided by electoral ward (smaller than

---

[22]In a study of participatory accountability mechanisms in India generally, I found political will at the highest level to be the *sine qua non* for success. Without it, no mechanism could last very long (Blair 2018).

[23]West Bengal (also under a CPM government from 1977 to 2011) has made notable progress in decentralization and has undertaken serious efforts at poverty reduction, but it has not taken up PB as such. See Crook and Sverisson (2001), also Robinson (2007: 15) and Maiti and de Faria (2017: 21).

[24]This paragraph and the next two are based on Menon et al. (2013) and Jobst and Malherbe (2017).

administrative wards), where they prioritized and submitted their suggestions, which were then included in the corporation budget.

Thus there was citizen participation in making suggestions and then in prioritizing those suggestions in that first year, but after that the citizen role was restricted to the suggestion phase only. A special effort to incorporate slum neighborhoods was also dropped after the first year. Nor has there been any information released about the status or completion of proposed projects. Municipal Corporation allocations to PB continued at an average of just under 1.5% of the total budget over the next half-dozen years. So there was some continuity, but aside from citizen input with suggestions, no participation in budgetary prioritization, allocation or monitoring. In a word, no real accountability of state providers to citizens. These changes are similar to Porto Alegre's retrogression to a lesser role for the poor in that city's PB after the PT's loss of power in 2004.

Two major reasons behind the failure of PB in Pune seem reasonably clear. Most importantly, there was no real political backing at the mayoral level. One administration launched PB in 2006 and a newly elected one came into office the following year. The first set up a process that had promise, and the second one implemented it for a year but then diluted citizen input to the point that the project could scarcely be called PB. A second problem emerged from bureaucratic and lower level political obstacles. City officials kept publicity low (one newspaper ad once a year), the submission process was inconvenient, and no feedback or monitoring information was provided. For their part, the elected corporators (city council members), jealous of a funding process not fully under their control, showed little support for it.

During its campaign for the 2015 elections in the National Capital Territory of Delhi, the reformist Am Admi Party had promised to decentralize power to gram sabhas and their urban counterpart, Mohalla sabhas.[25] After its victory, the new government launched a PB pilot in 11 of Delhi's 70 legislative assembly constituencies. The next year things were scaled up to all 70. Each of the territory's 2972 sabhas—urban and rural—was allotted a budget of one million rupees to spend on infrastructural projects (from an approved list). Public meetings were held in each sabha, and projects were suggested and voted upon. After that, however, the sabhas (which otherwise were granted some governing power) were restricted from allocating funds through citizen participation. PB's life in Delhi proved to be short.

## 6 Meeting the Goals of Social Accountability (at Least in Kerala)

Returning to our earlier list of social accountability goals, it is evident from the Heller team's survey after PB's first decade in Kerala that (1) the quality of governance had improved, that (2) there was a definite increase in empowerment for women and

[25]This paragraph draws mainly on Samy (2017).

Dalits/Adivasis, and that (3) public service delivery had become better (Heller et al. 2007).

Regarding well-being (the fourth goal), evidence for a PB role has proven somewhat elusive, even in places like Brazil where it has been possible to do statistical analysis over reasonably long periods. In a study covering the country's 220 largest cities, Boulding and Wampler (2010) found that while PB contributed marginally to reducing poverty rates,[26] it had virtually no effect on well-being indicators such as literacy, life expectancy, or infant mortality. In contrast, a few years later, Gonçalves (2014) found in her analysis of 3651 municipal areas, that those adopting PB had reduced infant mortality significantly in comparison with non-PB areas. And in a more fine-grained analysis, Touchton and Wampler (2014) found that when length of time a city used PB was factored in, its use for 8 or more years was associated with a 19% reduction in infant mortality. Analyses of PB and well-being measures like literacy and life expectancy have yet to be published. In any case, though, the very nature of Kerala's "big bang" PB renders this kind of analysis impossible.

If democracy promotion is to have as its goal only that people should have the ability to participate in public decisions that affect themselves, i.e., that democracy is a developmental end in itself, then PB's track record in Kerala must be reckoned a success, considering its achievements in improving governance and empowering marginal constituencies. And if democratization's goals are also to include societal impact, i.e., that democracy is a means as well as an end, then there is evidence that PB can improve public service delivery by connecting citizens directly with state providers using the Bank's short route rather than relying only on the long route. PB may well also have a downstream impact on well-being, of the sort that we can see glimpses of in Brazil, but quantitatively this cannot be measured in Kerala, given the impossibility of randomized control testing.

## 7 Conclusion

Kerala's experience shows that the World Bank's short route can be an effective engine for social accountability with PB, but only if it has strong state support. Strong CSOs help also. So does a broad gauge competitive politics in which parties reach out for support from a wide range of classes and ethnic groups. And an educated populace (if literacy >90% and it can be assumed that most of the illiterate 10% are poor, a good number of the poor have access to knowledge about local govt). But political will at the top appears the essential ingredient.

Can PB be replicated elsewhere in India? The fact that despite several tryouts in Indian cities it has not taken hold anywhere does not augur well. A number of cities (and states) do have competitive politics, which can be argued as a necessary condition for successful PB, but clearly it is not a sufficient one. The same could be said of strong CSO presence. As for education, in the 2011 census, Pune's literacy

[26]The World Bank (2008) also reported a poverty reduction.

rate stood at 89.6%, close to Kerala's 93.9%, and many other cities have impressive literacy rates as well. What appears to be lacking is any real political will to take up PB, even on an experimental basis. Since the Janaagraha organization began ranking Indian cities for quality of governance in 2014, Pune has climbed from 8th place out of 20 cities to the number one (out of 23) rank in 2017 (Janaagraha 2018). So if Pune does not have the political interest or will to engage PB more seriously, one must suspect that other cities will not either.

Is PB then worth trying in other areas of India? Given that it has shown demonstrable evidence of advancing three of social accountability's four goals in Kerala, and that some progress on the fourth goal can be inferred from Brazilian evidence, it is indeed. But whether it will be attempted in coming years is another matter.

## References

Abers R (1998) From clientelism to cooperation: local government, participatory policy, and civil organizing in Porto Alegre, Brazil. Polit Soc 26(4):511–537

Abers R (2000) Inventing local democracy: grassroots politics in Brazil. Lynne Rienner Publishers, Boulder

Avritzer L (1999) Public deliberation at the local level: Participatory budgeting in Brazil. Paper presented at the Experiments for deliberative democracy conference, University of Wisconsin, Jan 2000. http://www.ssc.wisc.edu/~wright/avritzer.pdf

Baiocchi G (1999) Participation, activism, and politics: the Porto Alegre experiment and deliberative democratic theory. Department of Sociology, University of Wisconsin-Madison, USA. http://www.ssc.wisc.edu/~wright/Baiocchi.pdf

Baiocchi G, Ganuza E (2017) Popular democracy: the paradox of participation. Stanford University Press, Stanford

Blair H (1985) Participation, public policy, political economy and development in Bangladesh, 1958–1985. World Dev 13(12):1231–1247

Blair H (2000) Participation and accountability at the periphery: democratic local governance in six countries. World Dev 28(1):21–39

Blair H (2013) Participatory budgeting and local governance. In: Öjendal J, Dellnas A (eds) The imperative of good local governance: challenges for the next decade of decentralization. United Nations University Press, Tokyo, pp 145–178

Blair H (2018) Citizen participation and political accountability for public service delivery in India: Mapping the World Bank's routes. J South Asian Dev 13(1):1–28

Boulding C, Wampler B (2010) Voice, votes, and resources: evaluating the effect of participatory budgeting on well-being. World Dev 38(1):125–135

Chathukulam J, John MS (2002) Five years of participatory planning in Kerala: rhetoric and reality. Econ Polit Wkly 37(49):4917–4926

CIDADE (Centro de Assessoria e Estudos Urbanos) (2010) Data on participatory budgeting in Porto Alegre. http://ongcidade.org/site/php/op/opEN.php?acao=dados_op

Crook RC, Sverisson AS (2001) Decentralisation and poverty-alleviation in developing countries: a comparative analysis or, is West Bengal unique? IDS Working Paper 130 https://opendocs.ids.ac.uk/opendocs/handle/123456789/3904

De Sousa Santos B (1998) Participatory budgeting in Porto Alegre: toward a redistributive democracy. Polit Soc 26(4):461–510

Friedman V (2018) The power of the yellow vest. New York Times, 4 Dec

George S, Neunecker M (2013) Democratic decentralisation and participatory budgeting: the Kerala experience. In: Sintomer Y, Traub-Merz R, Zhang J, Herzberg C (eds) Participatory budgeting in Asia and Europe. Palgrave Macmillan, Basingstoke, pp 66–85
GOK (Government of Kerala) (2018) Economic review 2016. State Planning Board, Thiruvananthapuram, Kerala. Also various earlier years
Gonçalves S (2014) The effects of participatory budgeting on municipal institutions and infant mortality in Brazil. World Dev 53:94–110
Harilal KN (2013) Confronting bureaucratic capture: rethinking participatory planning methodology in Kerala. Econ Polit Wkly 48(36):52–60
Harilal KN, Eswaran KK (2016) Agrarian question and democratic decentralization in Kerala. Agrar South: J Polit Econ 5(2&3):292–324
Heller P (2012) Democracy, participatory politics and development: some comparative lessons from Brazil, India, and South Africa. Polity 44(4):643–665
Heller P, Harilal KN, Shubham Chaudhuri S (2007) Building local democracy: evaluating the impact of decentralization in Kerala, India. World Dev 35(4):626–648
Isaac TMM, Franke RW (2002) Local democracy and development: the Kerala people's campaign for decentralized planning. Rowman and Littlefield Publishers, Lanham
Janaagraha (2018) Annual survey of India's city-systems. Janaagraha Centre for Citizenship and Democracy, Bangalore, Karnataka. http://www.janaagraha.org/asics/report/ASICS-report-2017-fin.pdf
Jobst M, Malherbe M (2017) Participatory budgeting in Pune: 'my city, my money'? Uppsala University, Department of Business Studies. http://www.diva-portal.org/smash/get/diva2:1104320/FULLTEXT01.pdf
Joshi A (2007) Producing social accountability? The impact of service delivery reforms. IDS Bull 38(6):10–17
Keefer P, Khemani S (2005) Democracy, public expenditures, and the poor: Understanding political incentives for providing public services. World Bank Res Obs 20(1):1–27
Koonings K (2004) Strengthening citizenship in Brazil's democracy: local participatory governance in Porto Alegre. Bull Lat Am Res 23(1):79–99
Maiti S, de Faria JV (2017) Participatory planning processes in Indian cities: its challenges and opportunities. J Sustain Urban Plan Prog 2(1):18–32
Malena C, McNeil M (2010) Social accountability in Africa: an introduction. In: McNeil M, Malena C (eds) Demanding good governance: lessons from social accountability initiatives in Africa. The World Bank, Washington, DC, pp 1–10
Malena C, Foster R, Singh J (2004) Social accountability: an introduction to the concept and emerging practice (Social Development Papers No. 76). The World Bank, Washington, DC
Mehta MC (1997) Making the law work for the environment. Asia Pac J Environ Law 2:349–358
Melgar T (2014) A time of closure? Participatory budgeting in Porto Alegre, Brazil, after the Workers' Party era. J Lat Am Stud 46(1):121–149
Menon S, Madhale A, Amarnath (2013) Participatory budgeting in Pune: a critical review. Centre for Environment Education, Pune. http://www.vikalpsangam.org/static/media/uploads/Stories_PDFs/participatory_budgeting_pune_review_cee_2013.pdf
Montambeault F (2019) 'It was once a radical democratic proposal': theories of gradual institutional change in Brazilian participatory budgeting. Lat Am Polit Soc 61(1):29–53
PAC (Public Affairs Centre) (2018) Public Affairs Index 2018. http://pai.pacindia.org/
Pateman C (2012) Participatory democracy revisited. Perspect Polit 10(1):7–19
PEO (Programme Evaluation Organisation) (2006) Evaluation report on decentralised experience of Kerala. Planning Commission, Government of India, New Delhi. http://103.251.43.137/greenstone/collect/imgmater/index/assoc/HASH01c0.dir/doc.pdf
Rajesh K (2009) Participatory institutions and people's practices in India: an analysis of decentralization experiences in Kerala State. Munich Personal RePEc Archive, Munich. https://mpra.ub.uni-muenchen.de/29544/1/MPRA_paper_29544.pdf

Robinson M (2007) Does decentralisation improve equity and efficiency in public service delivery provision? IDS Bull 38(1):7–17

Samy P (2017) Participatory budgeting: a case of Delhi. Center for Budget Governance & Accountability, New Delhi. http://www.cbgaindia.org/wp-content/uploads/2017/11/PARTICIPATORY-BUDGETING.pdf

Sebastian J, Kumary A, Nair S (2014) Kerala finances, 2002–03 to 2011–12: an evaluation, submitted to Fourteenth Finance Commission, Government of India. Gulati Institute of Finance and Taxation, Thiruvananthapuram, Kerala

Sintomer Y, Herzberg C, Röcke A, Alves M (2008) Participatory budgeting in Europe: potentials and challenges. Int J Urban Reg Res 32(1):164–178

Sintomer Y, Herzberg C, Allegretti G, Röcke A (2013) Participatory budgeting worldwide—updated version. Dialog Global No. 25. Engagement Global GmbH—Service für Entwicklungsinitiativen, Bonn

Touchton M, Wampler B (2014) Improving social well-being through new democratic institutions. Comp Polit Stud 47(10):1442–1469

Viscusi G (2018) Why people in yellow vests are blocking French roads. Washington Post, 10 Dec

Wainwright H (2003) Reclaim the state: experiments in popular democracy. Verso, London

WDR 2004 (World Bank) (2003) World development report 2004: making services work for poor people. The World Bank and Oxford University Press, Washington, DC

World Bank (2008) Brazil: toward a more inclusive and effective participatory budget in Porto Alegre, vol I, Main Report, Report No. 40144-BR. The World Bank, Washington, DC

**Harry Blair** is currently Senior Research Scholar at Yale University and serves as Associate Chair and Director of Undergraduate Studies in the South Asia Studies program there. He has worked as an academic political scientist and as a consultant since earning his Ph.D. at Duke University in 1970, mostly in South Asia, but also in Southeast Asia, Latin America and Eastern Europe. His topics have included agriculture, forestry, rural development, public administration, civil society, decentralization, elections and political parties, and macro-level politics. He has taught at Bucknell, Cornell, Rutgers and Yale Universities, and has served as consultant with USAID, DfID, Ford Foundation, Swedish SIDA, UNDP and the World Bank.

# Public-Private Partnerships to Improve Urban Environmental Services

**Bharat Dahiya and Bradford Gentry**

**Abstract** Public-Private Partnerships (PPPs) are most usefully viewed as a tool, not a religion. The 2030 Agenda for Sustainable Development introduces 'partnership' as one of the five critical dimensions of sustainable development, and lays emphasis on encouraging and promoting effective public, public-private and civil society partnerships. Within this larger context, the purpose of this chapter is to offer some thoughts on: (i) The reasons PPPs have generated such interest in the urban environmental arena; (ii) A way to understanding PPPs; (iii) The key features of successful PPPs; (iv) The current trends in and debate over PPPs in the urban water sector; and (v) Ways to make the best use of PPPs to help improve urban water services. The fact that private capital flows have remained above the Official Development Assistance flows since 2005—except 2015 and 2016, has sustained the interest of many parties in searching for profitable and impactful investment opportunities in urban environmental services. PPPs often start with questions vis-à-vis their goals, strengths and weaknesses, structure and processes. Successful PPPs often feature *individual champions* who address the tensions at the heart of many partnership efforts; *partnership space* or the context in which PPPs are formed; and *optimized structures and processes* that would respond to different urban environmental problems. Our analysis of the data (1990–2013) obtained from the World Bank's Private Participation in Infrastructure Project Database shows that whilst the number of water projects steadily increased from 1990 to 2007, the private capital flows to urban water sector declined from US$37 billion during 1990–2000 to US$25 billion between 2001 and 2010. The reasons for this decline in international private investment are many and varied: (i) functioning of public sector and political systems; (ii) private sector and commercial realities; and (iii) opposition to private sector involvement. Moving forward will require action on at least two important and interrelated fronts:

B. Dahiya (✉)
Research Center for Integrated Sustainable Development, College of Interdisciplinary Studies, Thammasat University, 2 PraChan Road, Pra Nakorn, Bangkok 10200, Thailand
e-mail: bharatdahiya.tu@gmail.com

B. Gentry
Yale University School of Forestry & Environmental Studies and Management, Kroon Hall 119, New Haven, CT 06511, USA


S. Cheema (ed.), *Governance for Urban Services*, Advances in 21st Century Human Settlements,
https://doi.org/10.1007/978-981-15-2973-3_4

first is *addressing the mayors' dilemma* about the choice of PPPs to improve urban environmental services, and second is *assessing the performance of all partners*.

**Keywords** Public-private partnerships · Urban environment · Private capital flows · Partnership space · Infrastructure and services · Sustainable development · SDG 11 · Multi-stakeholder partnerships

## 1 Understanding the Enigma of PPPs for Urban Environments

Public-Private Partnerships or PPPs have long generated both interest and controversy as a way to help address urban environmental issues. Some believe that they are the best approach to improving performance while increasing operational efficiency and access to private capital. Others believe just as strongly that they are often an unjustified abdication of public responsibility that leads to higher prices and few or disparate improvements in service delivery. This debate[1] is particularly intense in the urban water sector, where examples can be found to support both views,[2] and the debate is often more about philosophy than performance.[3]

In September 2015, the Member States of the United Nations adopted the 2030 Agenda for Sustainable Development that introduces five critical dimensions of sustainable development; these include people, planet, prosperity, peace and partnership (United Nations 2015). Thus, partnerships are presented as one of the important 'means of implementation'. The 2030 Agenda for Sustainable Development includes 17 Sustainable Development Goals (SDGs); the 17th SDG aims to 'Strengthen the means of implementation and revitalize the Global Partnership for Sustainable Development' (United Nations 2015: 28). In recent years, it has been increasingly recognized that to be effective PPPs need to be based on multi-stakeholder partnerships and cooperation, including community engagement and civil society participation (African Development Bank 2001; Asian Development Bank 2009; Dahiya and Okitasari 2018; World Bank 2019a). In view of this, the SDG 17 includes Target 17.17 to 'Encourage and promote effective public, public-private and civil society partnerships, building on the experience and resourcing strategies of partnerships'. Given the current state of global urbanization (Haase et al. 2018; UN-HABITAT 2013, 2016; United Nations 2017), the 2030 Agenda for Sustainable Development included SDG

[1]See Asian Development Bank (2018), Geddes (2011), Graham and Marvin (2001), Lanjekar (2010), Medhekar (2014), Nundy and Baru (2008), O'Neill (2010), Siemiatycki (2011), Srinivasan (2006).

[2]For example, compare Hall et al. (2002) with PADCO, Inc. (2002). Also see Fischer (2011) and Rodriguez et al. (2012).

[3]For the evolving developments in the dynamic field of PPPs, see Bayliss and van Waeyenberge (2018), Boardman and Vining (2012), Cui et al. (2018), Hall (2015), Hodge and Greve (2017), Leigland (2018), Marx (2019), Medda et al. (2013), Patrinos et al. (2009), Petersen (2019), Roehrich et al. (2014), Warner (2012), World Economic Forum (2017).

11 that aims to 'Make Cities and Human Settlements inclusive, safe, resilient and sustainable' (United Nations 2019).

PPPs are most usefully viewed as a tool, not a religion. They come in a wide variety of shapes and sizes (see World Bank et al. 2014). They work in some cases, not in others. Within this larger context, the purpose of this chapter is to offer some thoughts on:

(1) The reasons PPPs have generated such interest in the urban environmental arena;
(2) A way to understanding PPPs;
(3) The key features of successful PPPs;
(4) The current trends in and debate over PPPs in the urban water sector; and
(5) Ways to make the best use of PPPs to help improve urban water services.

The material in this chapter is based on a distillation of information collected since 1996 by the United Nations Development Programme's 'Public-Private Partnerships for Service Delivery' Program (PPPSD),[4] the Hixon Center for Urban Ecology at the Yale School of Forestry & Environmental Studies,[5] the World Bank's Urban Environment Thematic Group (see World Bank 2019b), the Public-Private Infrastructure Advisory Facility[6] (see PPIAF 2018), the Organization for Economic Co-operation and Development (see OECD 2019a), and partners from around the world.[7] The overlapping goals of these efforts have been to collect, analyze and disseminate lessons learned from efforts to use PPPs to improve the delivery of urban environmental services.

As such, the chapter includes working hypotheses developed and tested over time across a variety of experiences and contexts. Given the diversity of partners and activities involved, rigorous quantitative analyses have not been a major part of this work. Rather, the focus has been on embracing the differences across existing PPPs while distilling out common themes around which successful new partnerships may be crafted.

Following this brief introduction, Sect. 2 looks into why PPPs are needed to improve urban environments. In Sect. 3, the question of what are PPPs, including their goals, strengths and weaknesses, structure, and processes is discussed. Section 4 elaborates the key features of successful PPPs, which can be grouped in three major areas, namely individual champions; partnership space; and optimized structures and processes. Section 5 examines the current trends in the urban water sector PPPs, including the three sets of reasons for the decline in international private sector

[4]PPPSD was earlier called Public-Private Partnerships for the Urban Environment (see UNDP 2019).

[5]See: https://hixon.yale.edu.

[6]See: https://ppiaf.org.

[7]Particularly including: Mario R. Delos Reyes at the University of the Philippines; Jamal Ansari at the School of Planning and Architecture in Delhi, India; the late Kwabena Darko at the World Bank's Distance Learning Center in Ghana; Olena Maslyukivska at the University of Kiev-Mohyla Academy in the Ukraine; Shi Han, Center for Environmentally Sustainable Technology Transfer in Beijing, China; M. Sohail at the Water and Engineering Development Center at Loughborough University in the UK; and Hubert Jenny at the Asian Development Bank, China Resident Mission.

investment, namely the functioning of public sector and political systems; private sector and commercial realities; and opposition to private sector involvement. The concluding section dwells on the question: how best use PPPs for urban water services? In doing so, it provides some thoughts on how to navigate the complicated field of PPPs.

## 2 Why PPPs to Improve Urban Environments?

If governments were meeting the need for better urban environments there would be no need to discuss PPPs. Unfortunately, many are not. The "water decades" of the 1970s and 1980s focused on improving access to drinking water and sanitation through government funding and operations. While water services were extended to many new users, the overall number of people without access to adequate services continued to expand (Gentry and Abuyuan 2000), as the latest statistics confirm. The current needs in the water sector are staggering. The latest data shows that between 1990 and 2015 although 1.56 billion people gained access to improved sanitation in urban areas worldwide, the actual number of urban inhabitants without sanitation increased from 547 million to 674 million—an increase of 127 million (WHO and UNICEF 2014, 2017). During this period (1990–2015), while 1.20 billion urban dwellers gained access to piped water supply on premises across the world, still 595 million rely on unimproved sources of drinking water—an increase of 481 million (WHO and UNICEF 2014, 2017). By 2025, estimated annual investments in water infrastructure of over US$1 trillion are needed in the 'Organization for Economic Co-operation and Development' (OECD) countries, as well as Brazil, Russia, India and China (Ashley and Cashman 2006). Thus, water will account for a major share of global infrastructure investment until 2025. The problem, however, is that governments are finding it extremely difficult to meet these targets acting alone.

In the 1990s, two financing trends combined to increase interest in PPPs (Fig. 1): First, the amount of Official Development Assistance (ODA) flowing to all recipient countries stagnated around US$70–90 billion per year, well below the US$125 billion target set at the Rio Earth Summit (United Nations 1992). Second, as a result of the push toward freer markets and privatization, the amount of private investment flowing to emerging markets skyrocketed and remained above both ODA and the Rio Target, except for a couple of years after the financial crises of the late 1990s. Tapping into these flows of international private capital became a goal of many efforts to build PPPs.

During the first decade of 21st century, since bottoming out in 2002, private capital flows increased reaching the first peak of US$310 billion in 2007. A systematic review of the World Bank's lending for improving urban environmental quality found that 17% (44 in number) of then active projects involved private sector participation and these were mainly in the water and sanitation sector (Bigio and Dahiya 2004). Despite an abrupt fall in 2008 due to the global economic crisis, private capital flows reached a second, even higher peak of US$326 billion in 2011. The third highest peak was

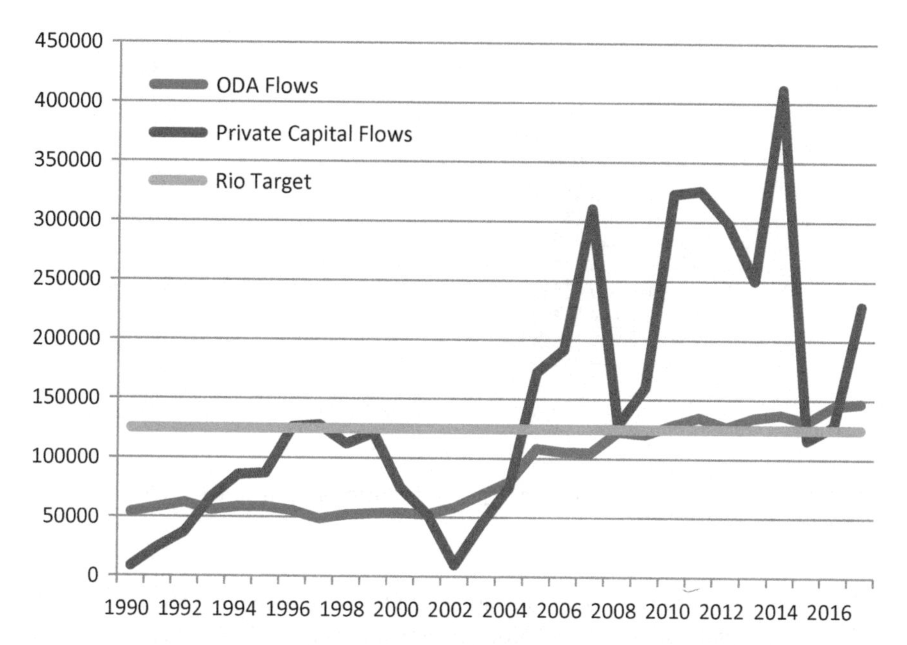


**Fig. 1** ODA and private capital flows to all recipient countries versus Rio target, 1990–2017 (US$ millions, current prices, 08/2019). *Source* Prepared by authors based on data obtained from OECD's 'Query Wizard for International Development Statistics' (see OECD 2019b: https://stats.oecd.org/qwids)

reached in 2014 when the private capital flows to all recipient countries recorded an investment of US$412 billion. The fact that these flows have remained above the ODA flows since 2005—except 2015 and 2016, has sustained the interest of many parties in searching for profitable and impactful investment opportunities in urban environmental services. A closer look at the breakdown of private flows during 2016–2017 shows that there was a big jump in 'lending to sovereigns' (Fig. 2); however, "these private flows are volatile and not a full substitute for aid" (Dag Hammarskjöld Foundation and United Nations Multi-Partner Trust Fund Office 2019:18; for details, see Kharas 2019). Moreover, during the same period (i.e. between 2016 and 2017), the cross-border private 'investments in infrastructure' declined marginally.

Further to the possibility of bringing additional private capital to bear, many supporters of PPPs argue that private participation can bring other benefits as well. In their view, the commercial incentives facing private companies make them more efficient than government providers. In addition, the creativity and drive of private firms is needed to develop new solutions to water delivery challenges. The expectation is that this combination will lead to more cost-effective improvements in services over time (Asian Development Bank 2008; World Bank 1997; World Bank et al. 2014).

While these arguments have traditionally been used to support the participation of large, multinational companies in PPPs (World Bank 2003), policy and advocacy attention has been expanded to include local, private companies. For example, in many cities, small firms—often informal—are the main suppliers to areas outside the

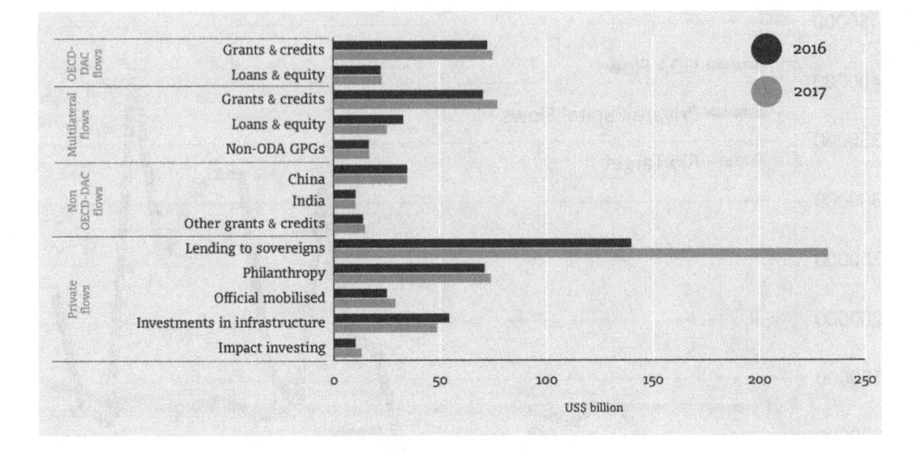


**Fig. 2** Broadly-defined international development contributions (current US$ billion, 09/2019). *Source* Kharas (2019: 71)

public water network (Angueletou-Marteau 2008; Development Workshop Angola 2009; Kjellén and McGranahan 2006; Pangare and Pangare 2008; Snell 1998; Solo 2003). They use low-cost methods in direct response to customer demand—but often outside the formal regulatory environment. Finding ways to integrate the activities of these decentralized providers with more centralized systems is one of the main PPP opportunities now being explored.

As a result, the interest in PPPs stems from the *fact* that many governments are having trouble addressing urban environmental issues and the *belief* that financial and operational benefits will flow from involving private firms. The questions then become: (a) how best to use PPPs to improve urban environments; and (b) how their performance compares to alternative approaches—such as reforming public sector provision.

## 3 What Are PPPs?

*The Economist* defines "public-private" as "[u]sing private firms to carry out aspects of government. This has become increasingly popular since the early 1980s as governments have tried to obtain some of the benefits of the private sector without going as far as full privatization".[8] Many people use the phrase 'public-private partnerships' in many different ways, covering a wide variety of relationships. At present, "… the definition of PPP lacks specificity and consistency, with different countries, sectors and projects conceptualising PPP differently in various political, economic

[8] *The Economist* does not include "partnerships" whilst defining "public-private". See: https://www.economist.com/economics-a-to-z/p#node-21529888 (accessed 3 September 2019).

and social environment contexts" (Wong et al. 2015: 262).[9] In order to be helpful to the efforts to improve urban environments, however, a definition of sufficient specificity to allow discussion and engagement should be used.

For purposes of this chapter, PPPs are defined as an *optimization tool*—a range of voluntary structures and processes for optimizing the application of public and private resources to meet shared goals[10] in a particular context. The individual components of this definition are described in more detail in the next section. In many ways, crafting a PPP is like traditional quilt making—taking the resources that are available and stitching them together into a whole that meets pressing local needs. For a quilt that means providing warmth—for an urban water PPP, it means providing more people with access to cleaner drinking water and healthier sanitation. Sometimes beautiful, sometimes not, but—if successful—extracting improved performance from the resources at hand.

As such, PPPs start with questions, not answers:

- What *goals* can the potential partners only achieve by acting together? If goals can be met acting alone, there is no need to take on the added burden of working with other parties.
- Where are the potential partners' *strengths and weaknesses* complementary? Every partner has strengths that can help make a partnership thrive—even the poorest slum dweller. The trick is matching assets with weaknesses to create more than any party can do alone.
- What PPP *structure* optimizes application of the available resources? A spectrum of possible PPP structures exists between fully public and fully private (as shown in Figs. 3 and 4 later). Finding the one that best fits the partners' goals in context is the key.
- What *processes* should be used to make the partnership effective? Partnerships are on-going relationships. They need to meet the needs of all partners—or they will not participate. When providing public services, they also need to be seen as legitimate.

Before turning to suggested answers to these questions, four *working assumptions* should be made explicit. First, and as stated above, PPPs should be seen as a tool, not a religion. They will not work for all problems in all situations. They are also really hard to do well. As such, they should not be viewed as a *silver bullet* solution for all urban environmental issues. They should be seen as one of a number of tools that might be used, and they should be evaluated in comparison to other options. Second, PPPs must focus on achieving a goal or vision that is shared by the partners, but which they cannot achieve acting alone. Every organization involved in a partnership has multiple loyalties—to their home organization and to the joint endeavor. Studies of such matrix organizations suggest that the best way to overcome these potentially conflicting loyalties is to have a clear target around which to organize the joint effort (Nicholson 1998). Third, PPPs should be seen as new structures and processes for action, but not new core values for the partners. Governments should protect the public interest.

[9]For a survey of definitions of PPPs, see Jomo et al. (2016).

[10]These could be shared or overlapping goals. Since a PPP is a voluntary arrangement, both sides need to conclude they will benefit from the partnership.

Businesses need to make a profit. Non-Governmental Organizations (NGOs) and civil society organizations should defend the interests of those whose voices are not heard. PPPs are best built in the areas where these core goals overlap—not on an expectation that potential partners will change their fundamental missions. Finally, PPPs are context specific—one size does not fit all. Goals, partners, structures, processes—all will vary from location to location, problem to problem. At the same time, however, successful partnerships appear to share some common themes. The next section suggests ways that these themes may help inform the crafting of PPPs.

## 4 How Think Through Successful PPPs?

A qualitative analysis of PPPs projects around the world brings forth several key features of successful partnerships, which can be grouped around three major areas: individual champions; partnership space; and optimized structures and processes. The working hypothesis is that attention to these key features while answering the questions posed above will increase the likelihood of developing a successful PPP.

### *4.1 Individual Champions*

Much of the discussion of PPPs focuses on arrangements between institutions: municipal government with multinational company and, maybe, various NGOs. While it is critical to understand these institutional relationships, focusing on institutions misses a key element of successful partnerships.

Partnerships start with individuals, not institutions.[11] The history of every successful partnership starts with a person who makes the partnership happen. These *individual champions* are the drivers behind the efforts to attract other individuals, who then bring their own organizations into the voluntary relationship.

Individual champions often have special personal skills that help address the tensions at the heart of many partnership efforts. On the one hand, they have the passion, realism and ability to inspire others that is required to bring and keep the partners together. On the other, they are willing to share leadership and—often most importantly—the credit for the partnership's successes. These potentially conflicting skills may well best be described in oxymoronic terms, such as *powerful humility* or *cynical idealism*. One hopes that many people have or can learn such skills. In practice, however, it seems that such individuals are fairly rare.[12]

[11] For a broader discussion on this subject, see Campbell (2012).

[12] It should be noted, however, that PPPs are established between organizations even though individuals may initiate them.

## 4.2 Partnership Space

While PPPs start with individuals, they need to scale up to address issues, such as urban environmental problems. Their ability to do so is constrained by the context—or *partnership space*—in which they are formed. While the precise features of the local context vary from place to place, understanding three main factors is key to building successful partnerships: shared reward; shared investment; and local constraints.

### 4.2.1 Shared Reward

As discussed above, since PPPs are voluntary arrangements, they are founded on each partner's belief that they will better achieve their own goals by working with others. PPPs are also built in areas where the potential partners' preexisting goals overlap, rather than by asking them to change their fundamental organizational missions. If potential partners do not have overlapping goals, there is no space in which to build a partnership. While detailed goals and missions vary across different organizations, starting from the archetypical goals of governmental, business and civil society organizations can help identify areas of potential overlap (Waddell 2002).

Presented in this way, the potential for a wide variety of PPPs becomes apparent (Table 1). As one illustration, PPPs in the urban water sector can fit:

- the governments' sustainable development goals of reducing child mortality, combating disease and sustaining the environment;
- businesses' goals of generating prosperity (UN-HABITAT 2013) and wealth, and expanding markets; and
- civil society's goals of improving the lives of the poor, reducing environmental impacts and, potentially, creating new economic opportunities.

While overlapping goals are a necessary prerequisite for PPPs, they are not sufficient. Many other factors also affect the partnership space available for any particular partnership effort.

### 4.2.2 Shared Investment

Another critical component is each partner's belief that the others bring something useful to the table—some asset(s) that will help make the partnership a success. It also involves recognition that each partner has one or more weaknesses that prevent it from achieving its goals by working alone.

Identifying and sharing information on potential partners' assets and weaknesses is one of the most difficult aspects of forming a PPP—and one where the specialized skills of individual champions are most important. While again the specifics will vary across partners, matrices of archetypical assets and weaknesses (Plummer and Waddell 2002) can be used to help think through potential combinations (Table 2).

**Table 1** Core goals: many areas of potential overlap

| Government (SDGs) | Business | Civil society |
|---|---|---|
| • SDG 1: no poverty<br>• SDG 2: zero hunger<br>• SDG 3: good health and well-being<br>• SDG 4: quality education<br>• SDG 5: gender equality<br>• SDG 6: clean water and sanitation<br>• SDG 7: affordable and clean energy<br>• SDG 8: decent work and economic growth<br>• SDG 9: industry, innovation and infrastructure<br>• SDG 10: reduced inequality<br>• SDG 11: sustainable cities and communities<br>• SDG 12: responsible consumption and production<br>• SDG 13: climate action<br>• SDG 14: life below water<br>• SDG 15: life on land<br>• SDG 16: peace and justice strong institutions<br>• SDG 17: partnerships to achieve the SDGs | • Generate new wealth<br>• Expand markets<br>• Develop new products<br>• Lower production/delivery costs<br>• Improve human resources<br>• Build reputation/brand<br>• Work with new partners<br>• Push for market frameworks/opportunities<br>• Embrace new and smart technologies<br>• Innovate<br>• Adapt to 4th industrial revolution | • Speak for those not represented<br>• Improve lives of the poor<br>• Create new economic opportunities<br>• Improve basic health care/education<br>• Reduce environmental impacts<br>• Strengthen local cultures/values<br>• Consolidate traditional knowledge<br>• Build on local wisdom, knowledge, expertise and experience |

*Source* Adapted by Authors from Waddell (2002) and United Nations (2015)

**Table 2** Core assets and weaknesses: room to optimize

| | Government | Business/private sector | Civil society |
|---|---|---|---|
| Assets | • Laws/regulation<br>• Enforcement<br>• Taxation revenues<br>• Policy knowledge<br>• Longevity<br>• Reputation | • Capital/financial assets<br>• Production systems<br>• Sector knowledge<br>• Innovation<br>• Reputation | • Inspirational/member assets<br>• Community bonds<br>• Community knowledge<br>• Reputation |
| Weaknesses | • Limited finances<br>• Inflexible rules<br>• Slow decisions<br>• Complexity of systems<br>• Desire to control | • Pull to monopoly<br>• Disregard for social and environmental externalities<br>• Short-term focus<br>• Inequality of outcomes<br>• Transactional focus<br>• Profit mindedness | • Narrow focus<br>• Amateurism<br>• Scarce materials<br>• Fragmentation<br>• Ideological parochialism |

*Source* Adapted from Plummer and Waddell (2002)

For example, in the urban water sector, a wide variety of useful combinations of assets and weaknesses are potentially available. They may include using:

- business capital and sectoral knowledge, plus time from members of civil society organizations, to expand water services to areas not now served by the government;
- government laws and enforcement, plus traditional wisdom, knowledge, expertise and experience of local communities and their grassroots and civil society organizations, to oversee business involvement in delivering water services; and
- centralized government authority and wide business reach to help support decentralized civil society efforts to improve local services.

Investment by all of the partners is also important for another reason—shared risk increases commitment. If each partner puts something it values into the partnership, then each partner has an additional incentive to make sure the partnership succeeds. As depicted in the matrix, such investments are not limited to money—time, knowledge, reputation, networks, linkages and other assets can also be just as important.

### 4.2.3 Within Local Constraints

Even with overlapping goals and complimentary assets/weaknesses, the space for any particular partnership is also defined by a variety of local constraints. Again, these constraints vary from place to place. They do, however, tend to fall into the following major categories:

- *Institutional*: Are the institutions, with which the individuals pushing the partnership are affiliated, likely to support the PPP?
- *Political*: Is the government willing, and does it have the capacity to support the proposed partnership? Are there opportunities for those potentially affected by the PPP to participate in the process?
- *Social*: Do potential users support the partnership? What is their willingness to contribute to its success and sustainability?
- *Economic*: What is the ability of users to pay for the proposed (improvement to) services? What is the scope of local capital sources? How much access to international capital markets is available?
- *Legal*: Do predictable investment frameworks exist? Does the government have the capacity and willingness to monitor and regulate private involvement in the provision of public services?
- *Environmental*: What quantities of environmental resources (such as water) are available? What is their quality?

The answers to these questions on local constraints, combined with those on the opportunities for shared reward and shared investment, define the partnership space potentially available in any particular location. Whether a successful PPP will be developed within the available partnership space then depends on the structures and processes applied.

## 4.3 Optimized Structures and Processes

Any discussion of the possibilities for a PPP grows out of a desire to see one or more tasks done better. Most urban environmental problems require a variety of responses. Some of these tasks are best done by governments, others by private parties, still others in some form of partnership arrangement.

As such, it can be helpful to think of the different PPP structures as falling along a spectrum from fully public to fully private according to the task to be done (Fig. 3). Doing so allows one to separate out different tasks (such as setting standards of performance or making capital investments) and then to consider a range of structures for achieving each of them involving greater or lesser degrees of public or private involvement. For example, in the urban water sector, PPPs tend to use one or more of the following structures (Fig. 4).

Note that this spectrum recognizes the options of having the government provide all water services (as in the Netherlands) or having the private sector provide all under government oversight (as in England and Wales). In between these extremes, a variety of voluntary arrangements combining public and private assets in pursuit of overlapping goals exist—PPPs. The major PPP types include:

- *Passive private investment in public provision*, such as municipal bonds. Widely used in the United States and a few other countries, such PPPs are not allowed or financially attractive at the municipal level in many countries.

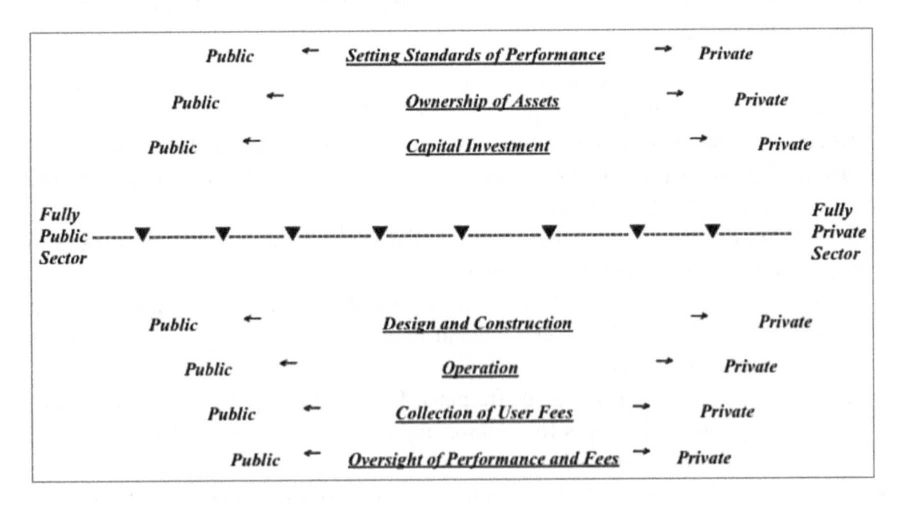


**Fig. 3** What structure: PPP task spectrum. *Source* Authors

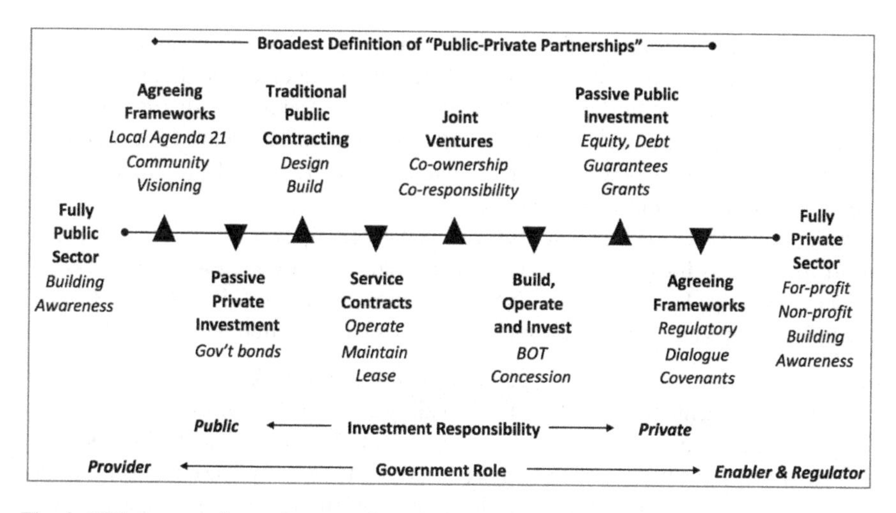


**Fig. 4** PPP structures for environmental services. *Source* Authors

- *Service or management contracts*, involving private operations within publicly controlled systems. Generally the easiest form of urban water PPP, allowing short-term arrangements tailored to government needs and not requiring substantial private investment.
- *Joint ventures*, involving combined public-private control of water systems. Directly combining public and private assets and weaknesses into a jointly owned company.
- *Private building, operation and investment* in publicly owned systems, such as concessions or various 'build, operate and transfer' (BOT) schemes. Long-term arrangements under which the government transfers both operational and investment responsibilities to the private sector while retaining ownership and oversight responsibilities (Gentry and Abuyuan 2000).

Choosing among these, or other, structures for any particular PPP requires a judgment as to which one best meets the goals of the partners in the local context (Table 3).

**Table 3** Optimizing goals across structures

| Goal | Fully public | Public $ and private operation | Joint operation and $ | Private operation and $ | Fully private |
|---|---|---|---|---|---|
| New $ | | | ✓ | ✓ | ✓ |
| Technical expertise | ✓ | ✓ | ✓ | ✓ | ✓ |
| Commercial incentives | | ✓ | ✓ | ✓ | ✓ |

*Source* Authors

As one example, if one of the key goals is to attract substantial amounts of new private investment in the water system, service and management contracts alone will not work. If municipalities are not allowed to borrow money directly (as is the case in many countries), municipal bonds will not work either. That leaves joint ventures, concessions, BOTs or other similar arrangements as the only PPP structures that might achieve the municipality's goals.

Finally, even if a structure that fits the local partnership space is chosen, it will still fail if the process for putting it in place is flawed. Lack of attention to process can kill a partnership. This is another place where the specialized skills of the individual champions come into play. They are the ones who have to find ways to:

(a) *Bring partners together*. In addition to the initial task of sparking people's interest in achieving a shared goal, scaling up a PPP requires sorting through the more formal rules and customs by which different organizations act. For example, since PPPs are new ways of working for many governments, they may well come into conflict with traditional rules on public procurement. More input by potential bidders on design criteria may be desirable. More disclosure to and involvement by users in the procurement process may be critical. Navigating the different traditions for how to *do business* that governments, companies and NGOs bring with them can be extremely difficult.

(b) *Keep partners together*. In addition to the possibility of receiving a shared reward and the risk of losing a shared investment, maintaining a PPP requires attention to broader relationship issues as well. For a start, the partners need to treat each other with respect—both their goals and their working methods. In addition, the partners need to be able to predict how the others will act and plan their own actions accordingly. Doing so requires a deep understanding of each other's underlying goals and methods. This combination of understanding, respecting and being able to predict provides a useful working definition of the *trust* one so frequently hears referenced in partnership discussions.

(c) *Build legitimacy*. When PPPs are being proposed to provide public services such as water, real concerns arise among users and those who believe that the government is the only legitimate provider of such services. Any effort to develop such a PPP needs to address these concerns throughout the entire process. This starts with the initial identification of goals and possible partners, continues through the development and operation of the partnership, and ends only when the PPP does. The key features of this effort are often: transparency in plans and performance; outreach to potentially affected groups; participation at different levels by circles of interested parties; and, ultimately, the PPP's performance against the stated goals.

(d) *Evolve the PPP over time*. PPPs exist to achieve a goal, and are not an end in themselves. Successful PPPs achieve their goals. Contexts change. As a result, PPPs also pose the question of what happens when its work is done or the context changes? Does it disband? Does it take on other tasks? While the answer varies across different partnerships, the implication is clear—any PPP must be prepared to evolve over time. Its participants and structure must anticipate and be ready to respond to change.

This approach to thinking through partnerships as a function of individual champions, partnership space, and optimized structures/processes offers a framework for building successful PPPs. It does not, however, answer the question of whether a PPP is the best tool to use to improve any particular urban environmental issue. The rest of this chapter explores the opportunities and limits of PPPs by describing some of the current trends in urban water sector partnerships and offering some thoughts on ways their effectiveness might be increased.

## 5 What Are the Current Trends in Urban Water Sector PPPs?

For as long as there have been cities, there have been both private and public providers of urban water services (Gentry and Abuyuan 2000). Private firms delivered most water and sanitation services in the United States and United Kingdom until concerns over their performance led to a gradual takeover by public providers in Victorian times. France has largely maintained its tradition of private operation of publicly owned water systems since mid-19th Century. Local private providers often serve rapidly expanding peri-urban areas of the mega-cities in the developing world.

The 1990s saw a huge increase in the number of and investments in private sector involvement in urban water services. The World Bank statistics shows that between 1991 and 2000, 219 new projects were initiated with total investments of US$37 billion (Fig. 5a, b). The year 1997 recorded the highest private sector investment in urban water sector, US$9.9 billion. Over that decade (1990s), for instance, huge, multi-year contracts to operate and expand existing water systems were negotiated in cities such as Buenos Aires, Jakarta, and Manila, and shares in traditionally publicly owned water companies were sold to private operating companies in Chile. In Asia, some of these projects in the 1990s had 'issues related to the allocation of risks between public and private partners, particularly foreign exchange risks' (Asian Development Bank 2009: 4). For instance, in Metro Manila in the Philippines, the state-owned Metropolitan Waterworks and Sewerage System (MWSS) was divided in 1997 into two PPP concessions. The PPP concession for the East Zone of Metro Manila was awarded to Manila Water Company, which 'provides water treatment, water distribution, sewerage and sanitation services to more than six million people in the East Zone, comprising a broad range of residential, semi-business, commercial and industrial customers' (Manila Water 2019). The PPP concession for the West Zone of Metro Manila was awarded to Maynilad Water Services, Inc. (Maynilad)—a consortium comprising Benpres Holdings Corporation, Suez Environment, Lyonnaise Asia Water Private Limited, and Metrobank. By the end of 1997, due to the Asian financial crisis and El Nino phenomenon, Maynilad could not meet its service and financial obligations; all of this resulted in a string of financial, legal and

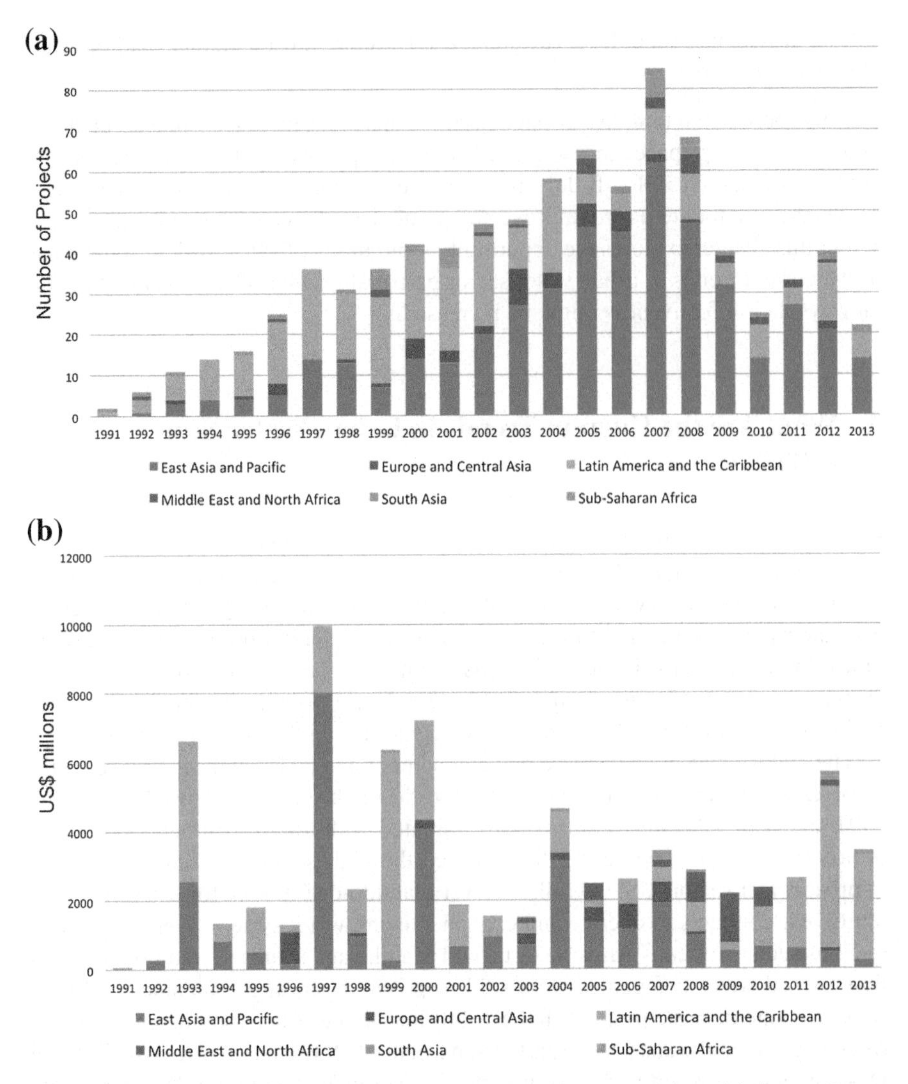


**Fig. 5** **a** Number of water projects, 1990–2013: by region and year of financial closure. *Source* Prepared by authors based on data obtained from the World Bank—Private Participation in Infrastructure (PPI) Project Database (http://ppi.worldbank.org/). **b** Investment in water projects, 1990–2013: by region and year of investment. *Source* Prepared by authors based on data obtained from the World Bank—Private Participation in Infrastructure (PPI) Project Database (http://ppi.worldbank.org/)

regulatory disputes between Maynilad and MWSS (Chiplunkar et al. 2008; Maynilad Water Services, Inc. 2017). In 2007, a joint venture between DMCI Holdings, Inc. (DMCI) and Metro Pacific Investments Corporation (MPIC)—or the DMCI-MPIC Water Company—won the competitive bidding and acquired 83.96% of Maynilad's shares. Thereafter, Maynilad slowly improved performance, and now provides safe and reliable water supply to over nine million people in the West Zone of Metro Manila (Maynilad Water Services, Inc. 2017).

During the past decade (2001–2010), private sector participation in urban water services increased in terms of the number of projects. Between 2001 and 2010, 533 new projects were initiated at the World Bank, compared to 219 during the 1990s. However, the private capital flows to urban water sector declined from US$37 billion during 1990–2000 to US$25 billion between 2001 and 2010. The leading recipients of private sector investments in urban water sector were East Asia and Pacific (mainly China), Latin America and the Caribbean, and Middle East and North Africa regions.[13] Private sector investments in the urban water sector further declined after the 2008 global economic crisis. This trend has continued until 2013, which recorded only 22 new projects with a total investment of US$3.4 billion.

The reasons for this decline in international private investment are many and varied, which can be grouped into at least three sets: (i) functioning of public sector and political systems; (ii) private sector and commercial realities; and (iii) opposition to private sector involvement.

## *5.1 Functioning of Public Sector and Political Systems*

(a) *Government Control of 'Public Service Units' (PSUs)*: Governments and political leaders often want to keep control of PSUs. For political, historical and economic factors, governments in developing countries perceive themselves to be "strong and dominant" institutions of governance. Governments do not want to lose the "sense of control" and, therefore, want to continue with their "strong and dominant" image.

(b) *Perception of Corruption*: The public sector has been a major provider of secure employment. Any (potential) move by governments towards private sector involvement in urban service provision—with attendant flexible employment arrangements, is seen as granting a favor to profit-making business enterprises. This problem is aggravated by examples of what some consider unhealthy PPPs, such as the privatization of water services in Greater Jakarta (see Argo and Laquian 2007; Hasibuan 2007; UN-HABITAT and UN-ESCAP 2010).

(c) *Time required for developing PPPs*: It takes a (very) long time to develop successful PPPs, as noted earlier. The long gestation period to design and implement PPPs is unhelpful on two accounts. First, it often goes beyond the usual 3–5 year

[13]Compared to these regions, private sector investments in water sector have remained low in South Asia and Sub-Saharan Africa (see Water and Sanitation Program 2011; World Bank 2014).

political tenure of (s)elected decision makers. If a PPP project takes longer than 3–5 years to design and implement, the likelihood of its getting support from the next/following administration cannot be guaranteed. Second, the PPP initiative taken by one (s)elected government is often not supported by the next government, especially if it belongs to another political party. It becomes politically difficult for the new government to support a PPP initiative started by a previous government, because the new government could be perceived as patronizing the *other political party's agenda* and/or its own weakness in finding new ideas to address urban service improvement.[14]

(d) *Question of Administrative Capacity*: National governments in many developing countries, let alone their urban local governments, lack the necessary administrative capacity to design and guide the development and implementation of PPPs. Moreover, they fall short of regulatory and monitoring capacity to ensure fairness, transparency and effectiveness in the implementation of PPPs (Dahiya 2014). This is despite the fact that many toolkits on PPPs are available, along with lessons learnt from past projects.[15] In short, one of the key factors for supporting successful PPP implementation is the 'public sector capacity to manage PPP appropriately' (Zen 2018).

### 5.2 Private Sector and Commercial Realities

The second set of factors behind the decline in PPPs relates to the private sector and commercial realities. Multinational water companies are pulling back from new investments in many markets. Many of their existing investments are in financial distress. Little new, international private money is going into rehabilitating water systems. Some of it goes back to the financial crises and resulting currency devaluations of the late 1990s. Many of the largest concession agreements were built around borrowing money in dollars and paying it back from water usage fees collected in local currencies. When the value of the local currencies declined (often by huge percentages) the effective cost of paying back the loans went up dramatically. In turn, this led to efforts to increase user fees—which resulted in increased public opposition (see below). Combining the large increases in user fees with, in many instances, much smaller improvements in performance compounded the public concern.

[14]For an example of problems associated with political intervention in transport sector in Bangkok, see Pongsiri (2012).

[15]See Asian Development Bank (2000, 2008), Delmon (2014), Mehta (2011), UNDP (2000, 2006), UN-ECE (2008), UN-ESCAP (2011), World Bank (1997, 2006, 2017a, 2017b), World Bank et al. (2014).

### 5.3 *Opposition to Private Sector Involvement*

Finally, activist coalitions and research networks have been formed to oppose any further private involvement in water services. Its members include those drawn from groups concerned about several different issues, including: globalization and the World Trade Organization (around international trade in water)[16]; particular efforts to bring in international water companies[17]; and the impact of doing so on public sector employees.[18] Much of the focus of these efforts is on reenergizing the option of keeping water services in public hands by reforming public utilities, not by replacing them with PPPs or private firms (Centre for Science and Environment 2011; Hall 2002, 2015).

Due to these factors, PPPs have not taken off as the dominant mechanism for improving urban environments in general, and in the urban water sector in particular. Yet, the need for more investment in the urban water sector remains. As such, more recent attention has been focused on opportunities for PPPs involving:

- *Public finance*, either ODA or government bonds, with private operational support (service or management contracts, joint ventures);
- *Local private capital*, from governmental pension funds or local capital markets where available; and
- *Links with local organizations*, where the centralized government system can tap into the decentralized activities of smaller firms and civil society organizations.

While these efforts are extremely useful, it seems unlikely that they will be sufficient to meet the need for capital. Ways need to be found to combine the available public and local funds with additional international private capital. In addition, it makes good sense to try and bring the efficiency and creativity of the private sector to bear on developing cost-effective ways to expand access to urban water services. As an optimization tool, PPPs should have a continuing role in the efforts to do so. This is because PPPs feature a performance-based management framework that combines strengthening the quality of public services and reducing the cost of their provision (Asian Development Bank 2014: v).

In a recent development, the World Bank's Public-Private-Partnership Legal Resource Center (PPPLRC) has started to examine the 'Key issues in PPPs for the Poor', and listed 10 key issues which include: (i) Monopoly of service providers; (ii) Prohibition on the delivery of service to informal settlements; (iii) Ad hoc solutions to services delivery; (iv) Unnecessary/inappropriate engineering and quality standards; (v) Judicial system; (vi) Regulator; (vii) Tariffs; (viii) Billing collection; (ix) High initial costs of setting up the service (e.g., cost of grid connection, solar panel

[16]Such as the Polaris Institute (http://www.polarisinstitute.org/) and Focus on the Global South (http://focusweb.org/).

[17]Such as the riots in 2000 protesting plans to privatize the water services in Cochabamba, Bolivia. Also see Focus on the Global South and Transnational Institute (2007).

[18]Such as the concerns raised by Public Services International and the research it funds through the Public Services International Research Unit (http://www.psiru.org/).

installation or setting up trash hauling); and (x) Community engagement through consultations (World Bank 2019c). At the time of writing this chapter, the World Bank was in the process of inviting comments and suggestions for links or material that could be included on the PPPLRC website (World Bank 2019c).

## 6 How Best Use PPPs for Urban Water Services?

The issues standing in the way of successful urban water PPPs are many and varied. Some are technical or financial in nature—what technologies provide the most cost effective solutions, how best to address the impact of currency fluctuations. Some raise basic questions about how much the different partners' goals overlap—network expansion versus poverty reduction versus making a profit. Others concern massive differences in initial information and expectations—estimated versus actual amounts needed to improve or expand existing water networks. Still others are rooted in fundamentally different working methods, such as the timing and basis for decision-making.

The most critical questions, however, concern how best to ensure that an urban water PPP actually serves the public interest. One of the key advantages of private sector involvement in the urban water sector, for instance, is "to clarify the true cost of the water, [and] to eliminate or quantify local government budget transfer and subsidies" (Jenny 2008). This begs the question: If local governments do not have the capacity to run a water system, do they have the capacity to oversee private sector involvement in the operation of one? Will private interests unfairly benefit from private involvement in water services? Are other tools (such as reforming public water utilities) more effective in meeting the public interest?

Such questions take us back to the beginning of the PPP analysis—what are we trying to accomplish and is a PPP a good tool for doing so in this location? While the full range of answers are beyond the scope of this paper, and will vary with each decision-maker and context, two areas seem especially important for further work. One has to do with the scope of the advice available to municipal mayors and officials. The other concerns the basis for assessing the performance of either public or private water services.

### *6.1 Addressing the Mayors' Dilemma*

When mayors consider their options for improving urban water services, they are often faced with the dilemma of making a choice of advisors—those who will tell them about PPPs or those who will tell them about reforming the public utility. Much of this appears to be driven by the donors supporting the advice—those in favor of private sector involvement and those against. Rare or non-existent are the advisors who can take a step back and effectively evaluate the public and private options against each other.

Thus, the choice between public or private is posed too early in the process. Efforts should be made to expand the pool of donors willing to support and advisors able to undertake a side-by-side comparison of both the options for public sector reform and for private sector participation. Doing so should help increase the legitimacy of the ultimate choice.

## 6.2 Assessing the Performance of All Partners

Any such efforts will also benefit from the second suggestion—taking a page from the sustainable forestry book and developing a system for certified water PPPs. All parties to PPPs have rights and responsibilities. All parties involved in urban water systems—providers, investors, users, overseers—also have rights and responsibilities, whether they are public or private. For example, while:

- *Providers* have the responsibility of meeting performance standards, they have the right to be given the resources to do so;
- *Investors* have the responsibility of conducting adequate due diligence before making their investment, they have the right to have the other parties to the investment contract fulfill their commitments;
- *Users* have the right to affordable water services, they also have an obligation to contribute to its provision; and
- *Overseers* have the right to collect data on system performance, they have the responsibility to enforce standards of performance and to do so in a fair and transparent manner.

Frequently, however, these rights and responsibilities are not understood, articulated or agreed among the parties to water services—whether they be public, private or PPPs. At a minimum, the result is inconsistent measures of performance across water systems and the methods for improving them. For PPPs, the results are often an inability to compare with public sector options and concerns about the legitimacy of the traditional PPP process.

When similar issues arose around timber harvesting, the Forest Stewardship Council was formed to develop standards for sustainable forestry and procedures for verifying that harvesting activities were meeting the standard[19] (see Forest Stewardship Council 2015, 2016). As a tripartite organization, Forest Stewardship Council includes representatives from governments, business and NGOs. Working together, Forest Stewardship Council members developed standards and verification procedures that enjoy a high degree of acceptance across parties in all three sectors. Similar initiatives have now been undertaken on marine resources, dams and other areas for which traditional policy-making processes have proven inadequate (Cashore 2002).

---

[19]See: https://ic.fsc.org/index.htm.

Certified water PPPs take this idea and expand it. Two steps are involved. First, a tripartite, global group would be assembled to develop standards and verification procedures for urban water systems. Many such standards already exist for providers of water services.[20] What is missing are: (i) the standards to be applied to the other parties in urban water systems (users, investors, overseers); and (ii) an articulation of all of these parties' rights to accompany their responsibilities. Second, the new standards and procedures could then be used in two ways: (i) to assess the performance of existing systems; and (ii) to inform the development of better solutions, including PPPs. Third-party verifiers could be used to evaluate the performance of existing systems—public or private—against the agreed standards and to suggest areas for improvement. On a new project, the articulated rights and responsibilities of the different parties could be used as the starting point for identifying areas for improvement and assigning responsibilities across different parties. In some cases, the result might be reform of the public system, in others the creation of a PPP or other private involvement. In either case, a widely accepted basis for tracking the performance of whatever approach is chosen would also then exist.

## 7 Conclusion

This chapter shows that PPPs are an important tool of governance for the delivery of infrastructure and services in general and that of urban environmental services in particular. If properly guided, international private capital flows are considered an important part of private sector contribution to help address social and environmental challenges, for example in water and sanitation, clean energy, and women and youth entrepreneurship. However, with regard the formulation and implementation of PPPs, some key challenges remain including: (i) understanding the local context, including administrative and technical capacity of urban local governments, (ii) figuring out whether a PPP or an improved public sector operation is the right solution for the delivery of services, and (iii) implementing a PPP in a proper manner, in case PPP is found to be the best institutional option to improve the provision of services. Moreover, performance over time is the key to addressing concerns about both inadequate urban environmental services and the potential role for urban environmental PPPs. In other words, reaching broad agreement on the goals of urban environmental services (such as urban water discussed in this chapter), evaluating both public and private options, and tracking performance against goals is necessary. Given the pressing need for large, new investments in expanding access to urban environmental services, and the potential for PPPs to help optimize the application of available public and private resources, they should have a major role to play in helping to meet urban environmental goals, emphasized as they are in local, sub-national, national, regional and global sustainable development agendas.

[20]Such as the WHO Guidelines for Drinking-water Quality (see WHO 2006).

As the institutional rhetoric on governance deepens across the world, the added value of PPPs—for the provision of infrastructure and services, ought to be understood *vis-à-vis* the ever-growing advocacy and operational practice of community engagement and civil society participation for sustainable development. Taking into account the on-the-ground realities with regard to the operational processes of governance, the conceptual and analytical work has already begun on the idea of 'public-private-people partnerships' (see Ng et al. 2013; Perjo et al. 2016). Thus, widening the conceptual and analytical focus beyond PPPs helps in looking into the emergent domain of 'multi-stakeholder partnerships', which are promoted as 'means of implementation' within the 2030 Agenda for Sustainable Development in general and under SDG 17 in particular (see Dahiya and Okitasari 2018; United Nations 2015). More research, policy and operational work needs to be done in order to develop a more refined understanding of PPPs to help them be even more effective tools of governance for the provision of infrastructure and services, within the larger context of growing advocacy for multi-stakeholder partnerships towards the implementation of the 2030 Agenda for Sustainable Development and beyond.

## References

African Development Bank (2001) Handbook on stakeholder consultation and participation in African Development Bank Operations. Environment and Sustainable Development Unit (OESU). Abidjan, Côte d'Ivoire: African Development Bank. Available at: https://www.afdb.org/fileadmin/uploads/afdb/Documents/Policy-Documents/Handbook%20on%20Stakeholder%20Consultaion.pdf

Angueletou-Marteau A (2008) Informal water suppliers meeting water needs in the peri-urban territories of Mumbai, an Indian perspective. Global changes and water resources: confronting the expanding and diversifying pressures: XIIIth world water congress, Sept, Montpellier, France

Argo T, Laquian A (2007) The privatization of water services: effects on the urban poor in Jakarta and Metro Manila. In: Laquian A, Tewari V, Hanley L (eds) The inclusive city: infrastructure and public services for the urban poor in Asia. Woodrow Wilson Center Press/Johns Hopkins University Press, Washington, DC/Baltimore, MD, pp 224–248

Ashley R, Cashman A (2006) The impacts of change on the long-term future demand for water sector infrastructure. Infrastructure to 2030: telecom, land transport, water and electricity. Organization for Economic Co-operation and Development, Paris, pp 241–347

Asian Development Bank (2000) Developing best practices for promoting private sector investment in infrastructure: water supply. Asian Development Bank, Mandaluyong

Asian Development Bank (2008) Public-private partnership handbook. Asian Development Bank, Manila

Asian Development Bank (2009) ADB assistance for public-private partnerships in infrastructure development—potential for more success. Impact Evaluation Department. Asian Development Bank, Mandaluyong

Asian Development Bank (2014) Public-private partnerships in urbanization in the People's Republic of China. Asian Development Bank, Mandaluyong

Asian Development Bank (2018) The office of public–private partnership. Asian Development Bank, Mandaluyong. Available at: https://www.adb.org/publications/office-of-public-private-partnership-flyer. Accessed 25 Aug 2019

Bayliss K, van Waeyenberge E (2018) Unpacking the public private partnership revival. J Dev Stud 54(4):577–593. https://doi.org/10.1080/00220388.2017.1303671

Bigio AG, Dahiya B (2004) Urban environment and infrastructure: toward livable cities. World Bank, Washington, DC

Boardman AE, Vining AR (2012) The political economy of public–private partnerships and analysis of their social value. Ann Public Coop Econ 83:117–141. https://doi.org/10.1111/j.1467-8292.2012.00457.x

Campbell T (2012) Beyond smart cities: how cities network, learn and innovate. Earthscan Publications, London

Cashore B (2002) Legitimacy and the privatization of environmental governance: how non-state market-driven (NSMD) governance systems gain rule-making authority. Governance 15(4):503–529

Centre for Science and Environment (2011) Water PPPs: are they here to stay? Available at: http://www.cseindia.org/node/2857. Accessed Oct 2014

Chiplunkar A, Dueñas MC, Flor M (2008) Maynilad on the mend: rebidding process infuses new life to a struggling concessionaire. Asian Development Bank, Mandaluyong. Available at: https://www.adb.org/sites/default/files/publication/29272/maynilad-mend.pdf. Accessed 19 Nov 2019

Cui C, Liu Y, Hope A, Wang J (2018) Review of studies on the public–private partnerships (PPP) for infrastructure projects. Int J Proj Manag 36(5):773–794. https://doi.org/10.1016/j.ijproman.2018.03.004

Dag Hammarskjöld Foundation and United Nations Multi-Partner Trust Fund Office (2019) Financing the UN development system: time for hard choices. Dag Hammarskjöld Foundation/United Nations Multi-Partner Trust Fund Office, Uppsala/New York

Dahiya B (2014) Southeast Asia and sustainable urbanization. Glob Asia 9(3):84–91

Dahiya B, Okitasari M (2018) Partnering for sustainable development: guidelines for multi-stakeholder partnerships to implement the 2030 agenda in Asia and the Pacific. United Nations University, Institute for the Advanced Study of Sustainability (UNU-IAS)/UN-ESCAP, Tokyo/Bangkok

Delmon VR (2014) Structuring private-sector participation (PSP) contracts for small scale water projects. Water and Sanitation Program, International Finance Corporation and the World Bank, Washington, DC

Development Workshop Angola (2009) The informal peri-urban water sector in Luanda. Report submitted to IDRC. Development Workshop Angola, Luanda

Fischer R (2011) The promise and peril of public-private partnerships: lessons from the chilean experience. International Growth Centre Working Paper 11/0483. London School of Economics, London

Focus on the Global South and Transnational Institute (2007) Water democracy: reclaiming public water in Asia—essay collection presented by the reclaiming public water network. Focus on the Global South/Transnational Institute, Bangkok/Amsterdam

Forest Stewardship Council (2015) FSC international standard: FSC principles and criteria for forest stewardship—FSC-STD-01-001 V5-2 EN. Forest Stewardship Council, Bonn

Forest Stewardship Council (2016) Standard setting in FSC—FSC-RP-standard setting V1-1. Forest Stewardship Council, Bonn

Geddes R (2011) The road to renewal: private investment in U.S. transportation infrastructure. American Enterprise Institute Press, Washington, DC

Gentry B, Abuyuan A (2000) Global trends in urban water supply and waste water financing and management: changing roles for the public and private sectors. Working paper CCNM/ENV(2000)36/FINAL. Organization for Economic Co-operation and Development, Paris

Graham S, Marvin S (2001) Splintering urbanism: networked infrastructures, technological mobilities and the urban condition. Routledge, London

Haase D, Guneralp B, Dahiya B, Bai X, Elmqvist T (2018) Global urbanization: perspectives and trends. In: Elmqvist T, Bai X, Frantzeskaki N, Griffith C, Maddox D, McPhearson T, Parnell S,

Romero-Lankao P, Simon D, Watkins M (eds) Urban planet: knowledge towards sustainable cities. Cambridge University Press, Cambridge, pp 19–44. https://doi.org/10.1017/9781316647554.003

Hall D (2002) Water in public hands. Public Services International Research Unit, Greenwich, United Kingdom (Working Paper)

Hall D (2015) Why public-private partnerships don't work: the many advantages of the public alternative. Public Services International Research Unit, University of Greenwich, United Kingdom. Available at: http://www.world-psi.org/sites/default/files/rapport_eng_56pages_a4_lr.pdf. Accessed Aug 2019

Hall D, Bayliss K, Lobina E (2002) Water privatization in Africa. Paper by the Public Services International Research Unit, University of Greenwich for the Municipal Services Project Conference, in Johannesburg, South Africa

Hasibuan D (2007) Jakarta water privatization: workers campaign to bring back water in public hands. Asian Development Bank Forum on Water Privatization, Kyoto

Hodge GA, Greve C (2017) On public–private partnership performance: a contemporary review. Public Works Manag Policy 22:55–78. https://doi.org/10.1177/1087724X16657830

Jenny H (2008) You said private sector participation? Asian Development Bank, Manila. Available at: http://www.adb.org/Water/Articles/2008/Private-Sector.asp#a1. Accessed Nov 2010

Jomo KS, Chowdhury A, Sharma K, Platz D (2016) Public-private partnerships and the 2030 agenda for sustainable development: fit for purpose? DESA Working Paper No. 148, ST/ESA/2016/DWP/148. United Nations, Department of Economics and Social Affairs, New York

Kharas H (2019) International financing of the sustainable development goals. In: Dag Hammarskjöld Foundation and United Nations Multi-Partner Trust Fund Office, Financing the UN Development System: Time for Hard Choices. Dag Hammarskjöld Foundation/United Nations Multi-Partner Trust Fund Office, Uppsala/New York, pp 71–73

Kjellén M, McGranahan G (2006) Informal water vendors and the urban poor. Human settlements working paper series water no. 3. International Institute for Environment and Development, London

Lanjekar P (2010) Public-private partnerships and urban water security: issues and prospects in Mumbai, India. Ritsumeikan J Asia Pacific Stud 27:167–176. Available at: https://en.apu.ac.jp/rcaps/uploads/fckeditor/publications/journal/RJAPS_V27_Prutha.pdf

Leigland J (2018) Public-private partnerships in developing countries: the emerging evidence-based critique. World Bank Res Obs 33(1):103–134. https://doi.org/10.1093/wbro/lkx008

Manila Water (2019) Business profile. Available at: https://www.manilawater.com/customer/about-us/our-company/business-profile. Accessed 21 Nov 2019

Marx A (2019) Public-private partnerships for sustainable development: exploring their design and its impact on effectiveness. Sustainability 11(4):1087. https://doi.org/10.3390/su11041087

Maynilad Water Services, Inc. (2017) History and transformation. Available at: http://www.mayniladwater.com.ph/company-history.php. Accessed 21 Nov 2019

Medda FR, Carbonaro G, Davis SL (2013) Public private partnerships in transportation: some insights from the European experience. IATSS Res 36(2):83–87. https://doi.org/10.1016/j.iatssr.2012.11.002

Medhekar, A. (2014) Public-private partnerships for inclusive development: role of private corporate sector in provision of healthcare services. Procedia - Soc Behav Sci 157:33–44. https://doi.org/10.1016/j.sbspro.2014.11.007

Mehta A (ed) (2011) Tool-kit for public-private partnerships in urban water supply for the State of Maharashtra, India. Asian Development Bank, Manila

Ng ST, Wong JMW, Wong KKW (2013) A public private people partnerships (P4) process framework for infrastructure development in Hong Kong. Cities 31:370–381. https://doi.org/10.1016/j.cities.2012.12.002

Nicholson N (1998) How hardwired is human behavior? Harvard Bus Rev 76(4):135–147

Nundy M, Baru RV (2008) Blurring of boundaries: public-private partnerships in health services in India. Econ Polit Wkly 43(4):62–71

O'Neill P (2010) Infrastructure financing and operation in the contemporary city. Geograph Res 48(1):3–12. https://doi.org/10.1111/j.1745-5871.2009.00606.x
OECD (2019a) Discover the OECD: together we create better policies for better lives. Organization of Economic Cooperation and Development, Paris. Available at: http://www.oecd.org/general/Key-information-about-the-OECD.pdf. Accessed 4 Aug 2019
OECD (2019b) QWIDS—Query wizard for international development statistics. Organization of Economic Cooperation and Development, Paris. Available at: https://stats.oecd.org/qwids/about.html
PADCO, Inc. (2002) A review of reports by private-sector-participation skeptics. Discussion Draft prepared for the Municipal Infrastructure Investment Unit, South Africa and USAID
Pangare G, Pangare V (2008) Informal water vendors and service providers in Uganda: the ground reality. The Water Dialogues. Available at: http://www.waterdialogues.org/documents/InformalWaterVendorsandServiceProvidersinUganda.pdf. Accessed Dec 2014
Patrinos HA, Barrera-Osorio F, Guáqueta J (2009) The role and impact of public-private partnerships in education. World Bank, Washington, DC
Perjo L, Fredricsson C, Oliveira e Costa S (2016) Public-private-people partnerships in urban planning. Working Paper (Deliverable 2.3.1 Potential and challenges of applying public-private-people partnership approach in urban planning), Baltic Urban Lab project, Interregg Central Baltic and European Union. Available at: https://www.nordregio.org/wp-content/uploads/2018/03/Public-Private-People-Partnership-in-urban-planning.pdf. Accessed 4 Sept 2019
Petersen OH (2019) Evaluating the costs, quality, and value for money of infrastructure public-private partnerships: a systematic literature review. Ann Public Coop Econ 90(2):227–244. https://doi.org/10.1111/apce.12243
Plummer J, Waddell S (2002) Building on the assets of potential partners. In: Plummer J (ed) Focusing partnerships: a sourcebook for municipal capacity building in public-private partnerships. Earthscan Publications, London, pp 67–124
Polaris Institute (2003) Global water grab. GATS ATTACK! Pamphlet Series, Ottawa
Pongsiri N (2012) Public-private partnerships and urban infrastructure development in Southeast Asia. In: Yap KS, Thuzar M (eds) Urbanization in Southeast Asia: issues & impacts. ISEAS Publishing and Centre for Liveable Cities, Singapore, pp 139–153
PPIAF (2018) Public-private infrastructure advisory facility—annual report 2018. Public-private infrastructure advisory facility. World Bank, Washington, DC
Rodriguez DJ, van den Berg C, McMahon A (2012) Investing in water infrastructure: capital, operations and maintenance. Water Partnership Program and the World Bank, Washington, DC
Roehrich JK, Lewis MA, George G (2014) Are public–private partnerships a healthy option? A systematic literature review. Soc Sci Med 113(July):110–119. https://doi.org/10.1016/j.socscimed.2014.03.037
Siemiatycki M (2011) Urban transportation public-private partnerships: drivers of uneven development? Environ Plan A 43(7):1707–1722. https://doi.org/10.1068/a43572
Snell S (1998) Water and sanitation for the urban poor—small scale providers: typology & profiles. United Nations Development Program—World Bank Water and Sanitation Program, Washington, DC
Solo TM (2003) Independent water entrepreneurs in Latin America: the other private sector in water services. World Bank, Washington, DC
Srinivasan K (2006) Public, private and voluntary agencies in solid waste management: a study in Chennai City. Econ Polit Wkly 41(22):2259–2268
UNDP (2000) Joint venture public-private partnerships for urban environmental services: report on UNDP/PPPUE's Project Development Facility (1995–1999). PPP working paper series vol II. United Nations Development Programme, New York
UNDP (2006) Global lessons learned in PPPUE local level initiatives in support of pro-poor partnerships for basic urban service provision. Institute for Housing and Urban Development Studies, Rotterdam

UNDP (2019) Public-private partnerships for service delivery. United Nations Development Programme. Available at: https://www.undp.org/content/undp/en/home/librarypage/capacity-building/pppsd.html

UN-ECE (2008) Guidebook on promoting good governance in public-private partnerships. United Nations Economic Commission for Europe, Geneva. Available at: https://www.unece.org/fileadmin/DAM/ceci/publications/ppp.pdf. Accessed Aug 2019

UN-ESCAP (2011) A guidebook on public-private partnership in infrastructure. Economic and Social Commission for Asia and the Pacific, Bangkok. Available at: https://www.unescap.org/resources/guidebook-public-private-partnership-infrastructure. Accessed Aug 2019

UN-HABITAT (2013) State of the world's cities 2012/2013: prosperity of cities. Routledge, New York

UN-HABITAT (2016) World cities report 2016—urbanization and development: emerging futures. UN-HABITAT, Nairobi

UN-HABITAT and UN-ESCAP (2010) The state of Asian cities 2010/11. UN-HABITAT, Fukuoka

United Nations (1992) Earth summit agenda 21, Chapter 33 (Finance). United Nations, New York

United Nations (2015) Transforming our world: the 2030 agenda for sustainable development. A/RES/70/1. Available at: https://sustainabledevelopment.un.org/post2015/transformingourworld/publication

United Nations (2017) The new urban agenda. A/RES/71/256. Habitat III and United Nations, New York

United Nations (2019) Sustainable development goal 11: make cities and human settlements inclusive, safe, resilient and sustainable. Available at: https://sustainabledevelopment.un.org/sdg11

Waddell S (2002) Core competences: a key force in business–government–civil society collaborations. J Corp Citizensh 7(Autumn):43–56. https://www.jstor.org/stable/jcorpciti.7.43

Warner ME (2012) Privatization and urban governance: the continuing challenges of efficiency, voice and integration. Cities 29(Suppl 2):S38–S43. https://doi.org/10.1016/j.cities.2012.06.007

Water and Sanitation Program (2011) Trends in private sector participation in the Indian water sector: a critical review. Water and Sanitation Program, New Delhi

WHO (2006) Guidelines for drinking-water quality: first addendum to third edition, vol 1, recommendations. Available at: https://www.who.int/water_sanitation_health/dwq/gdwq0506.pdf. Accessed 14 Aug 2019

WHO and UNICEF (2014) Progress on drinking water and sanitation—2014 update. WHO Press, Geneva

WHO and UNICEF (2017) Progress on drinking water, sanitation and hygiene: 2017 update and SDG baselines. World Health Organization (WHO) and the United Nations Children's Fund (UNICEF), Geneva. Licence: CC BY-NC-SA 3.0 IGO

Wong ELY, Yeoh E, Chau PYK, Yam CHK, Cheung AWL, Fung H (2015) How shall we examine and learn about public-private partnerships (PPPs) in the health sector? Realist evaluation of PPPs in Hong Kong. Soc Sci Med 147(December):261–269. https://doi.org/10.1016/j.socscimed.2015.11.012

World Bank (1997) Toolkits for private sector privatization in water and sanitation. World Bank, Washington, DC

World Bank (2003) Private participation in infrastructure: trends in developing countries in 1990–2001. World Bank, Washington, DC

World Bank (2006) Approaches to private participation in water services: a toolkit. Public-Private Infrastructure Advisory Facility and the World Bank, Washington, DC

World Bank (2014) Water PPPs in Africa. World Bank Group, Washington, DC

World Bank (2017a) Public-private partnerships: reference guide version 3. World Bank, Washington, DC

World Bank (2017b) Guidance on PPP contractual provisions. World Bank, Washington, DC

World Bank (2019a) A guide to community engagement for public-private partnerships: draft for discussion. International Finance Corporation, Washington, DC. Available at: https://www.commdev.org/wp-content/uploads/2019/06/PPP-Community-Engagement-Guide_FIN-for-WEB-1.pdf. Accessed 7 Sept 2019

World Bank (2019b) List of thematic groups and leaders. World Bank Group, Washington, DC. Available at: http://web.worldbank.org/WBSITE/EXTERNAL/WBI/0,,contentMDK:20212529~pagePK:209023~piPK:207535~theSitePK:213799,00.html. Accessed 4 Aug 2019

World Bank (2019c) Key issues in PPPs for the poor. Public-Private-Partnership Legal Resource Center. World Bank, Washington, DC. Available at: https://ppp.worldbank.org/public-private-partnership/key-issues-ppps-poor. Accessed 7 Sept 2019

World Bank, Asian Development Bank and Inter-American Development Bank (2014) Public-private partnerships: reference guide—version 2.0. World Bank, Asian Development Bank/Inter-American Development Bank, Washington, DC/Mandaluyong

World Economic Forum (2017) Harnessing public-private cooperation to deliver the new urban agenda. World Economic Forum, Davos

Zen F (2018) Public–private partnership development in Southeast Asia. ADB Economics Working Paper Series No. 553. Asian Development Bank, Mandaluyong

**Bharat Dahiya** is Director of Research Center for Integrated Sustainable Development at the College of Interdisciplinary Studies, Thammasat University, Bangkok, and Distinguished Professor at Urban Youth Academy, Seoul, Republic of Korea. He is Series Editor for the SCOPUS-indexed Springer book series, *Advances in 21st Century Human Settlements*. An award-winning urbanist, he combines research, policy analysis and development practice aimed at examining and tackling socio-economic, environmental and governance issues in the global urban context. Since early-1990s, Bharat's research and professional work has focused on sustainable cities and urbanization, strategic urban planning and development, urban environment and infrastructure, and urban resilience. Working with the World Bank, United Nations Human Settlements Programme (UN-HABITAT), the Asian Development Bank, United Nations Development Programme (UNDP), and United Nations University Institute for the Advanced Study of Sustainability (UNU-IAS), he initiated, led, managed and contributed to international projects on sustainable urban development in a number of countries. Bharat conceptualized and coordinated the preparation of United Nations' first-ever report on *The State of Asian Cities 2010/11* (UN-HABITAT and ESCAP, 2010). At the World Bank headquarters, he conducted the first-ever systematic review of the Bank's investments for improving urban liveability, published as a co-authored book, *Urban Environment and Infrastructure: Toward Livable Cities* (2004). For UNU-IAS and the Economic and Social Commission for Asia and the Pacific (ESCAP), he co-authored *Partnering for Sustainable Development: Guidelines for Multi-stakeholder Partnerships to Implement the 2030 Agenda in Asia and the Pacific* (UNU-IAS and ESCAP, 2018). More recently, he co-edited *New Urban Agenda for Asia-Pacific: Governance for Sustainable and Inclusive Cities* (Springer, 2020). He is a member of the International Advisory Board of the UN-HABITAT's *World Cities Report*. He serves on the editorial boards of *Cities: The International Journal of Urban Policy and Planning*, *Environment and Urbanization ASIA*, *Journal of Urban Culture Research*, *Jindal Journal of Public Policy*, and *National Geographical Journal of India*. Reuters, Inter Press Service, SciDev.Net, Nishi-Nippon, The Korean Economic Daily, China Daily, The Hindu, Deccan Herald, Bangkok Post, The Nation, UB Post, The Sunday Times, and Urban Gateway have quoted Bharat's work. He has held academic positions in Australia, Indonesia and Thailand.

Bharat completed his M.A. in Geography from Jawaharlal Nehru University, and Master of Planning from School of Planning and Architecture, both based in New Delhi. He holds a Ph.D. in Urban Governance, Planning and Environment from the University of Cambridge, United Kingdom.

**Brad Gentry** is the F. K. Weyerhaeuser Professor in the Practice at the Yale School of Forestry & Environmental Studies and the Yale School of Management, Senior Associate Dean for Professional Practice at the Yale School of Forestry & Environmental Studies, and a Director of the Yale Center for Business and the Environment. Trained as a biologist and a lawyer, his work focuses on strengthening the links between private investment and improved environmental performance. He has worked on land, water, energy, industrial and other projects in over 40 countries for private (GE, Suez Environnement, Working Lands Investment Partners), public (UNDP, World Bank, Secretariat for the Climate Change Convention, UNEP) and not-for-profit (Land Trust Alliance, The Trust for Public Land, the Elmina B. Sewall Foundation) organizations. He holds a B.A. from Swarthmore College and a JD from Harvard Law School.

# Gender Equality and Local Governance: Global Norms and Local Practices

**Annika Björkdahl and Lejla Somun-Krupalija**

**Abstract** From the global to the local level, gender inequality is the most persistent and entrenched challenge to development. The local level of governance is closest to citizens. Decisions taken at this level have the most direct effect on citizen's everyday lives, as improvements in living, working and leisure conditions depend on good local governance. This chapter focuses on local government. It attempts to describe in detail practical steps toward the localization and realization of SDG #5 at the local level by mapping how to improve, from a gender perspective, the analysis, monitoring, participation, decision-making and access to services to citizens at the local level in order to develop good local governments that serve all citizens. The aim of this chapter is to understand the processes of implementing SDG #5 with focus on urban local governance. We develop a theoretical framework to understand how global norms are translated into local practices. Moreover, we examine policy framework and practices in three municipalities in Bosnia and Herzegovina (BiH), Visoko, East Ilidža and Žepče to implement SDG #5, tools available and used for SDG implementation and actors involved in these processes at the municipal level. The research is based on fieldwork, qualitative text analysis, interviews and surveys in each of the three municipalities to map the use of tools to implement and mainstream gender equality into the decision-making processes, policies and practices at the local level.

**Keywords** Gender equality · Sustainable development goals · Bosnia-Herzegovina · Norms · Municipality · Gender mainstreaming

A. Björkdahl (✉) · L. Somun-Krupalija
Department of Political Science, Lund University, Box 52, 221 00 Lund, Sweden
e-mail: annika.bjorkdahl@svet.lu.se

L. Somun-Krupalija
e-mail: Leijla@mac.com


S. Cheema (ed.), *Governance for Urban Services*, Advances in 21st Century Human Settlements,
https://doi.org/10.1007/978-981-15-2973-3_5

## 1 Introduction

In September 2015, the 193 Member States of the United Nations adopted the 2030 Agenda for Sustainable Development. Comprised of 17 sustainable development goals (SDGs), 169 targets and 232 indicators, the Agenda tackles a broad range of global challenges, aiming to eradicate poverty, reduce multiple and intersecting inequalities, address climate change, end conflict and sustain peace. According to the Agenda "development will only be sustainable if its benefits accrue equally to both women and men; and women's rights will only become a reality if they are part of broader efforts to protect the planet and ensure that all people can live with dignity and respect" (UNWOMEN 2018: 14). Gender equality is a cross-cutting theme and, is, thus, relevant to the entire agenda and all of the 17 goals. But, SDG #5 focuses on achieving "gender equality and empower all women and girls". On this issue the global agenda for sustainable development builds on the commitments and norms contained in the Beijing Declaration and Platform for Action and the Convention on the Elimination of All Forms of Discrimination against Women (UN CEDAW).

From the global to the local level, gender inequality is the most persistent and entrenched challenge to development. The local level of governance is closest to citizens. Decisions taken at this level have the most direct effect on citizen's everyday lives, as any improvement to living, working and leisure conditions depends on good governance (Zelenev 2017: 1652–1655). Thus, this chapter focuses on local government, and attempts to describe in detail practical steps toward the localization and realization of SDG #5 at the local level. It does so by mapping how to improve, from a gender perspective, the analysis, monitoring, participation, decision-making and access to services to citizens at the local level, in order to develop good local governments that serve all citizens. The aim of this chapter is to understand the processes of implementing SDG #5 in urban local governance. Section 2 describes the theoretical framework used to understand how global norms are translated into local practices. Section 3 examines policy framework, practices, as well as tools available and used for SDG implementation and actors involved in these processes at the municipal level in Bosnia and Herzegovina (BiH). The final section presents main conclusions and recommendations.

We have employed a multi-method, comparative case study research design that combines ethnographic within-case analysis with cross-case comparison. It means that the depth of single cases can be captured, at the same time as it will be possible to find common factors and general conclusions (Gingrich and Fox 2002; Georg and Bennett 2005). A number of legal and policy documents has been analysed, including key gender equality laws, and various action plans for gender equality. In addition to qualitative text analysis of national, entity, canton and municipality documents relating to SDG #5, we have undertaken interviews with municipal decision-makers, policy-makers, civil servants, women organizations, advocates and other stakeholders. In addition, we have conducted surveys in each of the three municipalities to map the existence and use of tools to implement and mainstream gender equality into the decision-making processes, policies and practices at the local level. The survey was conducted with 34 participants, included 20 questions and the response rate was 100%. Moreover, we have invited relevant persons from the region to take part in focus groups to discuss specific SDG #5 targets and the best practices.

## 2 Towards a Theoretical Framework: Translating Global Ideas into Local Practices

The diffusion of global ideas and their effects on national and local policy and political practices and everyday behaviour are central research questions in international relations. Informed by constructivist approaches, International Relations scholars have focused on the role of ideas and norm, and what functions they have in international relations. In this literature, international norms are seen as social structures consisting of intersubjective understandings of appropriate behaviour in the international community (cf. Finnemore and Sikkink 1998: 891). They are typically defined as ideas of varying degrees of abstraction and specification with respect to fundamental values, organizing principles that resonate with many states and global actors and that have been incorporated into official policies, laws, treaties or agreements (cf. Krook and True 2010: 104). The general assumption has been that "good norms" travel and that international norm diffusion is a 'good thing' as good norms spread cooperative, liberal values throughout the international system, thereby socializing member of the international community into 'better' behaviour (cf. Risse et al. 1999; Finnemore and Sikkink 1998). "Gender" norms are among those thought to lead to better behaviour. For instance, gender balance in state decision-making, women's participation in public life and women empowerment are seen as "good things". Global norms shape the identity and interests of states through processes of norm adoption, internalization and socialization and then impact on state behaviour (Finnemore and Sikkink 1998; Björkdahl 2002; Sandholtz 2008; Krook and True 2010). These norm-takers i.e. the receiver of global norms are key to the complex processes of norm adoption. It also became increasingly clear that the national and local normative context affected norm adoption and the degree of 'normative fit' (Björkdahl 2002) between the international norms and local context. Amitav Acharya's (2004) influential article on norm localization was one of the first contributions that accounted for local agency in the norm diffusion process. Localization is defined by Acharya (2004: 245) as "the active construction (through discourse, framing, grafting, and cultural selection) of foreign ideas by local actors, which results in the former developing significant congruence with local beliefs and practices".

A crucial aspect of localization is that local actors actively choose and import the global norms. Hence, global norms are not imposed but local actors actually own the initiative of seeking change. In recognizing the ongoing constitution of norms, this approach confers an active role to local actors in identifying and giving meaning to the global norms and ideas and to construct a normative fit with the local normative context. Björkdahl (2012) argues that localization is a frictional process where global norms become local practices, but without fulfilling their 'promise of universality'. It is only when global norms assume concrete form as local institutions and processes that they move from one part of the world to another. In the practical implementation global ideas and policies may dissolve, unmake, and remake the global ideas and practice and how they work. Levitt and Merry (2007: 446) refer to such a process of appropriation and local adoption as vernacularization, as global 'ideas connect with locality, they take on some of the ideological and social attributes of the place, but also retain some of their original formulation'. Thus, processes of localization, vernacularization and translation highlighted the role of local actors (typically civil society organizations (CSOs), regional organizations, local elites and activists) and their efforts to reshape norms to create a normative match with the local normative context (Acharya 2004; Levitt and Merry 2007).

There are different types of outcome of norm diffusion such as adoption, adaptation, resistance or rejection. (Björkdahl et al. 2015). A successful outcome would be norm adaptation, which is thought to secure legitimacy in the local context. In this approach, the meaning of the global norm will change when it encounters the local context, whereas the local context persists. Local actors are viewed as strategically adapting global norms to regain local legitimacy. In both of these approaches, interaction between international and local actors remains undertheorized, since contributions seem to either ignore local agency, or assume that norms are localized only by local actors. We apply the notion of translation to describe the complex, conflictual and frictional processes through which global ideas are translated into local practices. Recent research has conceptualized translation as a process of interaction between global ideas and local practices and it reveals that translation takes place through patterns of power-laden, frictional interaction and feedback loops (Berger 2017; Zimmermann 2017; Björkdahl and Höglund 2013). It is also important to note that the adoption and adaptation of global ideas and policies may initiate rather than resolve contestation over its content as it generates new power-relations (Sandholtz 2008). Furthermore, norm diffusion and adaptation do not occur in a normative vacuum. Successful norm localization requires the displacement or silencing of competing, clashing and inconsistent norms in order to construct a normative match.

Given that such global ideas are constantly being contested and negotiated in heterogenous national and local contexts, they do not retain a single meaning, but develop many and shifting meanings (Wiener 2009) in response to the normative context and in different situations. Gender norms are particularly elusive in this sense because they consist "of two parts, 'gender' and 'equality,' that are each highly contested" (Krook and True 2010). A case in point is the UN CEDAW, which is not only the most ratified human rights treaty but also the one with the most state party reservations (True 2013).

Our theoretical framework brings to the fore: the process of localization; normative fit between global ideas and existing local normative framework in terms of legal and institutional framework; and local actors involved in translating the global ideas into local practices.

## 2.1 Gender Equality—Global Talk

The development of the Agenda 2030 and the 17 SDGs build on decades of norm advocacy work to promote gender equality and eliminate gender inequality and to change the global normative context. Thus, the development of norms and standards, international laws, declarations, agreements, UN resolutions and commitments, policies and strategic documents to address the challenge of establishing gender equality and the empowerment of women has been significant and meaningful. A fundamental point of departure for this development is *The United Nations Universal Declaration of Human Rights* and subsequent human rights instruments including: CEDAW, the Convention on the Rights of the Child, the Protocol to Prevent, Suppress and Punish Trafficking in Persons, Especially Women and Children, and Supplementing the United Nations Convention against Transnational Organised Crime. It also refers back to international landmark documents in the field of women's rights and status such as the declarations and action plans of the UN Conference on Women in Beijing and the UN Security Council Resolution 1325 on women, peace and security, and subsequent resolutions. It brings on board the 2030 Agenda for Sustainable Development in general but also emphasizes the SGD #5 that pertains to gender equality and empower all women and girls. It also refers back to the Rome Statute of the International Criminal Court, which for the first time in an international criminal law treaty defined and used, the term "gender" in 1998.

A number of conventions and agreements have been adopted at the regional level, such as the European Convention for the Protection of Human Rights and Fundamental Freedoms and the Istanbul Convention on combating violence against women These are also key background documents to guide the national institutional and legal framework in Bosnia and Herzegovina (BiH). Although not yet a EU candidate country, BiH laws on gender equality will have to harmonize with a number of gendered EU documents such the EU Action Plan for Gender Equality and Women's Empowerment through EU external relations 2016–2020, the EU Action Plan on Human Rights and Democracy 2015–2019, the Comprehensive approach to the EU implementation of the UN Security Council Resolution 1325 on women, peace and security, the European Commission's Strategy for Equality between Women and Men 2010–2015 and its successor, and relevant guidelines, and other documents.

In addition, United Nations Security Council Resolution 1325 (UNSCR 1325) was adopted in 2000 as a landmark document for the inclusion of women in peace and security. It requires women to be "represented in all decision-making processes" with regard to conflict prevention and conflict resolution and it brings to the fore the need to "involve women in all of the implementation mechanisms of the peace agreements". It highlights the importance of "women's equal participation and full involvement in all efforts for the maintenance of peace and promotion of peace and security" (UNSCR 1325). The resolution has had an important impact on BiH and it was one of the first states to adopt a National Action Plan for UNSCR 1325. Much of the motivation and driving forces behind the lobbying, development and finally adoption of the resolution was the events that took place during the Bosnian war in the 1990s. The resolution was adopted during the golden age of women's civic engagement in BiH and provided women's organisations in BiH with a tool to pressure the public as well as political parties and the national government to include more women in politics and policy-making through gender-quotas, gender equity legislation and gender agencies. Women's organisations in BiH attempted to use the moral and political authority of international acceptance of the resolution in order to exert pressure on local power-holders (Cockburn 2013).

## 2.2 *The Sustainable Development Goal #5*

The 2030 Global Agenda sets out a vision for the sustainable development of the world for the period 2015–2030. The SDGs encompass the unfinished business of the eight Millennium Development Goals (MDGs) to eradicate poverty, but the SDGs move beyond to break new ground in terms of linking peace, development and human rights (UN 2015; Güney Frahm 2018). In contrast to the MDGs, the implementation of the SDGs is a multi-dimensional process that goes beyond outcomes. It incorporates targets related to gender-based violence, harmful practices, unpaid care work, sexual and reproductive health and rights and legislative changes. All other 16 SDGs include gender mainstreaming in their targets, while SDG #5 is directed specifically on the promotion of gender equality and empowerment of women and girls (Bradshaw et al. 2017).

Women and girls represent half of the world's population and consequently half of its potential. Furthermore, women rights are human rights. While the world has achieved progress towards gender equality and women's empowerment under the MDGs, gender inequality still persists and stagnates social progress (Razavi 2016; Esquivel and Sweetman 2016). As of 2014, 143 countries have guaranteed equality between men and women in their Constitutions but 52 have yet to take this step. Advancing gender equality is critical to human development, a healthy society, reducing poverty, promoting education and to the well-being of women and girls as well as men and boys. Gender inequality affects women in many ways and in many places. It can start already at birth and follow them all their lives.

The SDG #5 is addressing key gender-related challenges societies around the world face. Child marriages affect girls more than boys and the notion of child marriages is misleading as UN statistics demonstrate that girls are vastly overrepresented in child marriages. Nearly 15 million girls per year marry before the age of 18. This means 37,000 girls marry everyday (UN.org). When girls marry young it effects their education. Although one third of the developing countries have achieved gender parity in primary education, girls still faces barriers to enter primary and secondary education in Sub-Saharan Africa, Oceania and Western Asia. Disadvantages in education translates into low skill, limited opportunities in the labour market, insecure position in the labour market, and low wages. On average women in the labour market still earns 24% less than men globally. In addition, about 35% of women experience sexual or intimate violence in relationships or sexual violence outside relationships. The UN estimates that 133 million women and girls have experienced female genital mutilation in some form and the vast majority of them live in Africa and the Middle East where such harmful practices are common (UN.org).

SDG #5 is worded more strongly than its MDG predecessor. It seeks to 'eliminate gender inequality' instead of 'promote gender equality'. There is a developing international institutional and legal framework in support of empowering women and move towards gender equality Yet, SDG #5 targets are not time-bound. Big promises are made, but there is no obligation to actually achieve the goal by a given date. In addition, the Agenda 2030 is silent about the tools and policies needed to attain the SDG #5 (Esquivel and Sweetman 2016).

## 2.3 Gender Mainstreaming

To ensure the implementation of the global policies on gender equality and the empowerment of women, the strategy of gender mainstreaming has been advocated by the UN and other international organisations. The UN defines gender mainstreaming as applying a gender perspective in all policies and programs so that, before decisions are taken, an analysis is made of the effects on women and men, respectively (United Nations 1995: 116). Gender mainstreaming is based on the assumption that gender equality cannot be achieved without considering the consequences of all policies on women as well as men, especially those that are disproportionately detrimental to women. The Beijing Platform for Action ratified by all state parties present at the 1995 Fourth UN World Conference on Women advocated mainstreaming as a new policy that involves working to 'promote a gender perspective in all legislation and policies' (paragraph 207, section d). Gender mainstreaming is clearly intended to change norms. Yet, the meaning of "gender" and "gender equality" are not well defined in these global policies or in the strategy of gender mainstreaming and this becomes a challenge when different levels of governance implement the norm of gender equality. Progress in mainstreaming gender can be monitored through three stages according to Moser and Annalise (2005); the adoption of the language of gender equality and gender mainstreaming; putting a gender mainstreaming policy in place; and the implementation of gender equality.

The implementation of SDG #5 is hampered by the fact that gender is a complex, multi-layered, and gender equality is a contested concept. There is little agreement beyond the basic definition of gender as the socially and culturally constructed identities of men and women and the prevalent subordination of women to men, which is key to gendered hierarchies (see Zalewski 2010). There is a difference between 'women' and 'gender' which needs to be acknowledged. Much like gender, the notion of gender equality is complex and it has multiple meanings. Gender equality is not a women's issue but refers to men as well as women. Women and men have equal rights, responsibilities and opportunities. It also means that the interests, needs and priorities of both women and men are taken into consideration when formulating policies and programs. Gender equality is one of the fundamental preconditions for the fulfilment of all human rights. Here equality between men and women i.e. gender equality means equal visibility, empowerment and participation of both men and women in all spheres of public and private life.

Formulating policies to respond to the goal of SDG #5 and its targets and funding them are the point at which the SDGs are 'translated' into concrete actions. Fukuda-Parr (2016) suggests that there is a danger that the SDGs and targets which contain most potential for transformation will be 'neglected in implementation through selectivity, simplification, and national adaptation'.

## 3 The Case of Bosnia and Herzegovina

BiH is one of the EU neighbourhood countries that is most advanced in its success in institutionalizing gender equity and equality (Hughson 2014: 8, 57). The institutionalization has taken place in the last nineteen years through the development of legal and policy frameworks, establishment of government gender mainstreaming mechanisms, and engendering government statistical institutions. Despite the advanced institutionalization of gender issues, the lives of women and girls have not significantly improved during this period. BiH has relatively greater capacity to empower women and girls and mainstream gender through its advanced gender equality frameworks. Yet one does not experience the effect of those frameworks and mechanisms at local level. BiH would thus be a good starting point of analysing how local governments could translate the SDG #5 Gender Equality into local policies and practice.

BiH comprises 145 municipalities. We have selected three municipalities Visoko Municipality, East Ilidža Municipality and Žepče Municipality for our study.[1] Given that the purpose of the study was to ascertain to what extent the global SDG #5 Gender Equality is being used in the local governance gender mainstreaming policies, it was decided to seek municipalities that already had some gender mainstreaming policies in place, and that were deemed to be gender champions in local governance for different reasons. The local government of these three municipalities have demonstrated political will to empower women and girls and mainstream gender issues through their participation in local, regional, and national campaigns, policy development processes and other activities.

[1] All three municipalities consist of urban and rural population, and its inhabitants represent one of the major ethnic groups of BiH (Bosniak, Serb and Croat respectively). Visoko is located in central BiH between Zenica and Sarajevo on Bosna River. It is organized into 24 local communities (communes). The municipality has 39,938 residents, out of which 11,205 live within the city limits, and it is one of the most densely populated areas in BiH. The vast majority living in Visoko is Bosniak (90%). The second selected municipality of the study is East Ilidža Municipality, which is a municipality of the city of Istočno Sarajevo, or East Sarajevo as it is more commonly known in English, located in RS. In the almost 28 $m^2$, there are 7649 women and 7069 men, 94% of which are living in urban areas across 73 streets. The third municipality in this comparative study is Žepče Municipality. It is located in a valley in the Zenica-Doboj Canton of the FBiH and the river Bosna flows through the town. Some 30,219 persons live in the municipality, among which only 5460 in urban areas, and the majority is Bosnian-Croats.

The Constitution of BiH and national legislation about gender ensure and protect equal representation of women and men before the law, regardless of their marital status or sex. The BiH Law on Gender Equality protects women engaged in political and public life. The BiH Agency for Gender Equality is the government body that was established by law and tasked with monitoring gender equality in BiH, and together with its two entity counterparts—Federation of BiH Gender Center and Republika Srpska Gender Centre—to monitor the compliance of national laws and policies with international human rights standards that concern gender equity and equality. We can see that there is a normative coherence between the global ideas inherent in SDG #5 and the existing BiH normative framework for gender equality.

The National Action Plans (NAPs) were established as means to realize the UNSCR 1325 agenda at the national level. They function as blueprints for how states prioritise, including which actors and areas governments deem important in the process of incorporating the goals of UNSCR 1325 into the national context. The potential impact of the NAPs however has proved to be dependent on how the government institutions choose to work with it. The BiH National Action Plan, adopted in 2010, stresses the security sector and also specifically points out co-operation with civil society actors as a specific goal. Sexual violence is present under two points, one concerning victims of the war, and the other concerning ongoing crimes of trafficking. From our fieldwork and interviews, the NAP seems to have been an effective tool for translating UNSCR 1325 into national policies and later on into local action plans. It is frequently referred to in interviews and focus groups as a useful tool for gender mainstreaming.

In addition to the NAP, BiH adopted its first Gender Action Plan (GAP) in 2006, and the third and current GAP was adopted in 2018. The GAPs suggest all institutions adopt two complementary approaches to strive for gender equality. The first approach, is the inclusion of a gender perspective in programs and policies, and the second is the development of programs that lead to the empowerment of women.

### *3.1 Institutional Mechanisms*

The following institutions, instruments and strategies are recognized as existing or potential tools to mainstream gender in general, and to implement SDG #5 in particular. Most of the tools are already recognized by the legal and policy frameworks and are grounded in the BiH Gender Equality Law or the GAP.

The municipal Committees on Gender Equality are one of the permanent bodies of the municipal council, a standard institutional mechanism for gender mainstreaming of the elected government, and they follow the model of such bodies in higher levels of government.[2] The purpose and mandate of these committees are to monitor,

[2]The Parliamentary Assembly of BiH has two chambers and one of them, the House of Representatives has had a Committee on Gender Equality since 2000 (https://www.parlament.ba/committee/read/21); at entity level, the Parliament of FBIH is also bicameral and its House of Representatives

report and cooperate on gender equality issues at local level and it stems from the BiH Gender Equality Law. Over 95% of the municipalities in BiH have established gender equality committees, however not all are "functional and their capacities need to be strengthened so that they will be able to implement the activities within their mandates" (Babić-Svetlin 2009). A small number of these committees are active in the implementation of the gender equality policy and legal framework, while some include members who do not have basic knowledge of this framework (OSCE 2009 and Miftari 2015).

The findings of this research came to similar conclusions, in that there is a discrepancy between the capacity and ability of the Committees on Gender Equality to mainstream gender at local level. The strength of the East Ilidža Gender Equality Committee was in the coordinated efforts of women working in decision-making or operational positions, as well as men who supported them in their work (in particular the Mayor who was in power when the Committee was established and who continued supporting it throughout his three mandates). However, the members are aware that the Committee does sometimes get engulfed in "humanitarian actions" of helping the needy, ill and impoverished members of society, which leaves little time for concrete gender mainstreaming of local governance.

The Gender Equality Committees in the other municipalities were relatively less active, even though they also had notable and distinguished women members who, in addition to being officially elected to the Municipal Councils, had decision-making positions in local governance. These Committees in Žepče and Visoko rarely met, and only in situations when there were items on the agenda that were specifically related to gender issues. If Municipal Gender Equality Committees will have an impact on gender mainstreaming and the implementation of SDG #5, they will need further support from entity Gender Centres, as well as civil society organizations that have experience in supporting institutional gender mainstreaming mechanisms.

The gender mainstreaming tool of appointing Gender Focal Points was one of the first to be utilized when the larger scale gender institutionalization began in BiH in 2000. At the local level, these positions were usually placed within the cabinet of the Mayor in order to ensure the gender focal point position within executive local authority was close to higher level decision-making. The next step was gender mainstreaming throughout the municipal departments. The result of these processes is that some municipalities do have some form of gender focal point, usually assigned to an existing position within a municipal department such as the case in East Ilidža municipality.

The Žepče Municipality is an example of how an entire unit within the municipality has been nominated as a gender focal point, namely the Žepče Development Agency. The Mayor delegates all incoming gender-related external contacts and cooperation to this important department. The Development Agency is in charge of preparing the Local Gender Action Plan (LGAP), monitoring its implementation and

has a Committee on Gender Equality, the National Assembly of RS has a Board for Equal Opportunities; at cantonal level in FBIH each of the ten cantonal assemblies has a Committee for Gender Equality.

reporting on it. However, although this Agency has adequate human resources, with experienced and gender sensitized staff and director, gender issues is just one in a myriad of tasks in their very long terms of reference such as overall municipal development, tourism development, rural development, investments, long-life learning, etc.

The SDG #5 as a tool has not yet been effectively used at the local level to mainstream gender. Although there are focal points at state level for all UN SDGs, instructions and guidelines have not reached local authorities. This becomes evident from our fieldwork in East Ilidža Municipality. Using the SDG #5 as a tool and ensuring the existing gender focal point within the municipality is also the focal person for SDG #5 were described by one interviewee as follows: "We still do not tag gender mainstreaming activities with the SDG #5 label, because there are no instructions from the higher levels of governance. However, now, before the forthcoming parliamentary elections [in October 2018] it could be used." (Vinka Berjan, East Ilidža, 16.05.18). Experience suggests that the gender focal point tool, as well as any other gender mainstreaming instrument or strategy can have an effect once it is adequately supported by decision makers, and when funding for their work and proposals is made available.

A GAP is used to track and monitor gender mainstreaming of municipal processes and policies to design outcomes for gender equality. Often such plan works through a cross-divisional approach that ensures that mainstreaming responsibilities are rooted in all departments of the municipality. The plan is based on the existing binding legal and policy national, regional and international frameworks as described above (please see above in description of normative match of SDG #5 targets and BiH legislation).

The LGAP are often the result of participatory engagements between the Gender Equality Team and members of staff, combined with baseline assessments of priority areas. The assessment presents key findings, recommendations for follow up action and targets for the municipality. Monitoring is expected to take place regularly.

Information on how many municipalities have LGAPs is not available, but as some research participants said, perhaps the more significant information would be to know how many LGAPs have funding and are being implemented. The LGAPs were developed for East Ilidža and Žepče, in relatively participatory manner, including gender—aware representatives of local authorities. The Žepče Development Agency believe that the introduction of SDG #5 "can help… improve the mechanism of coordination and monitoring of gender issues."

There is a plan to ensure that municipal budgets automatically have a budget line earmarked for LGAP in order to ensure it is implemented. The initiatives for gender responsive budgeting aim to create a direct link between social and economic policies by applying the tool of gender analysis to the way government budgets are formulated and implemented. BiH used the budgeting reform processes to introduce gender responsive budgeting and a strategy to advance gender equality (Avdagić and Hujić 2012). Although it seems the regular budgeting procedure and practice have been gender mainstreamed at entity level, this process has not reached local authorities (apart from a few pilot municipalities that have despite pioneering efforts failed to maintain gender responsive budgeting). Gender responsive budgeting is not an instrument applied in any of the studied municipalities, but in Žepče there is hope that it will be part of the future budgeting process reforms related to the introduction of "Program budgeting".

Gender balanced representation in decision-making institutions, bodies, public companies and other structures established and maintained by government is guaranteed by the BiH Gender Equality Law and the BiH Election Law, both of which call for a steady 40% representation of the less represented sex. Some local authorities have, like in Visoko Municipality, harmonized their legal frameworks with these laws, in order to guarantee gender equal representation.

> In the Statute [of Visoko Municipality] it is specified that it 'will ensure equal rights...including gender equality' and that the 'Municipal Council will respect the number of population according to the last census, in accordance with Election law making sure there is an equal partaking of both sexes' and 'when suggesting members for councils of municipal communities that the sexes are equally respected.' (Zekija Omerbegovic, Visoko, 26.12.17)

The current legally prescribed quota tools are recognized as a positive measure, and most research participants had knowledge of the quotas stipulated by the legal framework, but many of them found that these were not functioning in practice nor enabling the prescribed participation of women in decision-making.

> The [implementation of the] laws we have is questionable, because they are not being implemented. For instance, in our canton, all nominated ministers were men. One CSO complained about this inequality and reminded them of the legal obligation to include 40% of women in all executive government, but the Cantonal government ignored them. Thus, what use are laws if they are not implemented, if there are no sanctions for not implementing them? (Brigita Lovric, Žepče, 10.01.18)

The relatively low levels of women's participation in legislative bodies (municipal councils) and generally equal levels in executive bodies, in particular in public institutions, are more associated with stereotypical women's roles in social and health protection, and education sectors. Women's participation in decision-making is also dependent on the way they perceive politics in general, or the socially constructed expectation of who can and should be a politician and how s/he should behave.

Obstacles mentioned to women's political participation include: lack of awareness and support in political parties, and having to fight the "male mafia" and mutual recognition among women and their solidarity, as well as capacity-building of women who are elected or appointed to decision-making positions:

> Solidarity of women is important, it is interesting how women do not support women and do not vote for women...My experience in politics is that we do not ask each other for help and advice, but we ask our male colleagues, as if there is a prejudice that men know things better than women... (Berina Grahic, Žepče, 10.01.18)

Women's representation in decision-making is one of the key instruments for gender equality as it provides the critical mass of women's presence, provides them with access to resources and decision-making processes that have traditionally been reserved only for men. Although the quota mechanism cannot guarantee gender equality, when paired with political will, it is a useful reference tool for gender mainstreaming.

### 3.2 *Tools to Changing Perceptions and Behaviour*

Awareness-raising aims at showing how existing values and norms influence our picture of reality, perpetuate stereotypes and support mechanisms (re)producing inequality. It challenges values and gender norms by explaining how they influence and limit the opinions taken into consideration and decision-making. Besides that, awareness-raising aims at stimulating a general sensitivity to gender issues.

Research participants are recognizing that the awareness raising activities have created a palpable and visible change in BIH society. An important result of awareness-raising is that talking about violence against women is not a taboo anymore and that more cases are reported than before. Citizens are recognizing that violence against women and girls is not a private issue, but rather a problem that the entire society needs to get involved in.

> Has something changed? Yes, before we did not have the case of people reporting domestic violence, because both women and men thought that this violence was part of married life. People have become significantly aware and have access to information and knowledge [on domestic violence]. (Bosa Kalinic, East Ilidža, 25.12.17)

Awareness-raising is also recognized as a significant gender equality instrument and there are proposals and ideas on how it can be further developed to benefit future gender equality activities.

> To implement things, we need concrete examples. Gender equality is not men being discriminated against… We need to create conditions for people to understand gender equality. (Amila Koso, Visoko, 28.01.18)

Although awareness-raising on gender equality issues has had some impact during the last 20 years of intensive gender institutionalization processes, it still has not reached all the different parts of BiH in an equal way.

Gender training is neither a purpose in itself, nor a single tool to implement gender mainstreaming. It is often part of a set of other tools, instruments and strategies used to mainstream gender. Gender training should be a continuous long-term process, and also adapted to specific sectors and needs of the institution or organization which it is targeting.

Research participants were familiar with gender equality training, and had participated in various workshops. Gender training is also seen as one of the most important and effective tools for the future of gender mainstreaming and achieving gender equality.

Gender sensitive language is typical for languages that have a 'strong' male/female use of both nouns and verbs, such as the languages spoken in BiH. Although some countries are moving to introducing gender neutral titles for professions, in Slavic languages the use of male and female titles has been prominent and is part of the feminist and gender mainstreaming agendas. Instructions for using gender sensitive language were drafted by the Council of Ministers of BiH, and it is one of the most frequently adopted decisions of municipal councils (together with gender disaggregated data and equal representation of women and men). Although it might seem to be futile, or a trivial and insignificant move in the ocean of gender inequality issues, the use of gender sensitive language is perhaps another small step in the array of gender equality instruments and strategies, that like the quota, cannot assure quality of change, but at least does provide some visibility of change.

The issue of gender equality visibility was most evident in this research through the presentation of women in decision making positions as role models for society. The Mayor of Visoko, being one of the rare female Mayors in BiH, is most often seen as the role model and protagonist of an enabling environment for gender equality.

> When our Mayor was elected, as a woman Mayor, this also brought many changes when it came to the election of female councillor. (Seid Buric, Visoko, 06.10.17)

There are more women participating in political affairs of local communities and employment. The impact of this change was supported through institutionalization of gender issues and through the civil societies and their work of empowering women in every role and awareness-raising of the public.

Municipalities tend to employ the two most common means of promoting equal rights and opportunities in their programs: introducing quotas for women in municipal-funded projects and adopting affirmative measures to create equal opportunities for women. Municipalities also monitor the participation of women and men in their self-employment programs in the fields of entrepreneurship and agriculture. These are usually grants that local authorities provide to foster and develop these two self-employment sectors through start-ups, funding of equipment purchases, rental of office and industrial space and land, etc.

Women's access to ownership is still obstructed by traditional practice of male children inheriting all property, and in marriages the husbands being sole owners of property. This practice hinders women from accessing municipal grants, bank loans or participation in international development projects because they do not have bank guarantees in the form of proof of ownership. As one of the interviewee noted: "We educate people that there are laws, and that they can exercise their rights and then they go to enjoy their rights and they are rejected." (Melina Halilovic, Visoko, 26.12.17).

### *3.3 Tools to Mainstream Processes*

Gender disaggregated data or gender statistics "are defined as statistics that adequately reflect differences and inequalities in the situation of women and men in all areas of life".[3] The concept of disaggregating data by gender is a familiar gender mainstreaming tool to the municipalities, as it is a relatable and relatively simple concept.

Local authorities, like the Municipal Council of Žepče, have introduced local legal decisions that oblige their local authority to provide gender disaggregated data. Gender disaggregated data, as one of the tools for gender mainstreaming, is still a significant tool for policy development and assessment of women's access to resources

[3]UN Stat—Gender Statistics Manual, https://unstats.un.org/unsd/genderstatmanual/What-are-gender-stats.ashx.

and decision-making. Although it is frequently embraced by local authorities, it is not one easily put into practice.

Our mapping exercise shows that progress reports on the state of gender equality in municipalities is not taking place. The only regular reporting on an annual basis is on domestic violence, and some municipalities prepare an annual report and submit it to their respective ministries and Gender Centres. Furthermore, there are no guidelines or instructions on how to review municipal decisions, plans and other documents from a gender perspective. However, in the absence of gender analysis or gender checklists, some departments, using common sense and some incidental gender knowledge, tend to promote gender equality values in municipal work.

### *3.4 Cooperation for Gender Equality*

Cooperation with Gender Centres and the Agency of Gender Equality is crucial for local authorities as a guidance and instructions on how to implement existing instruments and strategies for gender equality at municipal level. Gender Centres can also be a mechanism to foster horizonal cooperation between municipalities on gender issues.

Sometimes local authorities are invited to contribute to state or entity-level strategy development processes, but the cooperation with other levels of government on gender equality is infrequent and irregular, with the exception of ministries that have the mandate to follow up on domestic violence. As one interview pointed out: "We never had any further interest or contact from any other level of institutions in BIH, we have never been invited to any gender institutions nor have we ever been told what they would expect from us" (Zekija Omerbegovic, Visoko, 26.12.17).

Cooperation with local CSO on gender equality issues takes place either through the municipal grants for CSOs, or by participating in CSO organized events. The organizations receiving municipal support for women's issues are usually streamlined into traditional women's role of cooking, handicrafts and heritage, as well as agriculture for household. Some research participants believe that municipal grants do provide opportunities for women activists as well. Women's organization Žene ženama and the Foundation CURE from Sarajevo supported researched municipalities in gender awareness, and the youth CSO Kult supported young women leaders.

Municipalities have received support from a number of donors for specific gender equality programs, or have participated in national programs funded by international donors, such as the UN Women and Swedish supported program on gender-based violence. Municipalities also expressed an interest in co-operating with international CSOs working on gender equality issues.

## 3.5 Gender Equality Tools in Practice

A survey on the extent to which these tools are utilized in the municipalities revealed that the three most prominent tools are Local Gender Action Plans, disaggregating municipal data by gender, and establishment of Municipal Council Committees for Gender Equality. The characteristics of these tools are that they tend to satisfy the *de jure* of gender equality, while to a great extent they are not making *de facto* changes in the everyday lives of women and girls. There is a lack of regular monitoring and reporting on the implementation of the Action Plan, Committees sometimes go through the whole four-year mandate without meeting, and although some gender disaggregated data is available, it rarely contributes to the municipal development planning processes. The gender mainstreaming champions in these municipalities, however, do find these tools effective in acquiring at least some means to mainstream gender (although not funded in municipal budgets) and find that these instruments contribute to gender awareness-raising and make gender equality more visible.

The tools least utilized or effective are women representation in decision-making, SDG #5 and gender responsive budgeting. Although quota systems for 40% of women's representation exist in the gender legal framework, women's political participation remains below 20% in municipal councils, and there are no mechanisms to ensure that at least 40% of women are nominated to decision-making positions in executive government. Two obstacles are that political parties do not invest in empowering women in politics, and politics still remains the bastion of male politicians. SDG #5 and gender responsive budgeting are tools that have not been introduced to local authorities.

## 3.6 Obstacles to Implementation of SDG #5 on Gender Equality

The overarching obstacle to the implementation of SDG #5 on gender equality at level of local authorities in BiH is that the 2030 Agenda for sustainable development has not been translated from state and entity levels to the municipal level. Only a few research participants in the municipalities we investigated had heard of the SDGs, or had some experience of them, while the vast majority of the interviewees became aware of the topic of SDGs and SDG #5 in particular through the interviews we conducted. The limited awareness of SDG #5 was compensated by a long-time engagement with gender equality issues through the GAP and the NAP.

Yet, there are several obstacles to mainstreaming gender and promoting gender equality. One such frequently recognized obstacle is the stereotypes about the role of women in society. One interviewee emphasized this in the following way: "Men are the biggest obstacle, men do not consider women worthy of those [higher] levels of government…this is part of their bringing up and their own personality…but not all men have the same views." (Slavica Govedarica, East Ilidža, 22.12.17).

Women were also accused of being an obstacle for gender equality, by male interviewees from East Ilidža and Žepče, and one female interviewee from Visoko. They argued that women need to be more active themselves (Žepče), and that there are not enough women making themselves available "to do the job" (East Ilidza), suggesting that women are not discriminated, but that they rather "have to seize the opportunities that are all already available and accessible to them" (Visoko).

To summarize, the key obstacles encountered by selected local authorities BiH in their advocacy for gender equality include: the lack of specific instruments and strategies for gender mainstreaming, inadequate political will, funding to implement related activities, and lack of awareness of the SDGs. Other obstacles are limited understanding of gender equality, patriarchal structures, lack of readiness of political parties to implement principles of gender equality, non-functioning Committees for Gender Equality, and of cooperation with Gender Centres at other government levels.

## 4 Conclusion

We find that that gender sensitive policies and gender equality are key elements of democratic governance, necessary for stable development and security in any community. Institutional mechanisms for gender equality are necessary to ensure that targets set out by SDG #5 will become part of the government programs and services. However, the danger in institutional mechanisms is that once they are established, they can become empty vessels that do not make concrete contributions to the improvement of the status of gender equality. In order to achieve the objectives of gender equality, a number of instruments and tools are necessary, including awareness-raising, gender mainstreaming know-how, support and political will, adequate funding, and sharing of knowledge and resources among institutional and organizational partners.

Gender equality is a concept that can be new to emerging local democracies and raising awareness can be a lengthy and challenging process. Societies with traditional and patriarchal norms pose the most difficult hurdles in promoting gender equality. Gender equality tools should, thus, be foreseen as ongoing continuous processes, and not one-off events and they are crucial in creating an enabling environment for institutional tools to function, but also as a resource for know-how and skills for gender mainstreaming.

Implementation of institutional mechanisms and legal and policy frameworks is an ongoing challenge. Data and information on the status of women and men, their access to resources and decision-making, and grassroot initiatives for gender equality are all inputs for development of municipal programs and strategies. Municipal authorities are in need of guidelines from BiH and gender mainstreaming institutions that, in addition to monitoring gender equality, are a resource on how gender is to be mainstreamed within government structures. Local authorities need to have open cooperation channels with CSOs, in particular women's organizations, that can be crucial in providing grass-roots information on gender equality, but also act as service providers for interventions that go beyond the human resource capacities of municipalities. Experiences on gender equality at local governance level demand sharing between national, regional and international frameworks, as best practices and lessons learned are models that can be replicated globally. Thus, cooperation with gender mechanisms at all levels of government, with CSOs, international donors, international CSOs as well as other partners are crucial.

## References


Acharya A (2004) How ideas spread: whose norms matter? Norm localization and institutional change in Asian regionalism. Int Org 58(2):239–275

Avdagić M, Hujić F (2012) Gender responsive budgeting as smart Economics: a comparative analysis between Bosnia and Herzegovina and republic of macedonia. J Econ Soc Stud 2(2)

Babić-Svetlin K (2009) Situation analysis: report on the status of gender Equality in Bosnia and Herzegovina. UNICEF, Sarajevo

Bennett A, George A (2005) Case studies and theory development in the social sciences. Boston: MIT Press

Berger T (2017) Global norms and local courts. Translating the rule of law in Bangladesh. Oxford: Oxford University Press

Björkdahl A (2002) From idea to norm. promoting conflict prevention. Lund University Press, Lund

Björkdahl A (2012) A gender-just peace: exploring the post-dayton peace process. J Peace Change 37(2):286–317

Björkdahl A, Höglund K (2013) Precarious peacebuilding: friction in global—local encounters. Peacebuilding 1(3)

Björkdahl A et al (eds) (2015) Importing EU norms? Conceptual framework and empirical findings. Springer, New York

Bradshaw S, Sylvia C, Brian L (2017) What we know, don't know and need to know for agenda 2030. Gend Place Culture 24(12):1667–1688. https://doi.org/10.1080/0966369X.2017.1395821

Cockburn C (2013) Against the odds—sustaining feminist momentum in post-war Bosnia-Herzegovina. Women's Stud Int For 37(1):26–35

Esquivel V, Sweetman C (2016) Gender and the sustainable development goals. Gend Dev 24(1):1–8. https://doi.org/10.1080/13552074.2016.1153318

Finnemore M, Sikkink K (1998) International norm dynamics and political change. Int Org 52(4):887–917

Fukuda-Parr S (2016) From the millennium development goals to the sustainable development goals: shifts in purpose, concept, and politics of global goal setting for development. Gend Dev 24(1):43–52

Gingrich A, Fox G (2002) Anthropology by comparison. Routledge, London

Güney Frahm I (2018) Agenda 2030: haunted by the ghost of the Third Way. J Dev Soc 34(1):56–76. https://doi.org/10.1177/0169796X17752418

Hughson M (2014) Gender equality profile for Bosnia and Herzegovina. Sarajevo: european commission. https://europa.ba/wp-content/uploads/2015/05/delegacijaEU_2014070314432045eng.pdf. Accessed 22 March 2017

Moser C, Annalise M (2005) Gender mainstreaming since Beijing. A review of success and limitation in international institutions. Gend Dev 13(2):11–22

OSCE (2009) The status and activities of municipal gender equality commissions in Bosnia and Herzegovina. OSCE Bosnia and Herzegovina Mission, Sarajevo

Krook M, True J (2010) Rethinking the life cycles of international norms: the united nations and the promotion of gender equality. Eur J Int Relat 18(1):103–127

Levitt P, Merry S (2007) Vernacularization on the ground: local uses of global women's rights in Peru, China, India and the United States. Glob Netw 9(4):441–461
Miftari E (2015) Gender equality in municipalities and cities of Bosnia and Herzegovina. Sarajevo Open Center and Foundation Cure, Sarajevo
Razavi S (2016) The 2030 agenda: challenges of implementation to attain gender equality and women's rights. Gend Dev 24(1):25–41
Risse T, Ropp SC, Kathryn S (eds) (1999) The power of human rights. Cambridge University Press, New York
Sandholtz W (2008) Dynamics of international norm change. Eur J Int Relat 14(1):101–131
True J (2013) Feminist problems with international norms: gender mainstreaming in global governance. In: Ann Tickner, Laura Sjhoberg (eds) Feminism and international relations. Conversations about the past. Present and future. London: Routledge
UN (2015) Transforming the world. The 2030 Agenda for Sustainable Development. https://sustainabledevelopment.un.org/post2015/transformingourworld. Accessed 8 May 2018
United Nations (1995) Gender mainstreaming. https://www.unwomen.org/en/how-we-work/un-system-coordination/gender-mainstreaming
UNSCR 1325 (2000) On women, peace and security. New York: United Nations. http://www.un.org/womenwatch/osagi/cdrom/documents/Background_Paper_Africa.pdf. Accessed 30 May 2018
UNWOMEN (2018) Turning promises into action: Gender Equality in the 2030 Agenda for Sustainable Development. UNWOMEN, New York
http://www.unwomen.org//media/headquarters/attachments/sections/library/publications/2018/sdg-report-summary-gender-equality-in-the-2030-agenda-for-sustainable-development-2018-en.pdf?la=en&vs=949. Accessed 6 May 2018
UN (2016) Gender equality: why it matters. https://www.un.org/sustainabledevelopment/wp-content/uploads/2018/09/Goal-5.pdf. Accessed 7 May 2018
Wiener A (2009) Enacting meaning-in-use: qualitative research on norms and international relations. Rev Int Stud 35(1):175–193
Zalewski Z (2010) I don't even know what gender is: a discussion of the connection between gender, gender mainstreaming and feminist theory. Rev Int Stud 36:3–27
Zelenev S (2017) Translating the 2030 agenda for sustainable development into local circumstances: principles and trade-offs. Int Soc Work 60(6):1652–1655
Zimmermann L (2017) Global norms with a local face? Rule-of-law promotion and norm translation. Cambridge: Cambridge University Press

**Annika Björkdahl** is Professor of Political Science at Lund University, Sweden. Her current research focuses on divided cities, peacebuilding and transitional justice, memory politics as well as the role of norms in international relations and pays particular attention to the gender dimension. Empirically, her research focuses on the Western Balkans, Cyprus and Northern Ireland. Björkdahl works with ethnographically inspired methods such as participant observations to provide a reading of material place and space, as well as interviews and focus groups to collect narrative and life-stories and to investigate silences. She has worked on conflict prevention issues at the Swedish Ministry for Foreign Affairs and with the United Nations, New York on developing an Early Warning System. Björkdahl was the manager of the EUFP7-project Just and Durable Peace 2009–2011. Among her recent publications are Rethinking Peacebuilding: The Quest for Just Peace in the Middle East and the Western Balkans (2013 Routledge), Divided

Cities (2015 Nordic Academic Press), Friction in Peacebuilding: Global and local encounters in post-conflict societies (2016 Routledge), and Spatializing Peace and Conflict (2016 Palgrave) and she has published articles in journal such as Peace and Change, Human Rights Review, Journal of European Public Policy, Millennium, and Security Dialogue. She is currently the Editor in Chief of Cooperation and Conflict.

**Lejla Somun-Krupalija** is Grant Manager for Western Balkans with the Kvinna till Kvinna Foundation in Sweden. Previously she was a researcher and a capacity building consultant in the fields of women's empowerment, gender mainstreaming, social inclusion and disability rights. She holds a post-graduate diploma (M.Sc.) in Forced Migration from the University of Oxford. She worked with UN Women, DFID, ILO, UNDP and other multilateral and bilateral agencies (I am not different: Children with Disabilities in B-H 2017, Baseline Study on the Prevention and Elimination of Violence Against Women in B-H 2014, Mapping Violence Against Women and Girls Support Services National Report 2015, Meeting EC Standards on National and Regional Gender Mainstreaming in the West Balkans Region, co-author 2009). She was councillor and Vice-Chair of Sarajevo City Council, thus extending her experience in applying the human-rights based approach in the legal and policy frameworks of B-H.

# Developing Capacities for Inclusive and Innovative Urban Governance

**Adriana Alberti and Mariastefania Senese**

**Abstract** Many cities throughout the world face multidimensional problems that need to be addressed to realize the Sustainable Development Goals (SDGs). With more than 50% of the world's population that currently lives in cities and a growing trend of urbanization expected to continue in the coming decades, cities must be well equipped to face multiple challenges. Developing countries, in particular, face serious problems, as cities are expected to overgrow. With already large populations living in slums and deplorable conditions, many of the challenges, such as freshwater supplies, sewage, and public health, will affect cities the most. Cities and urban local governments play, therefore, an essential role in the implementation efforts of the 2030 Agenda for Sustainable Development. Cities are critical in ensuring access to basic services, engaging people in decisions that affect their lives, creating opportunities for prosperity and well-being for all, especially for the urban poor, and protecting the environment. The implementation of the 2030 Agenda requires active policy interventions, innovative solutions, and new mindsets to overcome current challenges and ensure effective, inclusive, and accountable governance institutions at all levels. This chapter presents the global context of the 2030 Agenda for sustainable development (Sect. 1), urban trends and challenges facing cities (Sect. 2), critical role of local governments and cities in localizing the Agenda (Sect. 3) and a holistic approach to developing inclusive and innovative capacities for urban governance with examples from around the world (Sect. 4). The chapter concludes with key recommendations on how cities can build capacities to implement the 2030 Agenda for Sustainable Development effectively.

The views expressed herein are those of the authors and do not necessarily reflect the views of the United Nations.

A. Alberti (✉) · M. Senese
Division for Public Institutions and Digital Government, UN Department of Economic and Social Affairs, United Nations, 2 UN Plaza, New York, NY 10017, USA
e-mail: alberti@un.org

*Present Address:*
A. Alberti
301 West 115 St., New York, NY 10026, USA



S. Cheema (ed.), *Governance for Urban Services*,
Advances in 21st Century Human Settlements,
https://doi.org/10.1007/978-981-15-2973-3_6

**Keywords** Sustainable development goals · Capacities · Innovation · Urban governance · Transformational leadership · Changing mindsets · Smart cities

## 1 The Global Context of the 2030 Agenda for Sustainable Development

We are at a critical juncture in history as we are witnessing complex and interdependent social, economic, and environmental challenges, which are posing significant threats to the sustainability of our planet.[1]These problems are "not accidents of nature or the results of phenomena beyond our control. They result from actions and omissions of people—public institutions, the private sector, and others charged with protecting human rights and upholding human dignity".[2]

To address these social, economic, and environmental issues, in 2015, all 193 Member States of the United Nations endorsed an ambitious global compact for a better world. The 2030 Agenda for Sustainable Development is a transformative action plan for the planet, people, and prosperity. Its overarching goal is to eradicate poverty in all its forms, everywhere by 2030. It is a universal Agenda; all Member States have agreed to implement its 17 Sustainable Development Goals (SDGs).

Central to the Agenda is the principle of leaving no one behind. In an increasingly unequal world, the Agenda calls on all countries to step up their efforts to overcome the challenges faced by vulnerable groups, including people living in poverty, persons with disabilities, youth, indigenous people, and migrants. It also highlights the importance of addressing the challenges of countries in special situations, including African countries, Least Developed Countries (LDCs), Landlocked Least Developed Countries (LLDCs), and Small Island Developing States (SIDS).

The Agenda encourages governments worldwide to ensure responsive, inclusive, participatory, and representative decision-making at all levels (Goal 16.7). Indeed, the ambitious global goals outlined in the Agenda can only be realized through the participation of all people and partnerships among all sectors in society. Given the indivisible nature of the SDGs, the Agenda calls for integration and a holistic approach in pursuing prosperity and development for all, balancing the economic, social, and environmental dimensions of sustainable development.

Since 2015, governments from around the world have shown commitment to implementing the Agenda. In 2019, four years after the adoption of the Agenda, out of 193 UN Member States, 158 (7 for the second time) have already completed

---

[1]In his remarks at the Development Committee the UN SG, Antonio Guterres, emphasized that "we face a set of megatrends that are changing the context of our efforts—such as climate change, urbanization, migration, demographic changes, and the rapid technological change—including artificial intelligence—driving the Fourth Industrial Revolution". https://www.un.org/sg/en/content/sg/speeches/2018-10-13/remarks-development-committee.

[2]https://www.un.org/disabilities/documents/reports/SG_Synthesis_Report_Road_to_Dignity_by_2030.pdf.

their Voluntary National Reviews (VNR) of their implementation of the Agenda.[3] The VNRs are a critical mechanism that can facilitate the sharing of knowledge on challenges and lessons learned on how to make progress in the implementation of the SDGs. In 2018, cities started to report on their efforts to achieve the SDGs. In 2019, the Institute for Global Environmental Strategies (IGES) launched an online platform[4] that features local governments' progress on SDG implementation. The platform aims to share lessons learned on SDG achievement of local governments as a complement to the VNR process. "More than 20 cities around the world joined a declaration to advance the SDGs and to commit to tracking their progress through Voluntary Local Reviews (VLR). U.S. cities, including Pittsburgh and New York, talked about how they are moving forward on the SDGs in the United States. Furthermore, local and regional governments emphasized the need to localize the SDGs and pledged to leave no one, no place, and no territory behind."[5] The VNRs and VLRs so far presented at the High-level Political Forum (HLPF) by the United Nations Member States have shown that leaving no one behind is critical. They have also highlighted that transformation and integration require re-thinking the role of government and making public institutions more effective, accountable, and inclusive at all levels, in line with Goal 16 on peace, justice, and strong institutions.

## 2 Urban Trends and Challenges Facing Cities

Several "global megatrends," highlighted by the UN Secretary-General, Mr. António Guterres, will affect cities, as they grow, and urbanization expands. Such trends include "population growth and movements of people, climate change, food insecurity, and water scarcity".[6]

The New Urban Agenda highlights that "we are still far from adequately addressing these and other existing and emerging challenges, and there is a need to take advantage of the opportunities presented by urbanization as an engine of sustained and inclusive economic growth, social and cultural development, and environmental protection, and of its potential contributions to the achievement of transformative and sustainable development."[7]

Cities are currently facing many challenges, which are likely to increase over the next decades. The United Nations Report on the World's Cities highlighted that in 2018, the urban population rate is estimated at 55.3% of the world's population. However, by the year 2030, the number of people living in cities is expected to rise

[3] Voluntary National Reviews Database, https://sustainabledevelopment.un.org/vnrs/.

[4] https://iges.or.jp/en/projects/vlr.

[5] https://unfoundation.org/blog/post/3-takeaways-sdg-summit/.

[6] http://www.un.org/sustainabledevelopment/blog/2017/01/secretary-generals-remarks-at-the-world-economic-forum/.

[7] A/RES/71/256: New urban Agenda: http://www.un.org/en/development/desa/population/migration/generalassembly/docs/globalcompact/A_RES_71_256.pdf.

to 60% of the total global population. That would mean that cities with at least half a million inhabitants would house one-third of the world population. Implementing the 2030 Agenda, especially SDG 11 on inclusive, safe, resilient, and sustainable cities, requires a deep understanding of these urbanization trends and the challenges that they bring.

In 2018, the number of cities with at least 1 million inhabitants had risen to 548*, and by 2030, the number is expected to increase again to 706 cities. Cities with more than 10 million inhabitants, or "megacities," are projected to rise from 33 cities in 2018 to 43 in 2030. In 2018, 48 cities had populations between 5 and 10 million, and by 2030 this number is expected to increase to 66 cities worldwide. Out of these current 48 cities, ten will become megacities by 2030. Twenty-eight additional cities will house more than 5 million people between 2018 and 2030, of which 13 are in Asia and 10 in Africa. "It is expected that in 2050, 68% of the global population will live in urban areas" (World Urbanization Prospects 2018).

While cities have enabled people to advance socially and economically, there are growing urban challenges. These include the urban poor, the impact of climate change, insufficient funds to provide essential services, healthcare, and education, and declining infrastructure, which must be adequately addressed by relevant stakeholders and urban actors.[8] A pressing concern is the erosion of trust in representative government institutions both at national and local levels linked to their inability or perceived inability to effectively tackle problems related to the economy, social integration, and environmental protection.

Issues related to social integration and migrants have been on the rise in many cities. The World Happiness Report has devoted its 2018 edition to happiness and migration. A worrying concern that cities face today, which is related to social integration, is rising inequality and increases in levels of poverty. Eight hundred twenty-eight million people live in slums, and this number keeps growing. In some of the most advanced cities in the world, homelessness is a persisting human tragedy. Deplorable living conditions in poor and overcrowded urban settlements can pose severe threats to people's health conditions. Also, cities need to counteract health inequalities, which in towns and metropolitan areas are rooted in "differences in social status, income, ethnicity, gender, disability or sexual orientation."[9]

The levels of urban energy consumption and pollution are also worrying. While only 3% of the Earth's land is urbanized, cities account for 70% of energy consumption and carbon emissions. Many cities are also vulnerable to climate change and natural disasters due to their high concentration of people, which makes it crucial to build urban resilience to avoid human, social, and economic losses.[10] Today's urban contexts are increasing the vulnerability of the territories and the exposure to the impacts of climate change. Green infrastructures and nature-based solutions[11]

---

[8] http://www.un.org/sustainabledevelopment/cities/.

[9] https://www.who.int/sdhconference/background/news/facts/en/.

[10] https://www.un.org/sustainabledevelopment/wp-content/uploads/2016/08/16-00055K_Why-it-Matters_Goal-11_Cities_2p.pdf.

[11] Nature-based Solutions (NbS) are defined by International Union for Conservation of Nature (UCN) as "actions to protect, sustainably manage, and restore natural or modified ecosystems, that

provide significant benefits for urban resilience, the wellbeing of a community, and sustainable development. UN HABITAT's Strategic plan 2014–2019 also points to the "immense challenge for cities to create decent jobs and livelihoods for their people, including youth and women".

Another challenge that cities face, especially in the least developed countries, is a lack of or weak digital connectivity. The latter allows people to take advantage of new technologies and access digital services as well as participate in policy-making through online platforms. The demand for online services remains, however, conditioned by the widespread use of the Internet.

With more than 50% of the world's population living in cities and a growing trend of urbanization expected to continue in the coming decades, cities must be well equipped to face these challenges. Developing countries will likely face increasing problems as cities are expected to grow very rapidly. With already large populations living in slums and very poor conditions, many of the challenges, such as fresh water supplies, sewage, and public health, will affect cities the most.

Cities and urban local governments play an essential role in the implementation efforts of Agenda 2030. Cities need to ensure access to basic services, engage people in decisions that affect their lives, create opportunities for prosperity and well-being for all, especially for the urban poor, and protect the environment. Engagement of people is critical for broad-based ownership, commitment, and accountability of the 2030 Agenda among all sectors of society.

Many national governments are mobilizing and engaging local institutions and stakeholders in a dialogue on how to adapt the 2030 Agenda to the national and sub-national contexts. So far, the VNRs[12] have outlined several modalities, which involve local and regional governments. Also, many municipalities have started to submit VLRs to the HLPF. The VLRs could accompany the VNRs and highlight sustainable development achievement at the local level. The VLR process can involve citizens in the SDGs review process, "which contributes to strengthening accountability and making governance more inclusive".[13]

**Box 1 Towards Voluntary Local Reviews**

Toyama City, Shimokawa Town, Hokkaido, and Kitakyushu City, Fukuoka (Japan). The VNRs, prepared with the Strategic Research Fund of the Institute for Global Environmental Strategies (IGES), present the activities of the three Japanese local governments "engaged in advanced initiatives related to the

address societal challenges effectively and adaptively, simultaneously providing human well-being and biodiversity benefits".

[12]Voluntary National Reviews Database, https://sustainabledevelopment.un.org/vnrs/.

[13]https://sdg.iisd.org/news/iges-launches-voluntary-local-review-platform-for-sub-national-sdg-follow-up/.

SDGs".[14] In 2018, the three cities were selected by the Japanese government as "SDGs Future Cities." Toyama City has promoted a "compact city planning based on a polycentric transport network, to accurately respond to changing times and social demands and become a sustainable added-value creation city."[15] Shimokawa Town, Hokkaido, established its "Shimokawa Vision 2030: The Shimokawa Challenge: Connecting people and nature with the future," which aims to pursue sustainability through "cyclical forest management that maximizes the use of its forest resources".[16] Kitakyushu has established its SDGs vision to achieve the SDGs as "Fostering a trusted Green Growth City with true wealth and prosperity, contributing to the world." Kitakyushu has identified six priority goals and targets (SDGs of gender equality (Goal 5, target 5); Affordable and Clean Energy (Goal 7 target 2), Decent Work and Economic Growth (Goal 8, target 2 and 5); Industry, Innovation and Infrastructure (Goal 9, target 4), Responsible Consumption and Production (Goal 12, target 5), Partnerships to achieve the Goal (Goal 17. Target 7 and 17) to achieve the SDGs vision.[17]

In 2019, New York City[18] (United States) presented a VLR. The VLR addresses the SDGs of quality education (Goal 4), decent work and economic growth (Goal 8), reduced inequalities (Goal 10), climate action (Goal 13), peace, justice, and strong institutions (Goal 16), and promoting strategic partnerships for the SDGs (Goal 17) that were prioritized for the 2019 HLPF.[19]

Santana de Parnaíba (Brazil). The report was the result of a collaborative effort between municipal authorities, Gaia Education and the UNESCO Global Action Programme, as well as the private sector company Artesano. "The high engagement of civil servants, local business and civil society propelled the

[14]https://www.local2030.org/library/478/Toyama-City-the-Sustainable-Development-Goals-Report-Compact-City-Planning-based-on-Polycentric-Transport-Networks.pdf.

[15]Toyama City the Sustainable Development Goals Report—Compact City Planning based on Polycentric Transport Networks—2018-https://www.local2030.org/library/478/Toyama-City-the-Sustainable-Development-Goals-Report-Compact-City-Planning-based-on-Polycentric-Transport-Networks.pdf.

[16]Shimokawa Town the Sustainable Development Goals Report -The Shimokawa Challenge: Connecting people and nature with the future-, https://iges.or.jp/en/publication_documents/pub/policyreport/en/6571/Shimokawa_SDGsReport_EN_0713.pdf.

[17]Kitakyushu City the Sustainable Development Goals Report—Fostering a trusted Green Growth City with true wealth and prosperity, contributing to the world—2018; https://iges.or.jp/en/publication_documents/pub/policyreport/en/6569/Kitakyushu_SDGreport_EN_201810.pdf.

[18]2019 New York City's Implementation of the 2030 Agenda for Sustainable Development; https://www1.nyc.gov/assets/international/downloads/pdf/International-Affairs-VLR-2019.pdf.

[19]The 2019 VLR was based on OneNYC 2050 developed by the Mayor's Office of Climate Policy and Programs. In 2015, the city launched "One New York" a plan that aims to make New York city "the most resilient, equitable, and sustainable city in the world". In 2018, NYC announced it would become the first city in the world to present a review of its progress during the United Nations

creation of an SDG Commission by public decree with the task of holding the continuity of the SDGs implementation to 2030. The 'catalyser SDGs' for Santana de Parnaíba are—Peace, Justice and Strong Institutions (SDG 16), Quality Education (SDG 4), and Good Health and Well Being (SDG 3)."[20]

Bristol (United Kingdom). In 2019, the City of Bristol launched One City Plan to address systemic challenges and inequalities. The One City Plan is the result of extensive consultation and citizen engagement and enunciates a vision for making Bristol "a fair, healthy and sustainable city for all by 2050."[21] The plan integrates the Sustainable Development Goals (SDGs). The city of Bristol reports on the 17 SDGs by combining "a comprehensive review of statistical indicators with an extensive consultation exercise."[22]

Buenos Aires (Argentina). In 2019, the Review shows progress on the SDGs of Quality Education (Goal 4), Decent Work and Economic Growth (Goal 8); Reduced Inequalities (Goal 10); Climate Action (Goal 13); and Peace, Justice, and Strong Institutions (Goal 16), prioritized by the HLPF. The review shows also progress on SDG 5 (Gender Equality), which is a priority for the local government. "The commitment to the SDGs marks the roadmap towards a more sustainable and integrated city, where sustainable mobility is prioritized, which generates and drives its neighbours' talent, betting today on the jobs of the future, a city committed to climate action, diversity and gender equality."[23]

Los Angeles (United States). The report shows initiatives that are in place to address eight SDGs, starting with two priority goals for the city of Los Angeles, SDG 5 (Gender Equality) and SDG 11 (Sustainable Cities and Communities), followed by a summary of the goals prioritized for the 2019 HLPF, SDGs 4, 8, 10, 13, and 16.[24]

Local authorities, being closest to the people, are best placed to ensure that services are responsive to the needs of the most vulnerable groups[25] and that no one is left behind. The delivery of quality services—including education, health, water, and sanitation, and transportation—to the people will help to realize the SDGs. But given

---

during the HLPF. In its VLR NYC shows sustainable development achievements since 2015, using the SDG framework.

[20] Santana de Parnaíba 2030 Vision Connected to the Future; https://iges.or.jp/en/vlr/santana_de_parnaiba.

[21] Bristol and the SDGs A Voluntary Local review of progress 2019 https://www.bristol.ac.uk/media-library/sites/cabot-institute-2018/documents/BRISTOL%20AND%20THE%20SDGs.pdf.

[22] Idem.

[23] Voluntary Local Review Building a Sustainable and Inclusive Buenos Aires; https://iges.or.jp/sites/default/files/inline-files/buenos_aires_voluntary_local_review_1_0.pdf.

[24] Los Angeles Sustainable Development Goals, A Voluntary Local Review of Progress in 2019, https://sdg.lamayor.org/sites/g/files/wph1131/f/LA%27s_Voluntary_Local_Review_of_SDGs_2019.pdf.

[25] http://www.ose-france.org/categories/en/.

the many challenges that cities face today, new governance capacities to innovate and leave no one behind are needed both at the national and local levels if we are to achieve the SDGs by 2030.

## 3 A Holistic Approach to Developing Inclusive and Innovative Capacities for Urban Governance

Cities can represent powerful engines of growth and sources of prosperity (Duranton 2008),[26] but new governance models and capacities are required to seize these opportunities. Local governments need to rethink how they can engage all stakeholders in their decisions. They need to consider how they can better design and provide services. They need to examine how they can create green urban spaces that are conducive to people's interaction and how they can further mobilize funds and ideas to promote prosperity for all.

The success of the 2030 Agenda critically depends on a holistic approach to governance transformation at all levels. The Agenda envisages cities that are "participatory, promote civic engagement, engender a sense of belonging and ownership among all their inhabitants, prioritize safe, inclusive, accessible, green and quality public spaces that are friendly for families, and enhance social and intergenerational interactions".[27] For this to happen, local governments need to strengthen their capacities for inclusion, especially of the most vulnerable and innovation.

Partnerships, inclusion, and openness can help to mobilize new ideas, resources, including approaches to solving complex societal problems. The participation of people in the definition of priority areas, allocation of funds, as well as the design and delivery of services[28] is of critical importance to effectively addressing the needs of all segments of the population, particularly of the most vulnerable. Inclusive urban policies are essential to address inequality, promote employment opportunities, healthy lives, green urban infrastructure, and resilience to climate change.

**Box 2 City of Seoul and the Sustainable Development Goals**

The city of Seoul (Republic of Korea) has announced its plan to achieve the SDGs and has adopted its Sustainable Development Vision. "By 2030, Seoul plans to become a city that has a social security system suited for the city to satisfy the basic needs of vulnerable social groups, to make sure that all citizens

[26] http://siteresources.worldbank.org/EXTPREMNET/Resources/489960-1338997241035/Growth_Commission_Working_Paper_12_Cities_Engines_Growth_Prosperity_Developing_Countries.pdf.

[27] https://sustainabledevelopment.un.org/content/documents/17761NUAEnglish.pdf.

[28] http://workspace.unpan.org/sites/Internet/Documents/Aide-Memoire%20UNDESA%20Sessions%20Dubai%202013.pdf.

have access to safe and well-balanced food and receive quality education at a reasonable cost. Another goal is to reduce the concentration of fine dust to 70% of that in 2016 and the generation of greenhouse gases to 40% of that in 2005 to reinforce Seoul as a city that pre-emptively responds to climate change."[29]

At the local level, innovations are needed to equip local governments with new solutions to increasingly interdependent social, economic, and environmental issues. Innovation can bring about positive change to address complex challenges. It can refer to new policies and programs, new approaches, and new processes. It can involve the incorporation of new elements, a new combination of existing elements, or a significant change or departure from the traditional way of doing things.[30] It is essential to bear in mind that innovation is not an end in itself, but rather an instrument to improve services for the benefit of all. As the economist, Schumpeter once stated: "Innovation is mankind's effort to endlessly pursue change for a better world"[31] (Schumpeter 1912). Innovation should be a context-specific and holistic process to transform public institutions at all levels to achieve sustainable development goals.

Local authorities need to build capacities for inclusion and innovation in five key governance dimensions: (1) transformational leadership, new competencies and changing mindsets of public servants at all levels; (2) institutional and organizational innovation for urban governance; (3) partnership building and people's engagement; (4) knowledge sharing and management; and (5) digital transformation for inclusive, safe, resilient and sustainable cities.

## 3.1 *Transformational Leadership, New Competencies and Changing Mindsets of Public Servants*

Without local governments' dedicated efforts to mobilize and build the capacity of public servants, who are at the front line of service delivery, it is unlikely that the SDGs will be realized. Capacities include transformational leadership, new competencies, and a change in belief systems, attitudes, and skill sets that reflect the principles of the 2030 Agenda.

One of the most significant challenges in implementing the 2030 Agenda is to ensure that the institutional arrangements being set up or revamped to achieve the SDGs are effective. The latter refers to whether or not institutions produce desired results in solving complex societal issues, such as poverty eradication, food security, and climate change, among others. Effective institutions are strong institutions that

[29]http://localizingthesdgs.org/story/view/199.

[30]Innovation in Governance, replicating what works, UN DESA, 2006 https://publicadministration.un.org/publications/content/PDFs/E-Library%20Archives/2006%20Innovations%20in%20Governance_Replicating%20What.%20Works.pdf.

[31]http://workspace.unpan.org/sites/Internet/Documents/UNPAN98987.pdf.

depend on the degree to which actors internalize institutional beliefs and values. Although new institutional frameworks can be designed relatively quickly, old informal behaviors and consolidated belief systems may persist over time. "It is easiest to amend the law, it is more difficult to transform institutions, and it is most difficult to change people's mentality and habits" (Regulski 2000). It is, therefore, crucial to align civil servants' belief systems, values, and behaviors with the underlying principles and values of the new institutions that are being established and with the principles of the 2030 Agenda, including transformation, integration, leaving no one behind and inter-generational equity.

Changing mindsets in support of effective, accountable, and inclusive institutions requires strong transformational leadership at all levels. Weak leadership and human resources capacity can negatively impact the effectiveness and responsiveness of cities in delivering services. Ineffective leadership can lead to the lack of a shared vision articulated in a clear policy framework.[32] It can also translate into the inability to promote strategic and participatory planning, implementation, monitoring, and evaluation of service delivery. Weak organizational culture and over-centralization of decisions can also stifle innovation through lack of knowledge sharing and management, top-down approaches as well as low transparency and accountability.

Transformational leadership[33] plays a vital role in shaping policy decisions in cities (Stimson et al. 2009; Marshall and Finch 2006; McKinsey & Co 1994). It can guide and influence strategic decisions, priorities, and objectives for economic, social, and environmental improvements in cities. It can steer the values and incentives of stakeholders by involving them in the development process. Transformational leaders can encourage innovative governance models by guiding changes in the organization and creating a shared vision (Bryant 2003). For example, based on the belief that water is a public right and vital common good, in 2010, the City of Paris (France) took back control over the city's water supply and eliminated the problem of service fragmentation among multiple actors.

**Box 3 Water from Paris**

"Until 2010, the public water in Paris was provided in a fragmented manner between four entities. This fragmented approach led to a lack of comprehensive vision for water resources management. Audits demonstrated that water costs for the community were higher by 25–30% as a result. From 1980 to 2010, the price of drinking water was increased by an average of 7% annually. Since 2007, the city has begun a process of taking over its public water service. In 2008, it created a public authority, Water from Paris, to take charge of the

[32] http://workspace.unpan.org/sites/Internet/Documents/Aide-Memoire%20UNDESA%20Sessions%20Dubai%202013.pdf.

[33] "If a country must be transformed to achieve sustainable development and embrace a good society, transformational leadership must be pervasive in the entire society i.e. in public, private, civil society sectors at local, national and community levels" (Kauzya 2017).

production and distribution of water. A contract between Water from Paris and the City of Paris is the instrument for steering and evaluating the municipal guardianship of authority. It sets out criteria that guarantee balanced management by strictly supervising the evolution of expenses, strategic investment choices, and the price of water. Revisable every five years, it provides regular (monthly and semi-annual) evaluations and an annual assessment by the city departments. The emergence of a new model in public water management services has changed the production and distribution of quality drinking water at the lowest price. Water from Paris is a guarantor of access to water as a vital and essential element for all and a committed actor in the long-term preservation of water as natural resource threatened by the effects of climate change. With the establishment of a single operator of public drinking water, a net annual savings of more than € 30 million is realized. From January 1, 2011, the water price dropped by 8%. It also implemented the "right to water" through the development of free public fountains in the public space and an energy-climate plan aimed at reducing the ecological footprint of the company's activities. Eau de Paris contributed to mitigating climate change through the "greening" of its infrastructure and development of non-potable water".[34]

One way of strengthening public servants' competencies is to ensure that public servants' codes of conduct feature the 2030 Agenda principles of leaving no one behind, integration, and transformation. But this is not enough. Socialization to these new values and beliefs through various mechanisms, including training and a change in organizational culture, are central to changing mindsets.

For institutions to be inclusive, public servants need to develop new attitudes, skills, and behaviors to foster multi-stakeholder dialogue and the empowerment of grass-roots organizations. They need to work closely together with local communities, including women and vulnerable groups, to help promote buy-in of national policies and provide more significant opportunities to deliver services and engage people to leave no one behind. Leaders and public servants need to be trained to be service-minded and concerned about the welfare of people, shifting their focus away from internal processes towards the impact their actions have on society, particularly the poorest. As the 2030 Agenda calls for responsive, inclusive, and participatory decision-making at all levels, public servants will need emotional intelligence and the capacity to analyze their own biases to engage with people. They also need to understand the context in which they operate and the causes of social vulnerabilities.

Most importantly, public servants must change the way they communicate and interact. Open communication and collaboration are essential in delivering public services. Communication needs to be built around the needs of citizens, and this requires new skills and talents.

---

[34]2017 UNPSA Winner. 2017 UNPSA Updated Fact: Sheets. http://workspace.unpan.org/sites/Internet/Documents/UNPAN97341.pdf.

Public servants also require capacities in forecasting and assessing trade-offs among policy options and understanding the impact of those policies on present and future generations. These capacities are needed for long term planning to ensure that decisions that are made today do not adversely affect the people, planet, and prosperity in the future.

As the 4th industrial revolution is fast progressing, public servants must also continuously upgrade their digital skills. New technologies are transforming the way governments operate and can significantly support innovation at the local level. Increasing digital skills among the population as well as access to infrastructure and Internet connection are also indispensable to close the digital divide. Likewise, it will be essential to integrate online and offline communication with communities, so public servants will need to be familiar with different means of communication. They must also be able to effectively respond to the many demands and expectations for fast, sustained, and personalized information that can arise with the use of new technologies.

Though not new, designing and managing citizens' charters so that people can evaluate the performance of services, or setting up community scorecards to assess performance in the delivery of services, also requires specific skills. Moreover, promoting a diversified workforce in the public sector, including people from different backgrounds, will enhance the ability of local government officials to better interact with all groups in society, particularly vulnerable groups. Accountable institutions call for enhanced transparency and integrity in government. Ensuring free access to public information through, for example, opening up government data, requires new skillsets in data mining and analytics. Skills in the collection of disaggregated data and statistical capacity, risk management, monitoring and evaluation of policies and programs for poverty eradication, among others, are also critical for greater transparency and accountability.

Changing mindsets in the public sector, however, is not enough. Raising awareness of the values and principles of the 2030 Agenda among the population is of crucial importance as well if the SGDs are to be realized.

**Box 4 Colombia Cambia Tu Mente…Construye Paz**

In Colombia, for example, the initiative "Cambia Tu Mente…Construye Paz (change your mind… build peace)" aims to build a dialogue among members of rival gangs in neighborhoods of Manizales affected by gang violence. "The municipality of Manizales was confronted with the presence of gangs and a high rate of homicides in some neighbourhoods which influenced violence caused by armed youth. This also had a direct impact on the inhabitants' general perception of security and peace. To make holistic decisions, the initiatives' goal was to establish a dialogue with the members of the respective groups—the initiative aims at finding the root cause of the conflict. Its objective was also to find suitable solutions in consultation with the respective member of the groups. It seeks to change the mentality of people who belong to gangs, to manage real and lasting opportunities for them with the support of other public and private entities, and to provide good education and employment. In 2015 Manizales was cataloged by the Colombian Cities Network as the city with the best perception of security. This perception is currently the highest of the decade and follows the implementation of the initiative, among other actions. While the nearby city of Armenia has a homicide rate of 40/10,000 inhabitants per year, Manizales has a rate of 17/10,000".[35]

## *3.2 Institutional and Organizational Innovation for Urban Governance*

Given the interconnectedness among the SDGs, integrated sectoral plans are needed to advance sustainable development. The latter calls for innovation at the institutional level to break silos and work across institutions to support policy integration, mainly through better coordination between the national and local levels of government. The implementation of the 2030 Agenda requires "whole-of-government approaches so that different agencies work together across portfolio boundaries to develop integrated responses to the issues of policy development, program management, and service delivery" (Ojo et al. 2011). Horizontal and vertical institutional arrangements are critical for policy coherence and access to services.

Political, administrative, and financial decentralization is essential to equip local and regional governments with vital resources, both human and financial, which are needed to implement their development strategies. If properly managed, the process of decentralization can facilitate greater participation of people in problems' analysis, needs' identification, planning, and implementation. In addition to responsive development interventions, there are obvious benefits of increased ownership and buy-in among citizens and local institutions.[36] Local governments provide the most innovative means of conceiving, implementing, monitoring, and evaluating poverty reduction interventions since they are closest to those who will benefit from social programs and services. The risk of mismatches between development programs and the actual needs on the ground, which is characteristic of centrally-designed development projects, is thus reduced.

With public sectors offering an increasing number of services, the focus is shifting from what kinds of services are provided to how they are provided. In many countries, a host of services are increasingly coordinated and customized to fit the needs of

---

[35]2018 UNPSA Winner—http://workspace.unpan.org/sites/Internet/Documents/2018%20%20Winners%20with%20short%20paragraph_rev%20OD_EN_Clean.docx.pdf.

[36]2012 United Nations Public Service Forum United Nations Public Service Day, United Nations Public Service Awards ceremony, Workshop 1, institutions and leadership capacities to innovate and engage citizens in service delivery, Aide Memoire.

people. The inter-connectedness among development issues increasingly requires governments to integrate and coordinate policies and decision-making processes for improving service delivery and better responding to the needs of people.

In more advanced cities, services are delivered through a single government website that serves as an entry point to access all available services. A single government website enables people to access government services and information, independently of which government entity provides them. Open and collaborative governance models can help to foster transparency and accountability in urban development and have a substantial impact on local governments as shown by many examples from around the world.

**Box 5 Singapore**

The case from Singapore on Digitalization Plan—Exploit Data and Analytics for smart Digital Planning shows that communication and coordination among policy areas and agencies at all levels are helping governments deliver "as one" to the benefit of its citizens. To make Singapore "a great city to live, work and play," Singapore's Urban Redevelopment Authority (URA) was established to guide planning and facilitating the physical development of Singapore. Singapore's URA takes a holistic approach to achieve sustainable development. It relies on large volumes of data and collaboration with economic, social, and development agencies. It also involves consultation with stakeholders in the private and public sectors. URA places importance on capacity building and in-house training programs designed to seed a culture of planning that is data-driven and evidence-based. "The program equips officers with new skills and techniques in data analytics, including learning new software. With these favorable results, more agencies have collaborated with URA, allowing them to strengthen the culture of digital planning and partnerships. It allowed stakeholders, agencies, and customers to be engaged, share data and integrate business processes and systems."[37] URA has had a critical role in making an inspiring change in Singapore's physical landscape since its independence. One example is the redevelopment of the financial district area, also known as the "Golden Shoe" area. This area had an impressive renovation: whereas in the past, it was characterized by slums, "rundown buildings, street-side vendors, and congested streets it now features modern complexes and office skyscrapers and restored conservation properties."[38]

[37] IDC's Digital Transformation (DX) Awards honors "Digitalization Plan—Exploit data and analytics for smart digital planning" https://www.idcdxawards.com/award/urban-redevelopment-authority-sg/.

[38] http://eresources.nlb.gov.sg/infopedia/articles/SIP_1569_2009-09-18.html.

**Box 6 Republic of Korea**

In the Republic of Korea, the provincial government of Chungcheongnam-do Province has established an online fiscal information system on its website to strengthen the disclosure of its revenues, budget, expenditure, and settlement information to the public. "Since the global economic crisis in 2008, the increase in social welfare spending has put pressure on the finances of national and local governments. The seriousness of the local fiscal crisis caused by various irregularities of public officials and the poor financial management of the heads of local governments came to highlight the importance of integrity and transparency of local fiscal management. The Chungcheongnam-do provincial government has strengthened the disclosure of budget status, revenues and expenditure, and payment settlement on the website. In the case of revenues and expenditures, a fiscal information disclosure system was established, and 15 primary local governments in the province were connected for the first time. Every detail of the annual revenue, as well as expenditure status, was disclosed. Concerning the expenditure, all contract methods, contents, and parties were disclosed. As a result, citizens can check the budget execution status online. Fiscal surveillance has expanded, and transparency and efficiency of fiscal spending have been maximized. The National Finance Law was amended in December 2014. Subsequently, in May 2015, the Local Finance Act was amended, and from November 2015, all local governments are mandated to disclose daily revenues and expenditures through the Internet." The system has seen 15 local governments in the province sign a business agreement by which they disclose their current state of revenues and expenditures, including information on all contract methods, contents, and parties. As a result, citizens can check the budget expenditure operations on the website daily.[39]

## 3.3 *Partnership Building and Citizen Engagement*

The implementation of innovative and sustainable policies in cities has resulted very often from the engagement and participation of people in decision-making processes. Participation is not just a tool for involving the people but also a mechanism for empowering them and a right in itself.[40] Participation can increase people's capacity to influence the processes of collective decisions. It is critical to "rebuild" a relationship of trust with public institutions and to renew social cohesion contracts between the people and the city.

[39]https://publicadministration.un.org/unpsa/database/Home/Winners/2018-winners/Tax-administration.

[40]Participation is a fundamental right. Please see the Universal Declaration of Human Rights https://www.un.org/en/universal-declaration-human-rights/.

In recent years, the construction of new governance models for social inclusion has fostered the proliferation of digital and participation mechanisms within cities. Digital technologies have significantly improved the possibility of people's participation. New digital participatory mechanisms can help to bridge the information gap between the people and the local government and can bring significant and innovative solutions for the wellbeing of all. People's engagement in local planning (e.g., neighborhood committees and assemblies, town hall meetings, referenda, e-participation) is a crucial element for the implementation of local strategies.

**Box 7 Open and inclusive governance in the City of Madrid, Spain**

The city of Madrid (Spain) established more open, transparent, participatory, and inclusive models of governance. "The Madrid City Council was faced with the challenge of establishing new models of governance that will lead to more open, transparent, participatory, and inclusive governments. Before the initiative's implementation, the Council lacked a channel as well as a platform for citizen participation and discussion. An online platform housed on the Madrid Government website was created to enhance citizen participation and discussion by providing a channel through which people can directly and individually raise their ideas and needs and propose public services needed to meet them. The proposals published on the platform have a space for discussion, which is open to all citizens to exchange views and discuss their needs. Since 2015, 362,702 users have registered and participated in more than 5000 debates, made more than 21,000 proposals, and generated more than 4 million votes. Participatory budgets are making it possible to create 517 new services and facilities that the people of Madrid have proposed. Citizens' consultations have made it possible for the population to decide on the equipment of their squares or the pedestrianization of streets. In 2016, there was a 22.6% increase in the number of respondents who believed that the City Council facilitates and appreciates citizen participation in its decisions."[41]

**Box 8 Fund My Community, Australia**

In Australia, "Fund My Community" involves the community in deciding how AUD$1 million can be used to improve the lives of disadvantaged, isolated or vulnerable South Australians. "Australia is one of the wealthiest nations. The average annual income is over $80,000 AUD (approximately 57,000 USD), and the population enjoys free education and health care. Despite this, sections of the population experience economic or social disadvantage. Over 6% of

[41]2018 UNPSA Winner http://workspace.unpan.org/sites/Internet/Documents/2018%20%20Winners%20with%20short%20paragraph_rev%20OD_EN_Clean.docx.pdf.

its population is living in poverty (<50% of median income), and 8.6% are without recognized educational attainment. Community Benefit SA (CBSA) is the public name of the Charitable and Social Welfare Fund, a grant program operated by the Government of South Australia. Many organizations received multiple grants, especially the larger organizations with access to professional grant writers. As such, the process was biased against smaller organizations. However, little impact was evident from the Government's large investment. Despite allocating nearly AUD 70 million (approximately 50 million USD) in the programs and services over 20 years, there was no change in the communities with the highest levels of disadvantage and no reduction in the percent of the population experiencing disadvantage. Fund My Community (https://yoursay.sa.gov.au/fmcrounds/fund-my-community-2017) is a capacity-building grant program that uses digital participatory budgeting to allocate AUD 1 million (approximately 700,000 USD) annually to improve the lives of disadvantaged, isolated or vulnerable South Australians. For six weeks in February and March of each year, not-for-profit community groups submit applications for funding between $10,000 and AUD 100,000 for projects or services that will improve the lives of disadvantaged, isolated or vulnerable South Australians. For six weeks in April and May, South Australians are invited to take part in what is called the 'community assessment', during which citizens review the applications and select the projects or services that they think will have the most significant impact. Fund My Community contributes to the SDGs at two levels. At the administrative level, it helps to achieve Goal 16—building effective, accountable, and inclusive institutions at all levels. The funding outcome is transparent, as the result of a public deliberation in which the decisions of all participants are given equal weight and aggregated to determine the overall outcome. The simplified application process reduces the administrative burden to community groups applying for funding. The Program directly contributes to a transparent, inclusive, accountable, and respectful public service. Besides, the Program contributes to achieving the SDGs associated with human development (Goals 1–5) through the impact of the grant funding. By requiring members of the public to allocate the funding to a broad range of population groups, the Program has benefitted some of the poorest and most vulnerable South Australians."[42]

[42]2017 UNPSA Winner, https://publicadministration.un.org/en/UNPSA.

To address inequalities, some cities have promoted sustainable urbanization strategies including the human rights-based approach[43] to power ahead SDG implementation as in Vienna (Austria) (UCLG 2018); or a 'Right to the City'[44] approach as in Mexico City (Mexico) (UCLG 2018). Other cities have implemented a right-to-housing approach as the 'Cities for adequate housing' initiative led by Barcelona[45] (Spain) while others have addressed the "needs of specific vulnerable groups, especially refugees, migrants, and asylum seekers."[46]

To ensure no one is left behind, as called for in the 2030 Agenda, local governments need to rely on clear and accurate data to make sure that all sections of the community are developing in tandem. The latter requires developing strong partnerships with other stakeholders such as civil society and the private sector—or local businesses—who can help drive innovation and provide much needed financial resources.

People can also help to make cities safer for women. In Seoul (Republic of Korea), women were part of setting up an initiative to make the city safer for women. Lights were installed in parking lots, and car services with women drivers were provided around the city.[47] In Brazil, "combining architecture, universal design, and social work provides accessible housing for children and youths with disabilities living in poor conditions in the city of Rio de Janeiro."[48]

Cities not only need to become livable, inclusive, and safe but also resilient. The concept of resilience refers to the act of withstanding or recovering quickly from physical, economic, and social challenges. People's participation in programs aimed at preventing and/or mitigating the impact of natural disasters can be crucial. In the Philippines, through the project NOAH—Nationwide Operational Assessment of Hazards, the government engages people to address the impacts of natural hazards, particularly extreme weather events, by providing high-quality information to support the country's Disaster Risk Reduction Strategies. Project NOAH provides more Filipinos in the countryside with access to timely, accurate, and comprehensive weather information and their possible impacts, including flooding, storm surge, landslide, tsunami, earthquake, nuclear radiation—empowering the citizenry to make

---

[43]"The human rights-based approach focuses on those who are most marginalized, excluded or discriminated against. This often requires an analysis of gender norms, different forms of discrimination and power imbalances to ensure that interventions reach the most marginalized segments of the population", please see United Nations Population Fund https://www.unfpa.org/human-rights-based-approach.

[44]The "Right to the City" and the New Urban Agenda; http://sdg.iisd.org/commentary/policy-briefs/the-right-to-the-city-and-the-new-urban-agenda/.

[45]https://www.uclg-cisdp.org/en/news/latest-news/right-housing-joining-voices-local-governments-special-session-15th-july.

[46]Local and Regional Governments' Report to the 2018 HLPF—https://www.uclg.org/sites/default/files/towards_the_localization_of_the_sdgs.pdf.

[47]UNPSA winner, 2015 http://workspace.unpan.org/sites/Internet/Documents/UNPSA%202015%20Winners%20List.docx.pdf.

[48]https://zeroproject.org/practice/at-the-nexus-of-architecture-and-social-work-in-rio/ and UN DESA/DSPD Forum: Disability and development—Disability Inclusion and Accessible Urban Development, Appendix.

informed decisions, as well as enhancing the efficiency and readiness of concerned agencies and local governments.[49]

Another innovative case is the community mapping exercise implemented in Dar Es Salaam (Tanzania). Since some parts of the city are prone to frequent flooding, many homes end up being abandoned and become a fertile breeding ground for disease. The location of such households was gathered for a community mapping exercise in Tanzania through OpenStreetMap (OSM) technologies. In August 2015, Dar es Salaam—especially Tandale—faced a rare cholera outbreak. The OSM-based maps helped in response to the outbreak by identifying the most affected areas, locating victims, and providing other critically important information about water points and sanitation.

People's participation can also be effective in preventing disease outbreaks and managing health programs. In Punjab (Pakistan), citizens were involved in monitoring and providing feedback by developing a smartphone application to track anti-dengue activities by the field staff. Each fieldworker was given a basic Android-based smartphone, and the application enabled the "worker to take a picture of a completed task, tag its GPS coordinates (geotagging) and upload it to the system's back-end dashboard".[50] The system time-stamped the incoming pictures and mapped them to the field worker's phone numbers. This data was then visualized on a Google map, enabling easy and reliable analysis of "where and when the prevention activities were performed, or not".[51] Using thousands of Android phones, this system was used by 27 different government departments and hundreds of field workers. By running statistical analyses of data on larvae reports and geo-tagged patient-locations, aligning them to outbreaks in the field, and tracking the status of patients suffering from the virus, the application was soon refined to build a state-of-the-art epidemic early warning system.[52]

In Queensland, Australia, GIS was used for public health prevention. It helped to inform the placement of water pumps in Queensland villages that were most infected by Guinea Worm to ensure a safe water supply.[53] GIS applications were also used to enhance community-based child welfare services, as well as to identify distribution points for culturally appropriate promotion materials about diabetes in a multicultural community. Some other GIS applications used by local Queensland governments include quantifying major hazards in a neighborhood, predicting injuries of pedestrian children, and analyzing disease policy and planning. These applications have been integrated into targeted interventions. The result was a reduced prevalence of guinea worm disease in villages where pumps were introduced. It also resulted in children in high child poverty areas receiving subsidized meals while at family daycare. Moreover, it led to a targeted and culturally sensitive diabetes program,

[49] Nationwide Operational Assessment of Hazards; https://center.noah.up.edu.ph/.

[50] Punjab's model of m-governance: https://tribune.com.pk/story/631041/punjabs-model-of-m-governance/.

[51] Idem.

[52] Idem.

[53] https://docplayer.net/16272413-Report-on-gis-and-public-health-spatial-applications.html.

screening programs to assess hazards in high-risk neighborhoods, which also reduced overall costs, and locating clusters in space and time of child pedestrian injuries and suggesting interventions.

### 3.4 *Knowledge Sharing and Management for Innovation, Transparency, and Accountability*

A new organizational culture, supportive of social innovation and a new mindset, which emphasizes creative and proactive thinking, is essential to improve service delivery. In the Future We Want, world leaders recognized that inter- and intra-disciplinary sharing of information "is essential to create the individual and organizational knowledge necessary for achieving an integrated approach to sustainable development".[54] A holistic knowledge management strategy that promotes awareness of the SDGs and knowledge sharing appears critical for transformative changes.

Sharing and disseminating knowledge through open government data can also increase civic participation and strengthen the transparency and accountability of the cities. In turn, this can create trust and lead to citizen empowerment, including the vulnerable groups.

Sharing knowledge through more inclusive practices is an important mechanism to promote a positive change since it provides fertile ground for the adaptation of good ideas and practices to local contexts. Experiences of sharing knowledge on urban policies can facilitate comparison, ensure greater motivation and accountability, improve the quality of decisions, and foster the achievement of shared solutions. It can allow the construction of a national urban agenda, starting from the local realities of cities. For example, the knowledge program of the World Bank Disaster Risk Management (DRM) Hub, Tokyo (Japan), helps reduce risks from natural hazards by improving warning and management of disasters at national, local, and community levels. With its knowledge program, the Disaster Risk Management Hub in Tokyo develops "knowledge products and organizes knowledge exchanges and capacity building to connect developing countries with DRM solutions and expertise."[55]

Sharing knowledge on innovations and successful practices can provide governments[56] with a set of options on how to achieve the SDGs. It is critical to have global hubs that allow for the exchange of innovations in public administration among countries. For example, the United Nations Public Administration Network (UNPAN) is a global network that connects relevant international, regional, sub-regional, and national institutions and experts worldwide working on effective governance for

[54] Implementation of General Assembly resolution 61/16 on the strengthening of the Economic and Social Council Report of the Secretary-General, Bridging the knowledge gap: using the Council for "thought leadership" (C22) https://undocs.org/pdf?symbol=en/A/67/736.

[55] http://www.worldbank.org/en/news/feature/2014/02/03/drmhubtokyo-knowledge-program.

[56] http://unpan1.un.org/intradoc/groups/public/documents/other/unpan030362.pdf.

sustainable development in line with Goal 16 of the 2030 Agenda for Sustainable Development.[57] Also, the United Nations Public Service Awards (UNPSA) Innovation Hub showcases innovations in public service delivery worldwide to advance the 2030 Agenda for Sustainable Development. The UNPSA is an annual recognition of excellence in public service. It rewards the creative achievements and contributions of public service institutions towards a more effective, participatory, and responsive public administration in countries worldwide.[58] Rewarding creativity and excellence of public sector institutions is also a way to accelerate the implementation of SDG 16 on more effective, inclusive, and accountable public institutions in all countries. Sharing innovative practices on how local governments addressed the challenges related to the realization of the SDGs will enhance their capacity to respond to the 2030 Agenda.[59]

### *3.5 Promoting Smart Cities That are Inclusive, Safe, Resilient and Sustainable*

Innovation and the opportunities offered by the digital development and data revolution of recent years can help to transform government services and accelerate the realization of the 2030 Agenda for Sustainable Development as long as the risks associated with the use of new technologies are appropriately managed. "A smart, sustainable city is an innovative city that uses information and communication technologies (ICTs) and other means to improve quality of life, efficiency of urban operations and services, and competitiveness, while ensuring that it meets the needs of present and future generations with respect to economic, social, environmental as well as cultural aspects".[60]

The global diffusion of ICT in government, as well as higher investments in telecommunication infrastructure together with capacity building in human capital, are contributing to better service delivery. New technologies, if properly used, can promote people's empowerment, participation, access to information, education, and networking possibilities for all social groups, particularly older persons, young people, persons with disabilities, and indigenous peoples. ICT can have a significant impact on local government transformation by bringing greater effectiveness, efficiency, timeliness, and quality, for broader access to services. Cities that invest in cutting-edge technologies and digital skills are likely to reap the benefits of the digital revolution.

[57] http://www.unpan.org/.

[58] https://publicadministration.un.org/en/UNPSA.

[59] The UNPSA has collected more than 250 winning cases of innovative projects that are transferrable to other contexts.

[60] ISO/IEC JTC 1, Information Technology, Smart Cities, Preliminary Report, 2014; https://www.iso.org/files/live/sites/isoorg/files/developing_standards/docs/en/smart_cities_report-jtc1.pdf and UNECE and ITU, October 2015, https://www.itu.int/en/ITU-T/ssc/Pages/info-ssc.aspx.

The potential of the new generation of connections for service innovation is high. It can improve education and health care, enable efficient and effective resource management such as water and air pollution, plastic wastes, energy use, and traffic and parking-space management. However, complexity, costs, harmonized platforms, and standards, as well as security, are the biggest challenges of new technologies, including the 5G-technology revolution.

Many Governments make broadband access a high priority item on their policy agenda. As ICT is increasingly facilitating the interaction between governments and the public, it is essential to improve their access and guarantee the development of broadband[61] networks, helping to bridge the digital divide. Access increasingly emerges as a critical enabler of economic growth, distance education, and improved medical treatment of people in remote areas where advanced health care is scarce. High-speed broadband access and higher bandwidth are necessary components of the digital government. According to the 2018 United Nations e-Government Survey, "although both fixed- and mobile-broadband subscriptions have increased significantly around the world, the proportion of people who do not have access continues to far outnumber those who do. Lack of access remains a particular problem in low-income countries wherein 2016; only 12 out of every 100 people were Internet users, according to the latest data available. The middle-income countries rated higher regarding having more Internet users—42 people per 100—although a majority of their populations remain offline".[62]

The evolution towards a "smart" dimension of cities, therefore, can support the implementation of sustainable policies. Cities are organizing themselves to respond intelligently to the needs of the community. The smart city deploys "intelligent urban systems at the service of socio-economic development and improvement of urban quality of life".[63] Yet, it is important to remember that technology should always be seen as a means to improve the life of people and that there are risks that need to be addressed, including security and privacy issues.

Smart cities go in parallel with Big data. Big Data can improve the quality of life of citizens; it is a valuable source of information which, if adequately exploited, allows visualizing and understanding critical issues of a community.

- In Australia, the city of Melbourne is facilitating innovations across the city by engaging the community (residents, workers, businesses, students, and visitors) to design, develop and test the best ways to live, work and play in Melbourne.[64] The visualization and activation of city data can effectively promote understanding of,

[61] "Broadband refers to telecommunications in which band of frequencies is available to transmit information. As a result, more information can be transmitted in a given amount of time" (UN World Public Sector Report 2003, page 4).

[62] 2018 E-Government Survey, https://publicadministration.un.org/egovkb/Portals/egovkb/Documents/un/2018-Survey/E-Government%20Survey%202018_FINAL%20for%20web.pdf.

[63] https://egov.unu.edu/research/smart-cities-for-sustainable-development.html#outline.

[64] https://data.melbourne.vic.gov.au/.

and participation in, its evolution process as a smart city.[65] In Chile, the city of Santiago has developed a "territorial intelligence mechanism to invest resources ethically according to the interests of each area within the city".[66] Through a "smart" urban regeneration strategy, the program "Corazones de Barrio" will improve the living conditions in poorer neighborhoods.[67]

- In the Netherlands, the City of Amsterdam has invested in public-private partnerships, which has allowed the city to be transformed into an open-source urban lab, where new solutions aimed at improving the quality of life for citizens are developed.[68]

A continually evolving amount of data is stored, analyzed, processed, to create predictive models that make a city smart and can potentially improve the quality of life of all citizens. Data can range from the quality of the air to the energy efficiency of buildings, from food safety to the intervention of public health, from "intelligent" differentiated collection to computerized public transport, from planned bike-sharing to urban land register.

Science and technology can offer essential tools to make cities more resilient to several hazards, including those associated with climate change. They can help to anticipate challenges, like natural disasters. Forecasting and early warning systems based on scientific evidence, i.e., predictive analytics, are an essential element for making evidence-based decisions. They provide competent authorities and organizations with valuable time and information to undertake the necessary preparatory measures and to decide and coordinate intervention actions. Local governments need to implement "tools for measuring and increasing resilience".[69] They need to adopt an approach that embraces several stakeholders, including the private sector, and international stakeholders to develop multi-pronged interventions ranging from prevention in the planning phase to the monitoring and evaluation, including indicators underlying resilience assessment processes.

## 4 Measures to Improve Innovative and Inclusive Urban Governance

Cities are critical in ensuring access to basic services, engaging people in decisions that affect their lives, creating opportunities for prosperity and well-being for all, especially for the urban poor, and protecting the environment. The implementation of the 2030 Agenda requires active policy interventions, innovative solutions, and new mindsets to overcome current challenges and ensure effective, inclusive, and

[65] http://www.smartcityexpo.com/en/awards.

[66] http://www.sesantiago.cl/santiago-destaca-en-cumbre-mundial-de-smart-cities-en-barcelona/.

[67] http://corporacionciudades.cl/proyectos/corazones-de-barrio/.

[68] https://amsterdamsmartcity.com/.

[69] https://www.unisdr.org/archive/49451.

accountable governance institutions at all levels. In sum, five key measures can help cities to promote innovative and inclusive urban governance and achieve the SDGs:

1. **Adopting a holistic, innovative, and inclusive approach to urban governance through institutional capacity-building**. A holistic approach requires looking at all governance institutions in an integrated manner and promoting collaboration across levels of government and agencies to solve multi-dimensional problems. Inclusive local governments are built on a culture of innovation that cares about the poorest and most vulnerable and enables people to participate in urban governance. Innovative service delivery demands new institutional and organizational arrangements and a holistic approach to service delivery to address the complexity of social, economic, and environmental issues.
2. **Building the capacities of public servants and changing mindsets**. Transformational leadership and new mindsets are needed for inclusive and innovative urban governance. Values, such as impartiality, integrity, equity, non-discrimination, and inclusion, need to be internalized by public servants. More empathic, people-centric public servants can also help to rebuild trust in a municipality that cares about its people.
3. **Building capacities to share, analyze and disseminate information**. Knowledge sharing and dissemination of information and data is essential for innovation. Data, when appropriately used, can improve the quality of decision-making processes and help to respond promptly to the needs of citizens by delivering better services and bringing value to the community. The free flow of data and information can help generate innovative ideas and sustain innovation processes. Harnessing data can help governments unlock innovation and transform societies. It can increase the participation of citizens, strengthen transparency and accountability processes.
4. **Investing in the use of ICTs in local government, as well as in telecommunication infrastructure, together with capacity building in human capital, based on a clear, shared vision of development**. New technologies, if properly used, can significantly improve public service delivery and help to inform and possibly engage citizens for better decision-making processes, particularly at the local level. ICTs can support local government transformation by bringing greater effectiveness, efficiency, timeliness, and quality, for broader access to services.
5. **Building partnerships and engaging citizens**. Open and collaborative governance models can help to foster transparency and accountability in urban development and have a substantial impact on local governments. Innovative cities are those where local governments work hand in hand with their communities to find solutions that are uniquely fit for the challenges they face.

## References

Bryant S (2003) The role of transformational and transactional leadership in creating, sharing and exploiting organizational knowledge. J Leadersh Organ Stud 9(4):32–44

Duranton G (2008) Cities: engines of growth and prosperity for developing countries? The International Bank for Reconstruction and Development/The World Bank

Kauzya J (2017) Developing transformational leadership capacity in Africa's public-sector institutions to implement the 2030 agenda and achieve the SDGs. In: Presentation during the 38th roundtable conference of the African association for public administration and management (AAPAM), El Jadida, Kingdom of Morocco, 6th–10th November 2017

Marshall A, Finch D (2006) City leadership: giving city-regions the power to grow. Centre for Cities, London

McKinsey & Co (1994) Lead local, compete global. McKinsey & Co, Boston, MA

Ojo A, Janowski T, Estevez E (2011) Whole-of-government approach to information technology strategy management: building a sustainable collaborative technology environment in government, information polity. Inf Polity 16, 243–260. https://doi.org/10.3233/ip-2011-0237. IOS Press

Regulski J (2000) Samorząd III Rzeczypospolitej. Koncepcje i realizacja [Local government of the Third Polish Republic. Concepts and implementation], Warszawa, Wydawnictwo Naukowe PWN, p 388

Schumpeter JA (1912) Theorie der Wirtschaftlichen Entwicklung, 1912

Stimson R, Stough R, Salazar M (2009) Leadership and institutions in regional endogenous development. Edward Elgar, Cheltenham

United Cities and Local Governments (2018) Towards the Localization of the SDGs. https://www.uclg.org/sites/default/files/towards_the_localization_of_the_sdgs.pdf

World Urbanization Prospects (2018), p 2. (Revision)

**Adriana Alberti** is Chief of the Programme Management and Capacity Development Unit in the Division for Public Institutions and Digital Government of the Department of Economic and Social Affairs at United Nations Headquarters in New York. She has thirty years of experience in leading the development and implementation of governance and public administration capacity development activities in numerous countries from across the globe. She coordinated for several years the UN Public Service Forum and Awards Programme. She was Chief Technical Advisor of the Programme on Innovation in the Mediterranean Region. Before joining the United Nations, she was Visiting Scholar at the Center for International Studies of Princeton University, Visiting Scholar at the Institute of Advanced Social Studies in Cordoba (Spain), and conducted research on democracy and judicial systems at the Institute of Judicial Administration, University of Birmingham (United Kingdom). She worked at the Center for Judicial Studies of the University of Bologna and was Professor of Comparative Politics and European Union at Syracuse University and Dickinson College in Italy. She has edited and contributed to numerous publications and technical papers. Dr. Alberti was awarded a number of fellowships, including from the Italian National Research Council, the European University Institute, the Government of Spain for the Salvador de Madariaga Research Grant, and a fellowship from Harvard University. Dr. Alberti holds a Ph.D. in Social and Political Sciences from the European University Institute; a master's degree in Political Science *magna cum laude* from the University of Bologna and an Executive Education certificate from Harvard University.

**Mariastefania Senese** is Senior Governance and Public Administration Officer of the Division for Public Institutions and Digital Government (DPIDG), in the Department of Economic and Social Affairs (UN DESA) at the United Nations Headquarters in New York. Since 2000, she has contributed to the formulation, organization and management of capacity development programs. In her current position, she participates in the formulation, organization and management of mandated programs of governance, public administration, innovation and capacity development. Previously, she coordinated the United Public Service Forum and Awards Programme. She implemented participatory and inclusive initiatives in Syria, Turkey, and in the Middle East and North African Region to address the challenges of restoring governance and building trust in government. She also worked for the UN DESA Office in Rome where she contributed to projects related to decentralized governance and human resources development. Before joining UN DESA, Dr. Senese was Professor of Public Management at the National School of Administration of Italy and research fellow at the Department of Public Administration of the University of Rome "Tor Vergata", where she taught both at the undergraduate and the post-graduate level. Dr. Senese received her Ph.D. in Economics and Management of Public Administration from the University of Rome in 2004.

# Local Governance and Access to Urban Services: Political and Social Inclusion in Indonesia

**Wilmar Salim and Martin Drenth**

**Abstract** This study examines relationships between local democracy and the barriers to political and social inclusion of marginalized communities in two cities in Indonesia: Bandung as an example of metropolitan city; and Surakarta to give the perspective of a middle-sized city. Since Indonesia has implemented decentralization reforms, local governments carry out basic service delivery. The central government primarily facilitates local government with funding and policies such as slum improvements and financial support for the poor. A central theme in both central government policies and local government programs is the empowerment of marginalized communities of both their mindset and skills to earn. The community perception of government performance is generally high, except for the aspects of the politicization of public services. A difference between Surakarta and Bandung is that the respondents in Bandung believe the city has high levels of corruption. Generally, the respondents are more satisfied with the service delivery by the different government levels within the city, compared to the provincial and central governments and NGOs. This seems to be related to the higher level of interaction with local institutions and their services that benefit the communities. Both cities have recently implemented services to improve public participation, accountability, transparency and access to urban services. The main factors that led to these innovations are inclusive leadership, a community approach, allowing citizens to voice their aspirations, and the smart city concept. Informing marginalized groups about these services could empower them and contribute to the success of Sustainable Development Goal (SDG) 11.

**Keywords** Slum dwellers · Satisfaction · Participation · Accountability · Transparency · Innovation factors · Barriers to engagement

---

W. Salim (✉)
School of Architecture, Planning and Policy Development, Institut Teknologi Bandung, Jl. Ganesha 10, Labtek IX-A, Bandung 40132, Indonesia

*Present Address:*
M. Drenth
Research Center for Infrastructure and Regional Development, Institut Teknologi Bandung, Bintaro Jaya Sektor 9, Jl. Rajawali IV Blok HD9 no 18, Tangerang Selatan 15224, Indonesia



S. Cheema (ed.), *Governance for Urban Services*, Advances in 21st Century Human Settlements,
https://doi.org/10.1007/978-981-15-2973-3_7

This chapter discusses two questions: (1) to what extent and how are marginalized groups engaged in mechanisms and processes of local democracy—including community organization and participation, accountability and transparency in local governance, and better access to basic urban services; and (2) what are the barriers (structural and institutional) to the political and social inclusion of the marginalized groups in Indonesian cities. The methodology involves an institutional analysis approach comprising a review of by-laws and policy documents; interviews with key informants; case studies of Bandung and Surakarta; and a survey of selected slum settlements to learn about the perceptions of the slum residents about local governance and access to urban services.

After a brief review of urbanization and decentralization policies, this chapter provides an overview of slum alleviation programs and institutional arrangements. This is followed by the presentation of case studies of two cities, including findings of the surveys in selected slums and other low-income settlements, and the good practices and barriers to political and social inclusion in local governance and access to services.

## 1 Urbanization and Decentralization Policies

Indonesia has an annual urbanization of 4.1%, the highest rate in Asia. In 2015, Indonesia's urban population was estimated at around 54% of the total population. By 2025, 68% of Indonesia's residents are projected to live in cities (World Bank 2016b). In addition to an increasing urban population, the number of cities has also grown significantly in the past three decades. Between 1990 and 2010, the number of cities of all sizes increased from 47 to 93; large cities (over 1 million inhabitants) from 5 to 11; medium-sized cities in (between half a million and 1 million habitants) from 5 to 15; and small cities (between 100 and 500 thousand inhabitants) from 34 to 57. These numbers suggest that the growth of cities occurs not only in big metropolitan cities, but all the more so in medium and small cities. It is believed that this trend is triggered by decentralization policies that were put in place in 1999 (Salim and Hudalah 2020).

After the fall of Suharto in 1998, Indonesia implemented decentralization policies that enable local governments to prepare programs and policies that are better adjusted to the local situation. The Indonesian decentralization policies and regulations include Law 22 of 1999 on Regional Government and Law 25 of 1999 on Revenue Sharing between the Central Government and the Regional Governments. The objective of this decentralization was to improve public service delivery; to increase community participation; to have accountable local governments; and to improve public welfare. Law 32 of 2004 on Regional Administration that replaced the Law 22 of 1999 altered the objective of decentralization as to have local governments that strive for increasing public welfare and services, as well as competitiveness. The law amended by Law 23 of 2014 on Regional Government with emphasis on the synergy between regional and national governments.

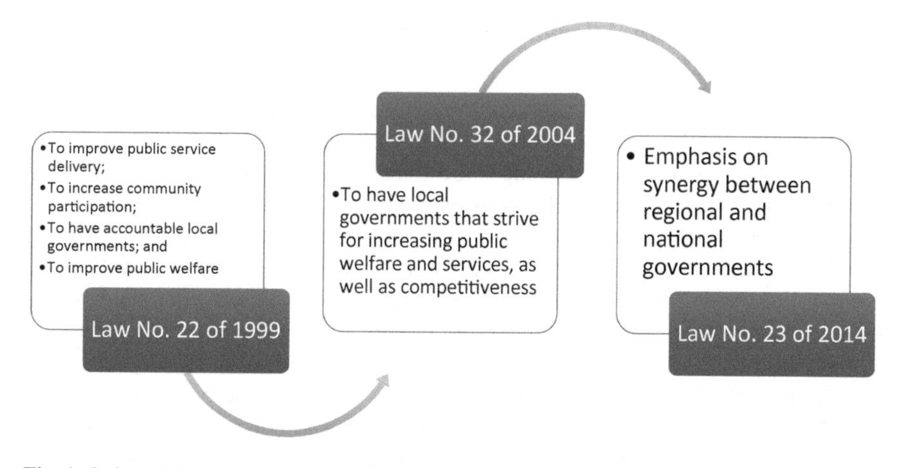


**Fig. 1** Indonesia's decentralization policy after "May 1998 Reformasi"

One of the implementing regulations of the law, Government Regulation No. 38 of 2007 on Division of Affairs between Governments, sets a clear division of responsibility for the central and local governments. For example, for education affairs, the central government is responsible for setting national policy, guidelines, standards, and criteria for running education at all levels and is also responsible for the management of higher education. Meanwhile, the local government is responsible for setting the operational plan, policy, and the management of pre-schools, schools, and non-formal education. Some mechanisms to increase transparency and accountability of the programs were introduced by the Ministry of Home Affairs, including clear and transparent funding mechanisms, a localization and bottom-up approach to the problem, and the use of independent parties in monitoring.

With regard to decentralization and the role of local governments, the National Strategy for Poverty Alleviation (NSPA) that was formulated in 2005 emphasizes the importance of the implementation of regional autonomy as part of the fulfillment of basic rights of the poor (Komite Penanggulangan Kemiskinan 2005). With increasing authorities and resources, regions are obligated to provide accessible, affordable and quality basic services for the poor. Decentralization will also offer ample space for the public to participate in the decision-making process. Despite the perceived benefits of decentralization towards poverty alleviation, NSPA experiences some challenges in its implementation, namely: weak coordination of central government's policies and programs; dependency of local governments toward financial resources from the center; local government's commitment in the budget allocation towards the fulfillment and protection of basic rights; the absence of minimum service standard for public service provision; and lastly the reformulation of public policy-making process to encourage openness and community participation.

## 1.1 National Slum Alleviation Programs

Indonesia has a long history of slum improvement programs. The Kampung Improvement Program (KIP) is considered as one of the most important and successful slum upgrading projects in the world. Some of the more recent programs include: (1) Urban Poverty Alleviation Project (UPP), between 1999 and 2005—a World Bank funded project aimed at alleviating poverty in urban areas; (2) Neighbourhood Upgrading and Shelter Sector Project (NUSSP) implemented from 2005 to 2010 in 32 cities by the Ministry of Public Works, which focused on both infrastructure and housing (funded by the Asian Development Bank) (ADB 2012); (3) National Program for Community Empowerment Urban or PNPM Urban, which is financially and technically supported by the World Bank to become a nation- wide program using the state budget, implemented between 2007 and 2014; and (4) Support for Self-help Housing Stimulus, focusing on housing rehabilitation, operated by the Ministry of Public Works and Public Housing. All these programs involved the allocation of small grants to communities, the cooperation of central and local governments, the communities themselves and in some cases public-private partnerships.

The current slum-related policies at the national level are three-fold (Minister of Public Works and Housing 2016): (1) restoration or upgrading, which is the reconstruction of slum housing and slums into decent housing and settlements through the improvement of houses, facilities, services and/or public utilities to restore their original functioning; (2) revitalization or urban renewal, which is the demolition and the overall rearrangement of houses, facilities, services and/or public utilities; and (3) resettlement, this is in order to create better housing and settlement conditions to protect the safety and security of communities by relocating them to new appropriate resettlement sites.

The above policies are implemented by the Ministry of Public Works and Public Housing through *Program Kota Tanpa Kumuh* (Cities without Slums), or KOTAKU as the national slum upgrading program implemented since 2015. It is a collaborative platform to improve the lives of the urban poor living in Indonesian cities. It aims to provide better water sources, sanitation, drainage, roads, and waste collection. It also attempts to build the local government's institutional capacity to prevent the creation of new slums (World Bank 2016a). The program aims to manage slum settlements and to support the 100-0-100-movement which refers to 100% universal access to safe water, 0% slum settlements and 100% access to safe sanitation. The program facilitates local governments and communities to build a collaborative platform by increasing the role of the local government and the communities. To date, the program is implemented in all 34 provinces, 269 districts/cities, and 11,067 villages nationwide, covering an area of 24,650 ha, or around 40% of the slum areas identified by the government.

The KOTAKU program involves building infrastructure and social and economic assistance for a more sustainable community life in slum settlements. Community organizations in (urban) villages have collected data on the baseline conditions for several slum indicators. After this, planning documents have been drafted based on an integration of community planning documents and those of the regencies/cities. This is done to prioritize decreasing slum settlements and to prevent new slum settlements from forming. The role of community organizations has been transformed from an orientation of measures against poverty towards measures against slums. The general goal of the program is to improve access to infrastructure and basic services in urban slum settlements in order to achieve urban settlements that are sustainable, productive, and suitable for living. This general goal has two implications: (1) increasing community access towards infrastructure and service facilities in urban slum settlements and (2) increasing the welfare of urban communities through the prevention of forming slum settlements and increasing the quality of the slum settlements, community based and with participation of the local government [Directorate General of Human Settlements of the Ministry of Public Works and Housing (n.d.)].

The KOTAKU program has only recently come into effect and the program is scheduled to end on December 31st 2021. Therefore, it is still too early to see the effectiveness of the program including the number of direct project beneficiaries. The current results of the program can be seen from the project development objective indicators. Good progress has been made for the following indicators: a functional task force for slum alleviation at the national level has been formed and functional task forces for slum alleviation at the local level *have* reached 76% (target of 90%). In addition, 56% (target of 90%) of local governments have completed Slum Improvement Action Plans (SIAPs) that have been approved by the mayor. Finally, 59% of urban villages have community settlement plans (CSPs) that have been consolidated with SIAPs (target of 90%) (World Bank 2017).

## *1.2 National Level Agencies and Institutions*

A single program involves many agencies; the involvement of these parties and multiple sectors is arranged with a vertical (central to regional) and a horizontal (between regional agencies or between work units) coordination system. The vertical coordination occurs in national programs (initiated by the central government) to be implemented in the regions. Meanwhile, the horizontal coordination occurs between local work units both in the implementation of national as well as local programs. This is because the implementation of a complex multi-sector and multi-actor program on the management of slum settlements cannot happen without the cooperation of many parties.

The national agencies directly involved in slum alleviation programs are as follows:

(1) The Directorate General of Human Settlements organizes the formulation and implementation of policies in the development of settlements; to support building arrangements; to develop the water supply system, the wastewater management system, and drainage system and waste disposal sites.

(2) The Ministry of Social Affairs has the task of organizing affairs in the areas of social rehabilitation, social security, social empowerment, social protection, and the handling of the poor to assist the president in organizing state governance and inclusiveness.
(3) The National Team for Accelerating Poverty Reduction was established as a cross-sector and cross-stakeholder coordinating body at the central government level headed by the vice president with the aim of aligning accelerated poverty reduction activities. Its main tasks are developing poverty reduction policies and programs; synergizing through the synchronization, harmonization, and integration of poverty alleviation programs in ministries/agencies; and supervising and controlling the implementation of poverty reduction programs and activities.

## 2 A Tale of Two Cities: Bandung and Surakarta

### *2.1 Bandung*

With a total area of 167.31 km consisting of 30 sub-districts and 151 urban villages, Bandung, in 2015, had 2,481,469 inhabitants and a population density of 14,831 inhabitants/km. In the coming years, the population is expected to continue to increase. Bandung's Human Development Index (HDI) significantly grew from 2011 to 2015. In 2015, the HDI reached 79.67 with a life expectancy component of 73.82 years, 13.62 years of expected schooling, mean years of schooling of 10.52 years, and expenditure per capita of IDR 15,608,850 per year. In the last five years, Bandung's HDI has consistently been higher than the national and provincial HDI.

In 2015 there were 445 slum locations in Bandung. The communities in slum settlements often face problems such as limited access to public services, particularly infrastructure. This is primarily caused by the fact that many slum settlements are constructed illegally and, therefore, are not included in formal frameworks of basic service provision such as water and electricity because the City Water Service and the State Electricity Company can only serve legal settlements. The Bandung City Government has issued the Bandung Urban Slum Settlement Plan (RKP-KP), which is assisted by the Directorate General of Human Settlements of the Ministry of Public Works and Housing, through one of its task forces in West Java Province (Development of Settlement Area and Building Arrangement) in 2015. The plan (RKP-KP) aims to achieve the target of 0% slums in 2019 through several programs.

In addition to those activities, the Bandung City Government also initiated several activities such as targeting various urban villages to improve partnerships with city government units; marketing the products of community groups and the sustainable operational funding for community self-help groups; "Kampung" Tamansari redevelopment; the division of government units working on slum area management; and creating various working groups such as on housing and settlements, water supply, and environmental sanitation. Besides the efforts of the Bandung City Government, there are several initiatives by the community, such as *Praksis* and *Komunitas Taboo: Kampung Kreatif (Taboo Community: Creative Kampung). Praksis* is an NGO in Bandung that focuses on participative planning by building bridges between various actors and guiding communities, marginal groups or minorities such as the poor in the planning process. The main mission of *Komunitas Taboo: Kampung Kreatif* is to develop a collective consciousness towards the understanding of democracy, i.e., the government process, management, community rights and the responsibilities of the country towards public service provision.

## 2.2 Surakarta

As an established old town, Surakarta is like a magnet that attracts residents from other districts to access economic opportunities and social services provided by the city. The urban population in 2015 according to official statistics was 557,606 people. Due to its attractiveness, the daytime population of Surakarta may reach around 2 million people (Aditama et al. 2016). The differences between daytime population and nighttime population is caused by the many residents from surrounding districts that come to the city to work, study, or for other activities. As an urbanizing city, Surakarta faces several social, cultural and economic problems. The main factor for this urbanization is the perception that a big city has all necessary amenities, including decent living conditions, job opportunities, educational facilities, and modern and better-maintained infrastructures.

During the leadership of Joko Widodo (currently the President of Indonesia) as mayor from 2005 until 2012, the city implemented various programs for different stakeholders, especially for the poor population, such as Surakarta Citizen's Health Maintenance and educational assistance for the citizens. These two programs focused on the poor populations that are not found in other districts surrounding Surakarta. Thus, both programs attracted many people to become residents of the city. Those who initially came to Surakarta for work gradually settled down in the city.

In 2015, there were almost 360 hectares of slum areas in 15 urban villages in the city. Several challenges in slum improvement programs include: about 50% of slum areas are considered as inadequate houses occupying state land and are not according to the land use plan; unclear legal status of the land, not all buildings have proper building construction permits and building use rights; density and irregularity of the buildings; inadequate sanitation, trash collection, and clean water; low awareness of the residents regarding environmental management; and disempowered low-income people.

The leadership of Joko Widodo and Vice Mayor F. X. Hadi Rudyatmo between 2005 and 2012 accommodated the marginalized communities through several initiatives such as resettlement from the riverbanks; improvement of houses not suitable for living; Surakarta Community Education Assistance; urban village development fund; a public complaint service with a website-based electronic system called Surakarta

Complaint Service Unit (ULAS); and urban village poverty reduction teams. Besides those programs, participative development planning is implemented through *Musrenbang*. This refers to the process of community discussion about local development needs. This is an annual process where residents meet to discuss issues that their communities face and where they decide upon the priorities for short-term improvements. The list of priorities is then handed over to the city planning department, which assigns resources. *Musrenbang* is a bottom-up approach in which communities can influence city budgets and neighborhood investments.

## 3 Findings of the Survey of Slum Settlements

This section presents the household survey data from the selected slum settlements in Bandung and Surakarta. Total of 120 respondents representing one household each in both cities were asked for their views on several topics. The study in Bandung involved 60 households in four community units (RW) in the urban village of Tamansari. Meanwhile in Surakarta, the survey involved 60 households in six community units in three urban villages.

### *3.1 Community Organization and Participation*

Television and cell phones are the most common and frequently used media of the respondents in Tamansari Urban Village, Bandung. Meanwhile, loudspeakers followed by newspapers, radio, and the internet are less used than television and cell phones. Both communication media are the most effective in disseminating news and information to the community in Tamansari. In Surakarta, television and cell phones are also widely used by the respondents; 58% of the respondents use cell phones often and 70% of the respondents use television. Television is used because it is more fulfilling to get audiovisual information compared to other media such as radio, which the respondents are abandoning. Only 13 households (22%) still rely on the radio as a communication medium. The internet and loudspeakers are rarely used; 75% of the respondents said they never use the internet and 90% of the respondents claimed to have never used loudspeakers. Loudspeakers are usually used to provide announcements of social activities for the community.

The involvement in community organizations is a sign of the inclusion and the participation of the respondents. In addition, it offers opportunities to influence the delivery of services and the overall access to these services for all residents of the study area. In Bandung, not all respondents are engaged in community institutions/organizations. Only a small proportion of the respondents (18%) is a member of community organizations and has family members who are active in community organizations (25%). Community institutions in the Tamansari area consist of youth organizations; Family Welfare Programs, the members of which are mothers; Qur'an recital groups; and administrators of neighborhood and community units.

In the case of community organizations, there is a quite striking difference between Surakarta and Bandung, where 65% of respondents in Surakarta declare to be involved in community organizations. In this case, most of the respondents claimed to be involved in urban village-based community organizations such as the Family Welfare Program and the Youth Center and various religious organizations. In Surakarta, the level of participation of respondents in social affairs and public affairs is high. This is influenced by the Javanese culture that has a strong tradition of communal work, instilling a culture of a sense of belonging and concern for the surrounding environment, and gathering with the community. This, of course, affects the pattern of social relations.

Related to the condition of local democracy in Bandung, embodied by the election of leaders/heads of local government, some respondents claimed to be often involved in the election or more than four times on average. This depends on the age of the respondent, the older the respondent the more frequently involved in elections the respondent will be. What needs to be emphasized is that there is still a small percentage of respondents, especially women, who say they have never been involved because they are represented by their husbands. There are also respondents who have never voted because they are newcomers in the area. In the process, the majority of respondents (75%) stated that family members actively and voluntarily choose their leader, provided they were old enough to vote (17 years or older). There is also the practice of leaders contacting the community to vote.

In Surakarta, 90% of respondents have followed the elections of neighborhood units/community units or community institutions more than twice. Only 7% of the respondents stated that they have only once participated in the democratic process at the neighborhood unit and community unit levels. The remaining 3% have never voted. When viewed from the quality of participation, the respondents' involvement is voluntary and proactive, meaning that people realize the importance of voting and using their rights as citizens. In Surakarta, 65% of the respondents vote actively and voluntary, while 35% of respondents claimed to vote because the head of the neighborhood unit, community unit or other institutions contacted them to vote.

In the community affairs in Bandung, respondents state that women are often involved (80%), especially when it comes to Family Welfare Program activities that involve mothers. Thus, most respondents consider the role of men and women to be equal in the community's affairs. However, a small proportion of respondents argue that women are rarely involved in certain cases, so men have a bigger role. An example is the physical work activities. There is also a small portion that views that the role of women is more active, this occurs mostly when the household head is a widow.

Women's participation in public affairs in Bandung does not only involve 'typical women activities', such as preparing food and cleaning up or other activities related to the Family Welfare Program. Rather, the involvement of women nowadays is much greater; women act as decision-makers in public affairs, serve in committees and organizations, and sometimes chair meetings or discussions. Nevertheless, the position of the chair in a committee or organization remains almost always occupied by men. Sometimes in the election of the chairman of a committee, there are female and male candidates but the majority of the citizens prefer men as leaders, while women can at most have a position as vice chairperson, secretary, treasurer, and division chairperson.

A point of interest is that when it comes to equality between men and women, the majority of the respondents answer that there is equality between men and women. However, in some cases, men consider women's position equal but women tend to withdraw themselves, lack confidence in their abilities, and decide to not participate outside family affairs. Some women still think that women's affairs only relate to the "kitchen" and "home", while men's affairs are much wider than that. Nevertheless, most citizens believe that women can play a bigger role. Currently, gender roles can even be reversed, with women earning a living and men staying at home to look after the household and the children, or both spouses work to earn a living.

In Surakarta, women also participate actively in public/community affairs, seen from the 87% of respondents stating that women are often actively involved through both the Family Welfare Program and other activities of the neighborhood and community units. Meanwhile, 7% states that women are rarely involved and 6% says women are almost never involved. Meanwhile, in terms of equality in male and female roles, 73% said that men and women play a comparable role, 22% stated a more active role for men and the remaining 5% considered the role of women more active.

Women's active participation in public/community affairs in Surakarta can be seen from 87% of respondents being actively involved through the Family Welfare Program or other activities of the neighborhood unit and community unit. Over time, women's participation in Surakarta has increased. In the 2000s, since participatory planning was implemented in the neighborhood unit, community unit, and urban village levels, regulatory developments have been implemented in the form of a 30% requirement for female participation in the planning process. Every planning level seeks to fulfill the rule. However, the role of women is generally in planning process is small because women are not active and the regulation of female participation is just a formality. In the decision-making process too, in many cases, women just agree with everything without expressing their own opinions.

There is a gap between the ideal and the existing condition of community involvement in local democratic processes and mechanisms, especially for marginalized groups or people living in slums. Many existing processes and mechanisms, in practice, rarely comply with procedures. The gap between the ideal conditions of marginalized groups can occur because of their limited assets; their limited social and physical assets cause marginalized groups to be often invisible and unheard so that their space of participation becomes smaller. These groups are often unidentified and unorganized. Without detecting such segments of marginalized groups, they do not enter into systems that can accommodate their needs. Therefore, the government needs to make efforts in identifying these segments so they can enter into the system.

### 3.2 Access to Vital Urban Services

The respondents in Bandung state that almost all basic services can be accessed except for sanitation in the form of wastewater and garbage, and health services. A small proportion of respondents lack access to sanitation services, especially wastewater, given the location of many houses along riverbanks. Due to the building density, there is no land for decent sewers, individual or communal. As a result, waste is discharged directly into the river. There are respondents who feel they lack easy and affordable access to health services. This may be because the respondents do not understand the procedure of registration and use of Indonesia's universal healthcare system facility.

The majority of the community in the study area in Bandung has access to clean water, independently obtained (individually or communally), from both shallow (3–15 m) and deep groundwater (>15 m) by digging wells. The water is machine-pumped, hand-pumped, or by the traditional way with pulleys and buckets. The only treatment before drinking the water and using it for cooking is by boiling it. In addition, a small part of the community is served by the piping network of the Local Water Supply Utility (a public and centralized system). However, this water must also be boiled before drinking it. Besides the provision of water by local public water utility and independently acquired water, drinking water refill depots sell bottled water or gallons of water as an alternative in meeting the needs of community drinking water.

The study area in Bandung also has communal restrooms managed by residents that are often used for bathing and washing clothes. These communal restrooms are provided by the Local Water Supply Utility PDAM Tirtawening, donor agencies, and aid programs from the Bandung City Government. The water quality from any of the sources mentioned before can be categorized as good, i.e., odorless, tasteless, and clear, although it has to be boiled before consuming it. In terms of continuity, clean water can be obtained at almost anytime and in a quantity that meets the basic needs of 120 L/person/day. However, in prolonged dry seasons, a common complaint is that it becomes more difficult to access water due to lower groundwater levels; water needs to be pumped from deeper than usual. Moreover, the groundwater slightly tastes and smells like metal but is still clear. In the rainy season, the water from the local water supplier is less clear and in the case of maintenance/draining of treatment plants, water supply is often very limited (6–12 h per day).

Although the Tamansari neighborhood in Bandung is located very close to the river, the residents do not use river water because it is heavily polluted and contaminated by domestic waste both from the upstream area and by residents who live along the riverbanks. The river is the backyard for houses and many residents in the Tamansari neighborhood dump garbage and domestic wastewater (both gray and black water) directly in the river. Only 25-50% of homes have family toilets and individual or communal septic tanks. The limited availability of land is a constraint for building septic tanks so the rest of the residents have toilets but wastewater flows directly into the river.

Besides wastewater, trash is often disposed of directly in the river, although the frequency of residents disposing of waste into the river decreases over time. The reduction of this bad habit is one positive impact of several movements initiated by communities concerned with the Cikapundung River Area. In addition, the Bandung City Government has also provided several schemes for waste management consisting of training and counseling related to recycling garbage, providing communal wastebaskets and taking garbage to trash collection sites located close to residential areas. This trash collection happens independently by individual citizens or paid officers. Residents who want their garbage to be collected from their home can pay the garbage collectors. The officers are usually part of the organization of the neighborhood unit or community unit (cleaning department) or are citizens who assist in collecting trash in return for contributions for their service.

The State Electricity Company supplies the entire Tamansari area. In some cases, especially in urban slums, there are some houses that are not registered at the State Electricity Company but still obtain electricity from neighbors or other houses. This may be done through an agreement of cost-sharing with households that have an electricity service or electricity can be stolen without the knowledge of households with the electricity service. Health services and educational services for the Tamansari community tend to be easily accessible considering that this area is located in the middle of an urban area with health and education facilities in various levels spread evenly. Many programs have been implemented to achieve the 9 years of compulsory basic education, such as the waiver of educational fees for elementary to junior high school, scholarship programs, and other facilities that assist citizens in accessing education.

In Surakarta, nearly all respondents stated that their needs for basic services such as healthcare, education, shelter, electricity, clean water, sanitation and garbage have been met. Only three respondents stated that they do not have access to clean water. These respondents live in Pasar Kliwon where water is only supplied at certain hours. The seven respondents who complained they do not have access to shelter also live in Pasar Kliwon, on the riverbanks. Their lack of access to shelter is because they are not eligible for help through the home improvement program because their houses are built on state land.

In general, clean water in Surakarta comes from the river, groundwater and mountains. The distribution of water sources is strongly influenced by the geographical typology of the area: the southern and western parts of Surakarta which have a sloping topography and are geographically close to Boyolali regency have access to mountain water, so that the water quality of the Local Water Supply Utility in Laweyan, Serengan and Banjarsari sub-districts is generally better. The central to eastern regions have varying access, some of which are supplied by the Local Water Supply Utility from Cokro Tulung (old connections), deep wells and treated river water. Meanwhile, the northern regions such as Mojosongo and parts of Banjarsari obtain water from deep wells (see Solo Kota Kita 2012, pp. 7–15).

In Semanggi Urban Village, including the survey locations, the community has a communal management system of a public hydrant that involves fees every time people take water, while water payments to the Local Water Supply Utility are carried out collectively. Meanwhile, the Local Water Supply Utility provides a master meter for the poor so they can have access to water with a cheaper basic cost. Clean water in Surakarta is usually used for washing, cleaning, flushing toilets, and cooking. The water needs to be boiled first to be able to be used as drinking water or for cooking. There is no tap water that can be consumed directly except for taps located in the vicinity of the Local Water Supply Utility office.

All respondents in the study locations in Surakarta stated that they have access to sanitation. In this case, the municipal government of Surakarta through the Public Works Agency provides sanitation in the slums, including in the three urban villages of study area: Semanggi, Mojosongo, and Sondakan. The provision of sanitation includes two programs, i.e. (1) the construction of communal wastewater treatment plants and connections to houses that do not have a waste management system, (2) communal restroom construction in densely populated areas where the community does not have a sanitation system in each house.

All of the respondents in Surakarta said they have access to health services. This shows that health services in Surakarta have reached all respondents in the slum locations. Health service is one of the most significant achievements of the Surakarta City Government with its Surakarta Community Health Care Program (PKMS). This program has been implemented since 2008, with the aim of providing health care insurance for the people of Surakarta, especially for the poor.

### 3.3 *The Expected Role of Local Government Dealing with Service Delivery*

It is the government's task to provide services that enhance the resilience of marginalized groups. The expectations of surveyed households in Bandung regarding public service provision are very high for the following aspects:

- Proper public announcements;
- Helping to build community-level shelters;
- Support for the elderly and people with disabilities;
- Cleaning and maintaining the shelters;
- Community-based preparedness;
- Developing more infrastructure for resilience; and
- Building appropriate infrastructure.

Meanwhile, the aspects with average or low expectations are:

- Mobilizing community members;
- Door-to-door awareness campaign; and
- Use of community radio.

In Surakarta, the respondents' expectations are high for the role of the city government in providing public services such as proper public announcements, mobilizing community members, and support for the elderly and people with disabilities. An average expectation level above 50% indicates that people expect the Surakarta City Government to continue to improve their public service provision. Meanwhile, as in Bandung, the lowest expectations are for the use of community radio.

Residents of cities generally expect their government to provide access to certain services. The marginalized groups in society depend on the government to provide these services, more so than the non-marginalized city dwellers. For instance, rich people can afford private hospitals, while poor people rely upon free healthcare provided by the government. Therefore, it is important to know which services these marginalized groups expect the government to provide. The respondents in Bandung have high or very high expectations from the government to provide access to livelihood, housing, recovery policy, health care, credit from the local government, and children's education. All six aspects are essential components in supporting their life. The answers from the respondents demonstrate that the six aspects are not fully achieved for most respondents, leading to a high expectation, with children's education being the highest.

In Surakarta, the community expectations of the role of the municipal governments in providing access to other services are also high. Over 50% of the respondents have high expectations for all six aspects. The highest expectation is for healthcare, followed by children's education. The respondents in Surakarta also state that social security programs such as healthcare and education support are very helpful for the community members. The respondents in Semanggi claim to benefit from the healthcare program, which offered cancer treatment with medication and surgery provided free of charge. The lowest expectation in Surakarta is for loans. It can be interpreted that the respondents in Surakarta do not really need this access to loans. Meanwhile, respondents in Bandung have high expectations of these loans. The main difference is the level of expectations between respondents where those in Bandung many respondents have very high expectations of the role of the municipal governments in providing access to services, while in Surakarta the respondents hardly have very high expectations.

## 3.4 Overall Assessment of Sense of Safety/Confidence

The sense of safety or confidence of the respondents in Bandung is generally on medium and (very) high levels. The sense of safety or confidence of the aspect of drinking water is the highest compared to other aspects, namely by 74% of respondents have high and very high levels of satisfaction. This is in line with findings in the previous discussion where 100% of the respondents state they have access to clean water. Although some of the water is obtained independently, the respondents do not find it difficult to acquire it and the availability of water is guaranteed at all times in quantity, quality, and continuity. In addition, the government's inability to provide piped water services to all communities is responded to by various programs

**Table 1** Sense of safety/confidence (in percentages)

| Aspect | Low | | Medium | | High | |
|---|---|---|---|---|---|---|
| | Bandung | Surakarta | Bandung | Surakarta | Bandung | Surakarta |
| Natural environment | 20 | 2 | 32 | 23 | 48 | 75 |
| Current living place | 13 | 0 | 35 | 22 | 52 | 78 |
| Drinking water | 5 | 18 | 21 | 15 | 74 | 67 |
| Community public safety | 15 | 3 | 27 | 10 | 58 | 87 |
| Future of the family | 17 | 0 | 33 | 17 | 50 | 83 |
| Government policy | 22 | 0 | 51 | 13 | 27 | 87 |
| Local government | 13 | 0 | 48 | 17 | 39 | 83 |
| Community leadership | 7 | 0 | 38 | 10 | 55 | 90 |

of procurement, repair, and maintenance of non-piped water infrastructure with various schemes by the central and local governments and donor agencies so that access to clean water is relatively easy (Table 1).

The respondents in Bandung assessed the aspects of public safety aspects, community leadership, current living place, future of the family, and the natural environment mostly at high and very high levels and only a small proportion values these aspects low and very low. This high assessment shows that the access to these aspects is good enough for most people. Whereas the above aspects are all valued almost 50% or higher, most respondents considered the aspect of local government and government policy in the neutral category while the remaining respondents show no significant difference between very low and very high assessments.

Most of the respondents in Surakarta reported high levels of safety or confidence towards all aspects asked; some even gave a very high rating. Nevertheless, there were also respondents who provided a very low assessment of the aspect of drinking water supply. As discussed before, one community lacks access to clean water, which is reflected in their sense of safety or confidence.

## *3.5 Overall Satisfaction with Organizational Performance*

The quality of the provision of services and help for marginalized groups depends on the responsible agencies/organizations at different levels. The following tables describe how the respondents in the study areas value their organizational performance. The results do not show a trend of very satisfied or very dissatisfied; overall, the respondents assess the performance in each level as satisfied or moderate.

The level of satisfaction about organizational performance in Bandung shows that most respondents are much more satisfied with the organizational performance of community/village leadership, local sub-district government and the city government

**Table 2** Satisfaction about organizational performance (in percentages)

| Organization | Low | | Medium | | High | |
|---|---|---|---|---|---|---|
| | Bandung | Surakarta | Bandung | Surakarta | Bandung | Surakarta |
| Central government | 33 | 15 | 40 | 22 | 27 | 63 |
| Provincial government | 20 | 13 | 51 | 54 | 29 | 33 |
| Municipal government | 18 | 1 | 30 | 15 | 52 | 84 |
| Sub-district government | 15 | 1 | 33 | 17 | 52 | 82 |
| Community/village leadership | 5 | 5 | 35 | 10 | 60 | 85 |
| NGOs | 32 | 5 | 38 | 93 | 30 | 2 |

compared to the performance of NGOs, the provincial government, and the central government. This indicates that the lower levels of government interact more directly with the community. The respondents tend to give higher assessments of satisfaction of an organization if it provides programs directly felt by them. This is in line with the responsibility of the leadership of villages, sub-districts, and municipal government who have a direct responsibility of serving the community. Decentralization leads to provincial and central government policies being implemented through the lower-level organizations in accordance with their respective authority so the central and provincial governments will rarely implement these programs or provide direct assistance to communities (Table 2).

In Surakarta, the level of community satisfaction of the performance of the organizations is diverse. The provincial government has the lowest level satisfaction and the respondents are most satisfied with the levels of government in the city and village. The performance of all these organizations shows that most citizens are highly satisfied with the municipal and village governments. This is because the community directly feels the public services provided by the municipal government. The high satisfaction with the municipal government may also be due to the limited knowledge of the community regarding the ownership of the program, in the sense that people generally know that the programs implemented in their areas are from the municipal or central government regardless of whether the program is conducted in cooperation with other parties. Generally, the public does not know about the provincial government programs.

Meanwhile, NGOs get the lowest assessment, as 5% of the respondents report low satisfaction and 93% of respondents were neither satisfied nor dissatisfied with the performance of NGOs in Surakarta. This is because the community does not know about the programs that are implemented by NGOs. During the interviews, many respondents stated that they were not even familiar with the term NGO. Many consider work related to public services the domain of the government rather than that of NGOs. In addition, the word NGO sometimes has a negative connotation of merely being a government's critic. This is due to the stigmatization of NGOs as anti-government movement in the New Order regime (see Antlov et al. 2005), a feeling that has been carried over till now.

### 3.6 Overall Assessment of the Institutional Features of Local Government

The previous tables described how respondents value the organizational performances. Meanwhile, the next table demonstrates how the respondents value the institutional features of the local government, as the government level that most influences the lives of city dwellers. In Bandung, the aspects of corruption, the politicization of service delivery, equitability of resource distribution, and level of accountability show medium and low levels of satisfaction. This demonstrates that the performance of these four aspects is insufficient in the eyes of the community. There are still many respondents who feel that resources are not evenly distributed or that there are imbalances on one side, there is a high level of politicization in public services, the level of corruption of the municipal government is high, and accountability is low in the eyes of some respondents. As an example of this, the previous Mayor of Bandung, Mr. Dada Rosada, was jailed due to a corruption case involving his staff and the misuse of social assistance fund. The recipients of the fund are mainly the poor who live in areas like in Tamansari. On the other hand, the aspects of community participation, transparency of activities, and gender sensitivity are rated high. Overall, the institutional features of the local government in Bandung are rated medium to high (Table 3).

Most respondents in Surakarta are satisfied with the performance of the local government, and some aspects stand out. Whereas the respondents in Bandung assessed the local government to have high levels of corruption, the perception of respondents in Surakarta is that corruption tends to be low. This proves the level of public confidence in the government in Surakarta. However, the perception of low corruption of local government does not necessarily mean that the rate of corruption in Surakarta really is low. Rather, this is related to public confidence in the leaders that represent the government, especially the mayor. From the time Joko Widodo was mayor to F. X. Hadi Rudyatmo, the people of Surakarta have high confidence in their leadership. This may result in an overall positive sentiment of the government's performance in terms of reducing the rate of corruption.

In addition, the practice of good public services in pockets of poverty and slums is one of the common parameters that encourage people to have high trust in the government. In Bandung, one of the four aspects of local government that the respondents consider problematic is the politicization of service delivery. This is also the aspect of local government that some respondents in Surakarta are dissatisfied with, even though most respondents still think this politicization is not a problem. Besides a high

**Table 3** Assessment of the institutional features of local government (in percentages)

| Institutional features | Low | | Medium | | High | |
|---|---|---|---|---|---|---|
| | Bandung | Surakarta | Bandung | Surakarta | Bandung | Surakarta |
| Corruption (clean government) | 65 | 3 | 22 | 12 | 13 | 85 |
| Quality of service delivery | 12 | 6 | 40 | 22 | 48 | 72 |
| Politicization of service delivery | 45 | 25 | 35 | 23 | 20 | 52 |
| Resource mobilization capacity | 13 | 3 | 42 | 59 | 45 | 38 |
| Resource management capacity | 17 | 3 | 46 | 59 | 37 | 38 |
| Equitability of resource distribution | 43 | 7 | 32 | 55 | 25 | 38 |
| Capacity to plan | 15 | 2 | 50 | 33 | 35 | 65 |
| Work efficiency | 15 | 3 | 50 | 59 | 35 | 38 |
| Adequacy of manpower | 25 | 0 | 45 | 55 | 30 | 45 |
| Transparency of activities | 20 | 0 | 23 | 55 | 57 | 45 |
| Level of accountability | 42 | 0 | 55 | 58 | 3 | 42 |
| Community participation | 2 | 0 | 33 | 27 | 65 | 73 |
| Level of trust by community members | 7 | 0 | 46 | 25 | 47 | 75 |
| Gender sensitivity | 28 | 5 | 17 | 42 | 55 | 53 |
| Responsiveness to special needs of marginalized groups | 18 | 2 | 40 | 40 | 42 | 58 |
| Overall image | 18 | 0 | 50 | 33 | 32 | 67 |

level of satisfaction with clean government (low level of corruption), other institutional features of Surakarta government rated high by respondents were the level of trust by community members, community participation, quality of service delivery, and the capacity to plan. Overall, the institutional features of local government of Surakarta are valued higher than Bandung.

## *3.7 Main Differences Between Bandung and Surakarta*

Most respondents consider that they have access to basic services even though many of these services are independently acquired rather than provided by the government. They generally have high expectations of the services the city should provide; the more important for their life the service is, the higher their expectations. However, the respondents seem to be satisfied with their basic services regardless that the quality of these services is not objectively sufficient. The scale of deficit in access to service is assessed through the access to vital services, support during disasters, and the expected role of city government. In Bandung, sanitation/waste disposal and health care are services that some respondents don't have access to, while in Surakarta it is clean water and shelter. In Bandung the type of support that respondents receive from the governments during disasters, which rarely happen, are mainly cash and food, while in Surakarta almost all basic necessities are provided by the governments. This is in line with what respondents in Bandung think of the capacity of city government to provide basic urban services (56% think that they are somewhat

**Table 4** Main differences between Bandung and Surakarta from the survey

| | Bandung (2.5 million pop.) | Surakarta (0.5 million pop.) |
|---|---|---|
| Respondents' ethnic composition | Varied, with Sundanese majority | Mainly Javanese |
| Household head education | Majority high-school graduates | Majority elementary-school graduates |
| Membership in community organization | Low (18%), which reflects a more heterogeneous society | High (65%), supported by strong tradition for communal work in Javanese culture |
| Final word in decision-making | Community leader | Community itself |
| Main deficit in urban services | Sanitation and waste disposal, which are unavailable<br>Health care | Shelter, as some respondents are living on state land illegally<br>Clean water |
| Highest expectation of access to services | Children's education | Health care |
| Highest sense of safety/confidence | Drinking water | Community leadership, public safety, government policy |
| Satisfaction toward city government | 52% satisfied | 84% satisfied |
| Satisfaction toward NGO | 30% satisfied | 2% satisfied |
| Top feature of government | Community participation (65%) | Clean government (85%) |
| Overall image of city government | 32% high | 67% high |

prepared), compared to respondents in Surakarta (83% think that they are somewhat prepared), In both cities the organizations in which respondents are more satisfied with are community/village leadership, subdistrict government, and city government. It suggests that the closer the government level to the public, the more satisfaction level they bring to the community. On the overall assessment of institutional features of the city government, Bandung respondents' satisfaction level is medium to high, while Surakarta respondents' is high to very high. Bandung's respondents have low satisfaction with regard to clean government, service delivery, equitability of resource distribution, and level of accountability. Furthermore, the high level of participation seems limited to local government levels (Table 4).

## 4 Good Practices and Innovations in Local Democracy

The study of local governance and access to urban services found a number of good practices and innovations dealing with three aspects of local democracy in slum settlements, i.e., participation, accountability, and transparency.

### *4.1 Participation*

**Musrenbang**—*Musrenbang* refers to the process of community discussion about local development needs. This bottom-up process is participatory in nature as it attempts to give communities a voice and a chance to influence the development planning that will be implemented. The *Musrenbang* process was introduced to replace Indonesia's former centralized and top-down government system. Law 25/2004 on the National Development Planning System institutionalized *Musrenbang* at all government levels and in the long-term, medium-term and yearly plans. It emphasizes the need to synchronize five planning approaches, i.e., political, participative, technocratic, bottom-up, and top-down in regional development planning. Since 2000, the communities in Surakarta have gotten used to the participative development process. Members of the public have been involved starting with the planning process at the community unit, neighborhood unit, and urban village level. Later, representatives of the community have also been involved with community discussions on the sub-district and city level. Originally, Surakarta was the pilot project for Musrenbang but because it is regulated in national legislation, it is replicated all over Indonesia albeit with various degrees of success (Kota Kita 2017).

**Regional Development and Empowerment Innovation Program (PIPPK)**—This program of the Bandung City Government was introduced in 2015. It aims to increase the community participation in the development process. It is implemented by the entire local government organization and through community institutions consisting of community units, the Family Welfare Program, youth organizations, and community empowerment institutions. PIPPK is formed based on the belief that dynamic changes occur within the community that can be implemented optimally through broad, bottom-up community participation, especially community empowerment in decision-making to solve problems.

Although steps were taken in recent years to streamline the development process in Bandung, the municipal government hopes that PIPPK can further refine the process. The program is expected to address various development problems at the local level by combining various existing programs and is expected to provide learning and empowerment for the community; the development process must be carried out through an innovative approach by all parties (collaboratively) and be the responsibility of all parties. PIPPK is carried out with the usual mechanisms and development processes, i.e., starting from the process of planning, implementation and controlling, evaluation and monitoring, and reporting and accountability. Its implementation requires direct participation and the involvement of the community at all stages of the development process, accommodated through community institutions in the scope of activities within community units (RW), the Family Welfare Program, youth organizations and community empowerment institutions.

**Active involvement of marginalized groups**—In Surakarta, the head of the Development Planning Agency ensures that all marginalized groups are included in the development planning process. He believes that all marginalized groups have already formed community groups. All known groups are involved in the development planning process, in cooperation with the government work unit related to their sector. The head of the Development Planning Agency actively asks community groups to be invited if they are not yet included in the development process. This good practice is easily replicable in other cities in Indonesia as it is not an institutional program but rather the mindset to actively involve each marginalized group. This is a way to empower these groups instead of making them fight for their own rights.

**Mider Projo and Sonjo Wargo Program**—Mider Projo is one of the programs that the current mayor of Surakarta carries out routinely every Friday morning. Meanwhile, Sonjo Wargo is a similar activity that the mayor does in the evening. Both activities involve meeting people in their own environment as to identify local needs and listen to their aspirations. The weekly activity of Mider Projo has been carried out in Surakarta since the time the current president Joko Widodo was Surakarta's mayor, when it was referred to as *blusukan.* This good practice is easily replicable as it does not require a large budget and it prevents the city administration from governing its citizens from an ivory tower.

## *4.2 Accountability*

**SAKIP Juara (Government Institution Performance Accountability System Champion Program)**—SAKIP Juara is one of Bandung's programs for creating responsible governance. Based on the basic principles of good governance and clean government, the government of Bandung seeks to start a paradigm shift from "Input-Oriented Government" to "Result-Oriented Government", based on the principle of accountability, meaning that the result of every program and activity of the state must be accountable to the public. The low Performance Accountability score of the previous Bandung City Government was identified to be caused by weak leadership commitment, resistance towards change, a low understanding of performance accountability and a low quality of performance data. Based on the identification of the four problems, the SAKIP Juara program encompasses five steps. The most innovative is the fourth step of developing the SILAKIP (SAKIP Online), BIRMS, and Bandung Command Center applications. Bandung's performance management system integrates financial and performance accountability by utilizing information technology.

***SAKIP Online or SILAKIP (Performance Accountability Reporting Information System)***—With SILAKIP/SAKIP online, performance reports of work units can be viewed both quarterly and yearly. SILAKIP is not only for government officials but is also accessible for the public to view the progress of the of work units in reaching their targets. The Ministry of Administrative and Bureaucratic Reform regularly audits the accountability of local governments. In 2015, Bandung was the only city in Indonesia to score an A on this. This is a remarkable improvement compared to 2013 when the city was ranked 400th out of all regencies and cities (Nurmatari 2016). This improvement shows how effective the programs of the Bandung City Government are in improving the accountability of the municipal government, which serves as an example for other cities in Indonesia to replicate.

***BIRMS (Bandung Integrated Resources Management System)***—BIRMS is an integrated bureaucratic information portal. Since 2013, all bureaucratic affairs are integrated into information technology. BIRMS is created so that leaders can make decisions quickly and accurately without the need for many meetings to see the progress of an activity. BIRMS is mostly intended to facilitate the performance of the city's work units but there is also some information that can be accessed by the public such as information related to public auctions, procurement, and direct appointments.

***Bandung Command Center***—This is the Bandung City Government's center for information management using information and communication technology. It was built as an effort for Bandung to become a leading and smart city. Some of its functions include monitoring the condition of the city (such as traffic congestion and street vendors) and viewing the bureaucratic performance in making decisions appropriately and quickly. Bandung Command Center is connected with city surveillance (CCTV) cameras installed in 80 strategic locations.

***Service for Public Complaint Handling***—In order to ensure that the government is accountable to the public, a public complaint handling system has been initiated by the central government. The LAPOR! application (meaning report in Indonesian, abbreviated for online Citizen Complaints Service) can be used through various media, i.e., SMS 1708, website (www.lapor.go.id), or Android and Blackberry apps. The LAPOR! website shows the status of the complaint and the way in which it is handled internally. This information is available to the public. The data on the complaints through the application are saved for analysis to be used in reports to leaders in the government. LAPOR! is actively used by Indonesian citizens; it receives around 800 reports each day. In 2013, the LAPOR! initiative was presented at the Open Government Partnership Summit in London.

Since 2014 the Municipality of Surakarta has also opened a public complaint service with a website-based electronic system called the *Unit Layanan Aduan Surakarta* (ULAS). This complaint system is established to improve government performance. Through a web- based complaint system, as well as through SMS, Twitter and Facebook, the Surakarta City Government is open 24/7 to community complaints. Through ULAS, Surakarta City Government increased the effectiveness of services to the community from initially receiving complaints through letters and SMS only to encompassing a wider range of communication media. Moreover, besides making complaints, the community can also monitor the follow-up directly and see how far the related government unit responds and acts to the complaint. Inevitably, the concerned government unit also has to be responsive and quick in following up on complaints. In this case, each government unit can monitor in real-time the performance of other units. Therefore, each unit is required to respond to complaints quickly and within a maximum of three days time.

### *4.3 Transparency*

**Websites of Government Agencies**—Transparency usually relates to participation and accountability as pillars of good governance. It ensures all citizens are entitled to obtain information about governance, public policies, and its formulation, implementation, and results. One of the things done to facilitate the public in obtaining budget planning information and its use is to provide information that is easily accessible through both written and electronic media. An example is the website of the Provincial Development Planning Agency (http://bappeda.jabarprov.go.id/). This website provides a lot of information for the residents of Bandung. It has an overview of different applications and public information. A different section of the website provides an overview of all the development plans that have been implemented. A nice feature of the website is the option to translate it from Indonesian to English and Sundanese, which is the native regional language of many residents of Bandung. Providing information on websites is a good way to offer transparency to citizens in general, as it can prevent people from having to travel to a government office to acquire information. Nevertheless, it is uncertain what the impact of this good practice is for the marginalized groups in society, considering the low use of the internet by the respondents.

**Presence of Government Agencies on Social Media**—The Provincial Regional Development Planning Agency Twitter account (@bappedajabar) is an example of the presence of government agencies on social media. Being active on Twitter is a good way for government agencies to increase their transparency and to provide information to the public in an easy way. Indonesians are very active on Twitter. In May 2016, Indonesia had 24.34 million active Twitter accounts, which is the third highest number worldwide. In 2016, Indonesian Twitter users sent 4.1 billion tweets

(Jakarta Globe 2017). The Provincial Regional Development Planning Agency uses its Twitter account to send important information to the public. An example of this is a tweet about a new regulation that would come into effect that toll roads now only accept e-toll instead of cash payments. This good practice is easily replicable, and requires only a small budget to implement.

**Bandung Planning Gallery (BPG)**—This gallery opened officially in early August 2017 and is the first city planning platform in Indonesia. BPG displays the urban development in the past, present, as well as future plans for the city. Various models of development are displayed, ranging from model of the city to three-dimensional maps of the Bandung basin. The main purpose of BPG is that every development plan is known by the community to offer transparency. According to the mayor of Bandung, BPG is developed in the spirit of open government. It is still too early to see the impact of the Bandung Planning Gallery. It is, however, a more accessible way for citizens to visualize plans, especially when they are not trained as planners, it can be hard to interpret maps on paper.

**Information Management and Documentation Office (PPID)**—The PPID office oversees access provision to public information for people who request it. The provision of information in Indonesia is based upon Law No. 14/2008 on Public Information Openness and Decision of the Minister of Communication and Information Technology No. 117/KEP/M.KOMINFO03/2010 on Information Management. The Information Management and Documentation Office was founded with the purpose of implementing a transparent and accountable information service in order to fulfill the rights of information applicants. The information provided in the framework of government transparency is very diverse and can include Regional Budget Management; Regional Planning; Procurement of goods and services; and Public Service. The public can access information in various ways: at a service desk; through the website ppid.bandung.go.id or through the websites of the respective work units; by sending a letter to the head of the Office of Communication and Information; by phone; by E-mail; and through social media. The office has various social media accounts, including Facebook, Twitter, Instagram, and a YouTube channel. As per November 2017, its website has been visited more than 13 million times.

**Social Media Account of the Former Mayor of Bandung**—In addition to the media and official portals of the Bandung City Government, data and information are also delivered through the social media accounts of the former mayor of Bandung. Through this account, he regularly made announcements and provided information on the progress, process, and development of plans. He conveyed his messages in an accessible way, often using Sundanese, the traditional language of Bandung. He also quickly responded to comments and suggestions by the public made in his social media account.

## 5 Factors that Led to the Innovations and Good Practices

There are many factors that underlie the aforementioned good practices and innovations in improving the access to basic services for marginalized groups as follows.

**Leadership**—Leadership is an important factor that underlies innovations in cities around the country. The current president, Joko Widodo has brought change in Surakarta during his time as mayor (2005–2012). The same is true for Bandung. A study by Ramdhani (2015) on the leadership of Ridwan Kamil, the former mayor of Bandung shows that he is one of the keys to the transformation of Bandung in the last 5 years. The former mayor of Jakarta, Basuki Tjahaja Purnama, is also seen as one of Indonesia's powerful leaders, according to the opinion leader survey of the University of Indonesia (Cahya 2016s).

**Communication style**—Good communication techniques are an important capital for the government of Bandung in interacting with the community. The use of social media such as Facebook, Twitter, Instagram, and YouTube channels is very effective in disseminating information widely. Languages ranging from Indonesian, English, to local language Sundanese, and style that is flexible but still elegant, humorous, and warm has become one of the main points in establishing communication between the government with its citizens.

**Community approach**—The direct approach to visit the community, either making Sunday a leader's routine to just eat and chat with citizens in Bandung, or Friday morning in Surakarta is an important way to provide communities with access to their leaders. During these visits the mayors provide routine information, socialize and listen to grievances and hopes of citizens to make them sympathetic to their leaders and government. This is a good way to increase people's willingness to participate to become part of and cooperative in the development and management of their city.

**Transformation of the Government Management Paradigm**—The government management paradigm that used to be oriented at input has been transformed to be oriented at output. Thus, various creative efforts have been undertaken to ensure the accountability of the government's performance with the principles of good and clean governance.

**Easiness for citizens to voice their aspirations to the government**—The Indonesian government makes it easier for citizens to report complaints. Systems such as LAPOR! and ULAS, by the local government of Surakarta, facilitate citizens in filing reports through various ways, such as SMS, websites, social media, or by phone.

**The smart city concept**—The implementation of the smart city concept in Bandung is implemented through incremental operational change which is based on the use of information technology. This concept integrates various fields. The smart city concept has a practical and efficient impact on urban management and services; it facilitates public services and shortens the time to solve problems as it is based on information technology. Bandung is considered one of the forerunners of the smart city concept in Indonesia. Jakarta has also applied the smart city concept in its Jakarta

Smart City project, which provides the city with a dedicated hub for the latest technology. It helps startups that focus on smart city technology with apps like Qlue and Trafi, and technology like smart lighting and ERP to shape the daily life of Jakarta's citizens. Another city in Indonesia well-known for its implementation of the smart city concept is Makassar. One of the latest smart city applications in Makassar is its smart CCTV, which serves as an example for other Indonesian cities (Winarko 2017).

## 6 Barriers to the Full Engagement of Marginalized Groups

The community groups living in slums encounter obstacles and challenges that often hamper the implementation of political and social inclusion and processes of local democracy. The experiences from slum management in Indonesia, especially in Bandung and Surakarta, show the major obstacles and challenges faced by various parties, either from the perspective of the community in the slums themselves or others as regulators and service providers, i.e., the government and other actors who act as mediators such as NGOs/Activists.

**Trust**—There is a lack of trust among various parties involved (between the community and the government, the community with NGOs, and NGOs with the government). Trust is often shaken by parties with political motivations. If people do not trust the government then no government effort will be effective. Conversely, if the community has high levels of trust then whatever the leaders say will be done and followed by the community. Participation and a good track record of leaders are believed to be tools that can increase trust. The survey results suggest that the level of trust of respondents in Surakarta toward the city government is generally higher than those in Bandung. According to the survey in Bandung, this is associated with the low level of satisfaction in regard to corruption, the politicization of service delivery, equitability of resource distribution, and levels of accountability. The trust in the government is also affected by inconsistencies in government policy. In one case, regulations are enforced strictly but in other cases, situations are left alone. An example is controlling illegal buildings on riverbanks by the government for violating regulations in one location but not acting in other locations.

**Political and personal interests**—The community is an object and political tool that is easily instigated by groups with certain interests. There are provocations from parties with certain political interests to make the community do not get involved in the planning process.

This is reflected in the level of satisfaction of respondents in both cities toward this institutional feature of politicization of service delivery, in which they felt that public service delivery has been politicized. Another example is the existence of third-party brokers in development projects. The brokers make it difficult for the government to convey its good faith to the communities. Often the government's mandate is not distributed as a whole because of these parties. The most common example is in the process of compensation for land. Communities often do not get their due right because a part of their share is taken by the brokers. Structurally, these brokers can come from either party, the local apparatus and from the residents' representatives.

**Process of participation**—Processes and mechanisms of community participation for the marginalized groups in slum areas are not working to fully utilize the potential contribution of communities. Their involvement is still constrained due to their limited assets, so their needs may not be accommodated. Often processes seem very instant and the government's engagement is only a final call or notification rather than a two-party dialogue and without inputs of the opinions and aspirations of the marginalized groups from the beginning of the planning process. Community involvement in formal planning process is still limited, even though community participation has been formalized in the national legislation regarding the development planning system and local government affairs. Despite the high level of participation in local election and community affairs, many respondents in Bandung acknowledge that not all communities are involved in the *Musrenbang* (Development Planning Discussions) forum. Some do not even know about the forum.

**The mindset of communities**—Communities often have their own priority and do not care about the process. Moreover, they often want to get quick solutions without considering the long-term consequences. Changing the patterns of community behavior and the process of adaptation to new things are challenging on its own. The limited amount of land and the ever- increasing urban populations necessitates non-conventional long-term slum handling solutions such as vertical dwelling, which many communities are not used to. Changes in behavioral patterns of life and in the outlook of people who are used to living in landed- houses are needed as they now must live in vertical dwellings, which require serious time and adaptation. It is a great challenge to educate the public, vis a vis the marginalized groups, about concepts of participation, accountability, transparency, and access to basic services.

**City budget allocation**—Decentralization policy has improved the fiscal capacity of local governments, especially municipal governments. This is shown by the case of Bandung, where almost 50% of the city budget originates from locally generated revenues, in which 40% is from local taxes and levies. The large percentage of locally generated revenues implies that it is in the city discretion on how to spend those revenues. However, from the expenditure information we learned that although 57% of the budget is spent for performing mandatory basic services, as mandated by the decentralization policy, based on expenditure account, the biggest expenditure proportion is actually for government employees expenditure (38%), followed by 33% for goods and services expenditure. Meanwhile, budget for capital expenditure is only 24%. Thus, although in terms of function the city has allocated its budget for the education (17.76%), health (16.44%), public works and spatial planning (11.48%), and housing and settlement (7.93%) sectors, we cannot tell how much the proportion is spent for actual development, compared to the routine expenditure for government employees. Greater allocation for routine, rather than capital expenditure is a nation-wide phenomenon.

**Data**—In Surakarta, many programs are aimed at poverty reduction. However, poverty alleviation is still constrained by chaotic data management because the municipal government does not have a master data set that can act as a joint reference. Each work unit has its own data for poverty alleviation programs. As a result, many programs are not on target or have double recipients for funding. Meanwhile, there is also national data. One of the respondents stated that the biggest problem of poverty alleviation in Surakarta is the ego of the sectoral work units that collect their own data. He pointed out that the Education Office has different data on school-aged children compared to the data from community-based poverty assessments. The difference is because the data of the Education office comes from schools, while the data from Urban Village Poverty Reduction Team comes from the community. Therefore, single data is important to be a common reference by all work units. Another problem related to data, as pointed out in the interviews with key informants, is a difference in attitude to poverty data between the local government of Surakarta and the regional higher tier government. Surakarta is innovative in compiling local data, which usually shows a higher rate of poverty. Meanwhile, the regional government is more focused on showing their accomplishments in reducing poverty rate and the number of poor people.

## 7 Conclusion

Since 1999 Indonesia has implemented many decentralization policies that enable local governments to prepare programs and policies that are better adjusted to the local situation. This has facilitated participatory planning in Indonesia, as local governments have more power to develop bottom-up participatory planning programs that involve local communities, NGOs, and other community organizations. At the same time, Indonesia is experiencing rapid rate of urbanization. More than half of its 260 millions populations live in urban areas. In addition, the number of cities has also grown significantly. The number of cities today (>100) is almost double than it was 20 years ago. As the poverty rate is still around 10% of population, cities are where the poor are living in slum areas and become one of the marginalized groups.

To answer the study questions, i.e. to what extent the marginalized groups are engaged in local democracy mechanism, and the barriers to social and political inclusion, the study found that the respondents in both cities actively participate in local democracy although they do not necessarily trust the government or NGOs. The governments in both cities attempt to reach the most marginalized groups by providing channels of participation through innovations and good practices that have led to a change in the behavior of the marginalized communities to be more open and to participate more actively in the programs of the government. The innovations and good practices are inseparable from several important underlying factors as discussed, i.e., leadership, communication style, a community approach, a transformation in the government management paradigm, the smart city concept, and easiness for citizens to tell the government their aspirations.

Although there are mechanisms to ensure the participation of marginalized groups, there are weaknesses in government institutions' capacity to reach out to these groups.

In Surakarta, one of the major constraints is the lack of urban administration data system to support the service program. There are differences in data on the poor, as well as in the beneficiaries of the program, which is sometimes also influenced by the political dynamics at the bottom level. The results from the study in Bandung show that some of the important barriers include a lack of trust in the government, in part by inconsistencies in the law enforcement. Despite the fact that many policies have been developed for a more inclusive city, in practice, these policies are not always as inclusive as how they were designed to be.

This leads to final point about its policy implications for the Indonesia i.e. coping with the barriers to inclusion for marginalized groups. The need is for government service models that emphasize the importance of intermediary groups or facilitators to link and build closeness with the marginalized groups and for collaboration between community leaders and social activists to reach these groups.

## References


Aditama K, Soedwiwahjono N, Putri RA (2016) Pola Perjalanan Penduduk Pinggiran Menuju Kota Surakarta Ditinjau dari Aspek Spasial dan Aspek Aspasial. In: Arsitektura, vol 14, no 1

Antlov H, Ibrahim R, van Tuijl P (2005) NGO governance and accountability in Indonesia; challenges in a newly democratic country. Available at: http://www.icnl.org/research/library/files/Indonesia/Peter_NGO%20accountability%20in%20Indonesia%20July%2005%20version.pdf. Accessed at 14 Jan 2018

Asian Development Bank (2012) The neighborhood upgrading and shelter sector project in Indonesia. https://www.adb.org/sites/default/files/publication/29879/neighborhood-upgrading. Accessed 23 Nov 2017

Directorate General of Human Settlements Ministry of Public Works and Public Housing Tentang Program Kota Tanpa Kumuh (KOTAKU). http://kotaku.pu.go.id/aboutdetil.asp?mid=1&catid=5. Accessed 10 Oct 2017

Government of Indonesia (2007) Government regulation no. 38/2007 on division of affairs between governments. Republik Indonesia, Jakarta

Government of Indonesia (2004) Law No. 25 of 2004 on national development planning system Republik Indonesia, Jakarta

Government of Indonesia (2004) Law No. 33 of 2004 on revenue sharing between the central government and the regional governments. Jakarta: Republik Indonesia

Government of Indonesia (2014) Law No. 23 of 2014 on Regional Government. Jakarta: Republik Indonesia, Jakarta

Jakarta Globe (2017) Indonesia fifth largest country in-terms of twitter users. In: Jakarta Globe. http://jakartaglobe.id/news/indonesia-fifth-largest-country-in-terms-of-twitter-users/. Accessed 10 Oct 2017

Komite Penanggulangan Kemiskinan (2005) Strategi Nasional Pengentasan Kemiskinan [National strategy for poverty alleviation]. Republik Indonesia, Jakarta

Kota Kita (2017) Improving the transparency, inclusivity and impact of participatory budgeting in Indonesia. http://www.makingallvoicescount.org/project/kota-kita/. Accessed 23 Nov 2017

Minister of Public Works and Housing Republic Indonesia (2016) Regulation of the minister of public works and housing 02/PRT/M/2016 on Increasing the quality of slum housing and slum settlements

Ministry of Social Affairs (2017) Program Keluarga Harapan. https://www.kemsos.go.id/program-keluarga-harapan. Accessed 7 Oct 2017

Nurmatari A (2016) Kota Bandung Raih Penghargaan Tertinggi Sistem SAKIP dari KemenpanRB. In: detikNews. https://news.detik.com/berita-jawa-barat/3140234/kota-bandung-raih-penghargaan-tertinggi-sistem-sakip-dari-kemenpanrb. Accessed 23 Nov 2017

Ramdhani LE (2015) Fenomena Kepemimpinan Fenomenal. Jurnal Borneo Administrator 11(3):268–296

Salim W, Hudalah D (2020) Urban governance challenges and reforms in Indonesia: towards a new urban agenda. In: Dahiya B, Das A (eds) New urban agenda in Asia-Pacific, Advances in 21st century human settlements

Solo Kota Kita (2012) City and community profile: Solo, Central Java, Indonesia. Kota Kita. www.kotakita.org. Accessed at 14 Jan 2018

Winarko W (2017) Three Indonesian smart cities to look up to. In: The Jakarta Post. http://www.thejakartapost.com/life/2017/06/09/three-indonesian-smart-cities-to-look-up-to.html. Accessed 23 Nov 2017

World Bank (2016a) Indonesia: Improving infrastructure for millions of urban poor. Press release, World Bank. http://www.worldbank.org/en/news/press-release/2016/07/12/indonesia-improving-infrastructure-for-millions-of-urban-poor. Accessed 10 Oct 2017

World Bank (2016b) Indonesia's urban story. World Bank. http://pubdocs.worldbank.org/en/452811465807212968/IDN-URBAN-ENGLISH.pdf. Accessed at 14 Jan 2018

World Bank (2017) Indonesia national slum upgrading project latest implementation status and results report. World Bank. http://projects.worldbank.org/P154782/?lang=en&tab=results. Accessed 23 Nov 2017

**Wilmar Salim** is a Senior Lecturer and former Chair of Graduate Program in Regional and City Planning, the School of Architecture, Planning and Policy Development, at Institut Teknologi Bandung (ITB) in Indonesia. He is recently appointed as the Head of Research Center for Infrastructure and Regional Development in his university. He is also affiliated with the ITB's Center for Climate Change as Senior Researcher. He earned a Ph.D. in Urban and Regional Planning from the University of Hawaii at Manoa. His areas of expertise include: plan and policy implementation and evaluation; decentralization in developing countries; local and metropolitan governance; poverty reduction strategies; and mainstreaming climate change adaptation into development planning. He has published journal articles and book chapters on Indonesia's urban and regional studies. He has worked as consultant for ADB, GIZ, JICA, UNDP, the World Bank, as well various Indonesian government agencies. He was a Visiting Associate Professor at the Graduate School of International Development and Cooperation of Hiroshima University in 2016. He is an active member of Indonesian Regional Science Association (IRSA) and Research Committee 21 Sociology of Urban and Regional Development of the International Sociological Association (ISA). He is now a Senior Fellow of Partnership for Australia-Indonesia Research (PAIR), an initiative of Australia-Indonesia Centre.

**Martin Drenth** graduated from the University of Groningen with a degree in Human Geography and Planning. During his bachelor studies, he was selected for the Challenge Program targeted at the top 5–10% of his faculty. His first international academic exposure was an Erasmus Program at the University of Reading. Since 2013, Martin has lived in Indonesia. He is a researcher affiliated with the Bandung Institute of Technology (ITB). He has been involved with the Climate Change Center and the Center for Research on Infrastructure and Regional Studies of ITB. Besides research, he has been a proofreader and translator at the Journal of City and Regional Planning. Most of his professional life has been spent in Indonesia where he was able to develop an academic network. He has varying research interests including inclusive planning, the water-sensitive city, climate change adaptation and resilience in an urban context. His main interest has always been on the Global South. His work in developing countries explores the effects of planning and how it disproportionately affects minority and marginal groups. During his bachelors, he studied the segregation among the three major ethnic groups in Paramaribo, Suriname. Martin's first research in Indonesia was for his master thesis when he assessed the possibility of applying brownfield redevelopment in Jakarta. He studied the regulatory and institutional mechanisms that affect planning. The study found that powerful market forces can hijack well-designed plans, which underlines the need to develop governance capacity and more effective development control.

# Political and Social Inclusion and Local Democracy in Indian Cities: Case Studies of Delhi and Bengaluru

**Debolina Kundu**

**Abstract** Growth of slums, urban poverty and inequality, deteriorating quality of the urban environment, unplanned growth, rapid periurbanisation and growth of sprawls, and deficiencies in the coverage of basic urban services have plagued Indian urbanisation. Through micro-level surveys in select slums and squatters in Delhi and Bangalore, this study attempts to identify the structural and institutional barriers to full engagement of marginalized groups in processes and mechanisms of local democracy, including participation in community based organizations in the provision of basic urban services. Results indicate two prominent categories of exclusion, viz, institutional including housing, land and basic services and structural including barriers to participation of women in economic, social and political activities. These manifest in various forms of inadequacies for different vulnerable groups. To elucidate, the migrants, living mostly in rented accommodation, have lower coverage to basic amenities as compared to the non-migrants. Similarly, the condition of the scheduled population is poor as compared to non-scheduled population. The condition is also poor for the low-income households. The slum dwellers from both cities perceive that providing support to elderly and disabled population, women and child safety, improving connectivity to slums and developing infrastructure and community based preparedness for disaster should be on high priority for the government. The slum dwellers showed low faith in local government as compared to NGOs, in whom they have higher confidence. An integrated approach to urban planning and management with proper targeting of beneficiaries may ensure inclusion of the disadvantaged sections of the society.

The paper is based on a research funded by Swedish International Centre for Local Democracy project titled "Political and Social Inclusion and local democracy in Asian cities: Case of India", and carried out by East West Centre in 2017 with data support from World Vision, India. The author is grateful to the research team for their contribution, especially Dr. Arvind Kumar Pandey, Ms. Pragya Sharma and Dr. Arpita Banerjee for their research support.

D. Kundu (✉)
National Institute of Urban Affairs, Core 4B, 1st Floor, India Habitat Centre, Lodhi Road, New Delhi 110003, India
e-mail: dkundu@niua.org



S. Cheema (ed.), *Governance for Urban Services*, Advances in 21st Century Human Settlements,
https://doi.org/10.1007/978-981-15-2973-3_8

# 1 Introduction

The globe is more urban now. Urbanization, which is strongly correlated with economic development, is emerging as a strong transformative force reshaping the world's regions. It is impacting both urban and rural landscapes, giving rise to peri-urbanisation while bringing prosperity to many urban regions. Concomitantly, urbanization has also opened up new forms of deprivation, unsustainability, polarization and divergence in development and incomes between urban and rural areas and within urban areas, resulting in exclusionary urbanisation (Kundu 2016).

Indian urbanization has also been exclusionary in nature. The country, after urbanizing very rapidly in the post-independence period in the decade of the fifties registered a decline in urban growth rates since the 80s, when it reached the peak of 3.8% growth rate. Despite its contribution of 52.58% in the national GDP,[1] urbanization in India has been essentially unplanned. It has been accompanied by growth of slums, poverty and inequity, deteriorating quality of the environment, unplanned expansion, rapid peri-urbanization, and deficiencies in the coverage of basic urban services. Urban India currently is mired in crime and violence, congestion, exposure to pollution and disparity in the access to education and health.

The consistent decline in the growth rate of urban population over the past two decades of the last century led to the Tenth Plan expressed concern over 'the moderate pace of urbanisation'. The Eleventh Plan admitted that 'the degree of urbanisation in India is one of the lowest in the world' and considered planned urbanisation through new growth centres in the form of small and medium towns its major challenge. The Approach Paper to the Twelfth Plan also recognised the need to promote spatially-balanced urbanisation.

The approach papers to the 11th and 12th Five Year Plans (FYP) have accepted the fact that migrants are the most vulnerable among the poor. Recent literature (Kundu and Saraswati 2012; Kundu 2003, 2009, Kundu and Samanta 2011; Bhagat 2012) shows that cities are increasingly becoming exclusionary. The policy environment with regard to poor migrants is not conducive. Many city Master Plans aim to keep migrants at bay; rural development and agriculture policies aim to control out-migration; and migration is viewed as a socially and politically destabilizing process. In the last few decades, demolition and eviction drives of slums have become a routine practice in a bid to make Indian cities world class. This scenario has made the lives of slum dwellers more vulnerable.

Since the 1990s, attempts have been made both by the central as well as state governments to make a few large cities more attractive. Macro level programs adopted at state and city levels have helped in pushing out the slums and squatters from the better off areas of the city. This has accelerated the process of segmentation and accentuated intracity disparity (Kundu 2009). This is reflected in social, political and economic spheres. Exclusionary forces have affected adversely the marginalized communities in various forms. Two prominent categories of exclusions are discernible: institutional inequality including housing, land and basic services and structural inequality

[1] Source: Central Statistical Organisation, National Accounts Statistics, 2016.

including barriers to participation of women in economic, social and political activities. These manifest in various forms of inadequacies for migrants, youth and elderly to lead a decent lifestyle.

At the same time, as a signatory to the New Urban Agenda and Sustainable Development Goals (SDGs) 2030, the government is committed to leave no one behind in the process of development. This commitment is reflected in various Missions and policy statements of the government. It, therefore, becomes pertinent for the government and other stakeholders to address the existing forms of exclusion and barriers to political and social equity. In addition, it becomes important to ensure full participation of the marginalized communities including women, youth, migrants and ethnic minorities in the processes of local democracy in cities and towns to ensure equitable access to basic amenities.

In reality, however, elite capture of urban governance, forces of globalization and labor market restrictions have resulted in limited access for the poor and vulnerable to housing and basic services in urban India. The resident welfare associations that came up in the big cities in the last few decades as parallel agencies to wards committees,[2] are increasingly responsible for the operation and maintenance of services (Kundu 2011). Further, globalization has resulted in selective big city bias in government interventions. Exclusionary urbanization has restricted the entry of poor migrants to cities because of their low levels of skill and education coupled with high cost of living in cities. Given this background, the current research examines the determinants of political and social inclusion of the marginalized groups, including scheduled castes and scheduled tribes, women, poor migrants and minorities (Muslims) in selected cities.

Following the introduction, Sect. 2 presents an overview of government policies and programs to improve the coverage to urban services. Section 3 describes the scale of macro level deficits in the coverage of housing and basic services including water and sanitation in urban India. Section 4 presents micro-level results based on surveys in select slums and squatter settlements in two cities and the structural and institutional barriers to full engagement of marginalized groups in processes and mechanisms of local democracy, including coverage of basic urban services and participation in community based organizations. Section 5 examines structural and institutional factors that have influenced access to services in urban India. Section 6 presents conclusions The scope of the study is pan India with reference to the institutional structure and policy interventions. However, primary survey is limited to two cities, viz, Delhi and Bengaluru. Here, two slums in each city with differential tenure status have been selected for detailed analysis with regard to inclusion of poor and marginalized communities in the access to basic amenities, education and health facilities. The study is based on both macro and micro level analysis taking the slums of Delhi and Bengaluru as case studies.

Delhi or National Capital Territory (NCT) has the highest population density of 11,297 people per square kilometer as per 2011 Census of India. It covers an area of

[2]The Wards Committees are supposed to involve the citizens in local governance, as a mandate of the 74th CAA, 1992.

1484 km$^2$. The Delhi Urban Agglomeration (UA) consists of 16.34 million people while the Municipal Corporation of Delhi area accounts for more than 11 million people (Census 2011). The other city is Bengaluru (erstwhile Bangalore); the state capital of Karnataka which covers an area of 741 km$^2$. (Census 2011). Most of the city lies in the Bengaluru urban district. Urban population of Karnataka is more than 22.1 million of which Bengaluru is home to around more than 9 million urban population (Census 2011).

## 2 Government Policies and Programs

Since independence, various programs related to urban development have been launched, specifically under various five year plans. However, until recently, India did not have a national urban policy to guide the process of urbanisation in a planned manner. In recent years, India has moved from a 'business-as-usual approach' to paying systematic attention to urbanisation and its challenges. Policy makers and planners believe that Indian cities need a robust framework of urban policy that can guide its current development trajectory. The National Urban Policy Framework (NUDF), 2018 comes exactly three decades after the National Commission on Urbanisation (1988), which came up with a roadmap on tackling growth of metropolitan cities by promoting growth of small and mid-sized cities, which could act as counter-magnets. The idea of the NUDF is to have a fluid plan that can be tweaked at regular intervals, instead of one rigid Master Plan that no agency follows or implements, which is at the root of the urban mess.

It may be mentioned here that the Government of India's policies on slums have undergone a paradigm shift in recent decades. In the decades of the seventies and eighties, the government emphasized 'no slum cities', which implied forceful resettlement or rehabilitation of the slum dwellers. Removing slums from central areas and transportation nodes of cities meant that the new settlements where those slum dwellers were relocated, remained on the outskirts of the city, far from their workplace, thus further worsening the welfare of the slum dwellers. With this realisation, the government started focusing on slum upgrading and slum rehabilitation programmes. The government focused on infrastructural development of slums through schemes such as Environmental Improvement of Urban Slums (1972) and Sites and Services schemes (1980), and the Community Development Programme (1988). Gradually, slum upgradation policies shifted their focus from providing and improving infrastructure in the slums to improving the quality of life of slum dwellers. At present, the government has adopted the role of a facilitator in housing the poor instead of being a direct provider of housing. Moreover, the private sector is encouraged to provide housing for the poor.

Initiatives taken by the government for housing the poor over the last 60 years include: Integrated Subsidized Housing Scheme (1952) for industrial workers and economically weaker sections; Low Income Group Housing Scheme (1956); Slum Improvement/Clearance Scheme (initiated in 1956 and discontinued in 1972 at the national level); Environmental Improvement of Urban Slums (1972); National Slum Development Programme (1996), Scheme for Housing and Shelter Upgradation (SHASHU as part of Nehru Rozgar Yojna, introduced in 1989 and discontinued in 1997); Night shelters (1988–1989); Two Million Housing Programmes, Valmiki Ambedkar Awas Yojana launched in 2001–2002 (VAMBAY); Jawaharlal Nehru National Urban Renewal Mission (JNNURM); Rajiv Awas Yojana (RAY); Rajiv Rinn Yojana (RRY); and the latest Pradhan Mantri Awas Yojana (PMAY). In addition, various ministries have had their own programs targeted towards their area of work.[3] In the 1990s, the national five-year plans adopted an inclusive agenda and launched several programs as "missions" involving private sector participation—with clearly defined objectives, scopes, timelines, milestones, as well as measurable outcomes and service levels. These programs, however, demonstrated a "big-city bias". Some of the major initiatives of the government in the recent past are discussed below.

**Jawaharlal Nehru Urban Renewal Mission (JNNURM)** launched in 2005 in mission mode, is the first program having an exclusively urban focus with four sub-missions—(i) Urban Infrastructure and Governance (UIG) with a focus on infrastructural development covering 65 mission cities (ii) Urban Infrastructure Development of Small and Medium Towns (UIDSSMT), which aimed at planned urban infrastructural improvement in all towns/cities except 65 mission cities (iii) Basic Services to the Urban Poor (BSUP) with a focus on integrated development of slums covering the same 65 cities as UIG and (iv) the Sub-Mission for Integrated Housing and Slum Development Programme (IHSDP), which aimed at holistic slum development, in all cities/towns except those covered under BSUP.

**Affordable Housing in Partnership (AHP)**—Realizing that mere effort of the government would be insufficient to address the housing shortage, the scheme of Affordable Housing in Partnership (AHP) was introduced in 2009. AHP sought to promote various kinds of public-private partnerships such as government with the private sector, cooperative sector, financial services sector, state parastatals, ULBs etc. to create affordable housing stock.

**Rajiv Awas Yojana (RAY)**—RAY, again a mission mode program, was launched in 2011. It envisaged a 'slum-free India' by encouraging states to tackle the problem of slums in a definitive manner and called for a multi-pronged approach focused on bringing existing slums within the formal system, redressing the failures of the formal system that lie behind the creation of slums.

**Pradhan Mantri Awas Yojana (PMAY)**—PMAY was launched in June 2015 for the period 2015–2022. Under this scheme, the central government provides financial and technical assistance to cities for in situ rehabilitation of existing slum dwellers using land as a resource through private participation; credit linked subsidy to both poor and non-poor; affordable housing in partnership and subsidy for beneficiary-led individual house construction or enhancement. The mission provides flexibility to the states for choosing the best options amongst four verticals of the mission to meet the demand of housing in their states.

---

[3]Nation Resource Centre, School of Planning and Architecture, New Delhi (2009), "Affordable Housing for Urban Poor".

## 3 Macro Level Analysis

A macro level analysis shows that 17.4% of the urban population lived in slums in 2011 where housing conditions and infrastructure facilities were far from satisfactory. Nearly 2.9% of the urban houses were in a dilapidated condition (Census 2011). The urban housing shortage in the country was 18.78 million in 2012 as per the estimates of the Technical Group on Urban Housing Shortage (2012–2017). Of this, 0.53 million households were homeless and mainly dominated by single male migrants. Further, 70.6% of the households in urban India were covered with tap water,[4] 19% households either had no toilet within their premises or defecated in the open, and 13% households had no bathing facilities within the home. These conditions were worse in the slums. Moreover, access to basic amenities was not uniform across the states and urban centres—economically developed states and metropolitan cities had better infrastructural facilities as compared to less developed states and non-metropolitan cities. Disparities in the coverage of basic amenities were also noted by caste and class affiliation of urban dwellers and across migrant and non-migrant households. Current evidence from national level official data suggests a declining rural to urban migration. With exclusionary migration becoming a dominant phenomena, the rural people find it increasingly difficult to gain foothold in cities. Urbanisation in the neo-liberal regime has rendered cities less affordable for the poor in terms of living and access to basic amenities.

In view of the above, in the subsequent section, an analysis is done mainly by four different groups, (1) male-female headed households, (2) migrant-non-migrants, (3) social groups and (4) income groups to examine the marginalities in housing and access to basic services in urban areas at the macro level.

### *3.1 Male-Female Headed Households*

Female headship in India is often an outcome of death or ailment of the male members of a household and its share is more or less nominal. Widowed, separated and unmarried single household women are economically poorer and live in vulnerable conditions. A high percentage of such women draw income from informal sector work; characterised by job insecurity, low and irregular wages and poor working conditions. The results from the two rounds of NSS show that although the gap between female and male headed households living in *pucca*[5] houses reduced during 2002–2012; still the percentage of female headed households living in *pucca*

[4]Water from unsafe sources includes uncovered wells, springs, rivers/canals, tanks/ponds/lakes.

[5]National Sample Survey of India defines a pucca house as 'a housing unit which has walls and roof made of materials such as cement, concrete, oven burnt bricks, hollow cement/ash bricks, stone, stone blocks, jack boards (cement plastered reeds), iron, zinc or other metal sheets, timber, tiles, slate, corrugated iron, asbestos cement sheet, veneer, plywood, artificial wood of synthetic material and polyvinyl chloride (PVC) material'.

houses was less as compared to male headed households in both periods. The ownership pattern of the households show that percentage share of the households living in rented houses was slightly higher for female headed households as compared to males. For female headed households, this percentage share slightly increased during 2002–12 reflecting higher vulnerability for the former in terms of housing structure and ownership of houses in 2012.

There is a slight difference between male and female headed households in the access to tap water as main source of drinking water. The percentage of households having access to tap water was slightly higher for male headed households in 2002. However, in 2012, this gap narrowed down with 69% households in both categories having tap water as the main source of drinking water. One striking result in 2012 was an increment in the share of bottled water. It shows that during 2002–12, the dependency on bottled water increased both for all households. In 2012, the share of households having own toilet increased for both and the gap between these two remained same. The percentage share of households having underground drainage facility almost doubled during 2002–2012 and the gap between them came down. It is evident from the above analyses that in comparison to male headed households, the conditions of housing and basic amenities were slightly poor for female headed households in 2002, although it improved in 2012.

### *3.2 Migrant and Non-migrant Households*

Migrants from poor background face multiple vulnerabilities. They face social disconnect from the urban society as well as 'sedentary bias' in access to housing and basic amenities. The results from NSS show that the percentage of households living in *pucca* houses increased during 2002–12 both for migrant and non-migrant households. Interestingly, the percentage of migrant households living in *pucca* houses was higher as compared to non-migrant households both in 2002 and 2012. However, the ownership pattern show a different result. More than 60% non-migrants owned houses in 2002 and 2012 as well. In comparison, only less than 15% migrant households owned houses, and a majority of them lived in rented accommodation (73.9% in 2002 and 80.8% in 2012).

The condition of basic services was also poor among migrant households. In 2002, only 66.8% migrant households had access to tap water as the main source of drinking water as compared to 74% non-migrant households. However, this share further declined to 60% for migrant households in 2012 as compared to 69.5% for non-migrant households in 2012. A higher dependency on bottled water is seen among migrant households as compared to non-migrants. The access to toilet facilities among migrant households during 2002–2012 show a slight improvement. Half of the migrant households in 2012 had separate toilets in their houses. Also, in comparison to non-migrant households, the condition of migrant households with access to toilet facilities was poor in both 2002 and 2012 as more than 40% migrant households in 2002 and 2012 used shared toilets. This share was almost half among non-migrants households. In terms of access to drainage facilities, the condition of migrant households was slightly better than non-migrants. In 2002, a total of 37.9% migrant households had access to underground drainage facility as compared to 28.8% non-migrant households. In 2012, the situation improved for both categories, as half of the migrant households and 45% non-migrant households had access to underground drainage facility. The access to open *pucca*[6] *and* open *katcha*[7] drains and no-drainage categories declined during 2002–2012 for both types of households with corresponding increments in the share of households with access to underground drainage facilities.

### 3.3 Social Groups

In India, people from certain castes were historically deprived from possession of land and social capital and, therefore, a majority of the population from these castes are still poor and vulnerable. This is reflected even today in the pattern of urban housing and access to basic amenities. The results from NSS show that although the percentage of households living in pucca houses was very high across social groups both in 2002 and 2012, but in comparison to others (general category) the percentage of households living in concrete households was low among Other Backward Castes (OBCs), Scheduled Castes (SCs) and Scheduled Tribes (STs). During 2002–12, this share increased across social groups. The highest increment was found in STs and SCs. The ownership pattern shows that the SCs households in urban India had the highest percentage share of owned houses in 2002 followed by OBCs, others and STs. However, this pattern reversed in 2012 with households from other category owning the highest percentage of houses followed by SCs, OBCs and STs. Households living in rented houses were highest among STs and OBCs in 2012 as compared to others and SCs, reflecting their vulnerable situation.

The condition of basic amenities among STs was very poor as compared to other social groups. The results show that tap was the main source of drinking water across social groups but its access was highest among other categories followed by OBCs, SCs and STs in 2002. In 2012, the share declined slightly with increment in the other category. The analysis shows that dependency on bottled water increased across social groups but the OBCs and SCs had higher dependency on bottled water as compared to others and STs. The percentage share of households having access to own or separate toilet in their households was highest among others followed by OBCs, SCs and STs in 2002. Although the share increased across social groups in 2012, but households from other category still had the highest share as compared to other social groups. In 2002, the percentage share of households who used shared toilets was highest among STs followed by OBCs. The percentage share of households that did not have toilets in their houses was also highest among STs followed by SCs and

[6]Open drains made of materials like bricks, stones and cement concrete, etc.

[7]Ordinary channels cut through the ground to allow water to pass.

OBCs. In comparison to these social groups, the share of households with no toilet facilities was least among the others category. Despite the reduction in the share of households with no toilets facilities across social groups in 2012, a significant percentage of SC (19.6%) and ST (17.2%) households had no toilet facility in their houses. The pattern of drainage facility across social groups shows that despite a decline in the percentage share of the households with no drainage facility during 2002–12, nearly 20% households from SCs and STs groups had no drainage facility in 2012. In 2012, the share of the households with access to underground drainage facility improved significantly across social groups and this improvement was more among SCs and STs as compared to other social groups. It is evident from the analysis that the condition of housing and other basic amenities among SC and ST households in urban India is poor as compared to OBCs and other categories.

### *3.4 Income Groups*

The income group wise condition of housing in urban India shows that the percentage share of households living in concrete houses was highest in higher income groups. Also, the share of households living in pucca houses increased during 2002–2012. However, there is no strong relationship between ownership pattern of housing and income level of the households as the percentage share of the households living rented accommodation was highest among high income groups (Q4 and Q5) as compared to lower and middle income groups both in 2002 and 2012. Correspondingly, the percentage share of the households living in owned houses were high in lower and middle income groups as compared to high income groups.

The condition of basic services was much better in the households from higher income groups, as with increasing level of income groups, the share of households using tap water as the main source of drinking water increased. Likewise, with increasing income level, the share of households having own latrine/separate latrine in households increased during 2002–2012. The condition of drainage facilities was also better in the households from higher income groups. In contrast, the share of households having no drainage facilities was highest among lower income groups. It is apparent from the analyses that there is persistent inequality in the condition of housing and basic amenities across income groups, as households from higher income groups had better housing and basic services as compared to middle and lower income groups.

## 4 Micro Level Results of Surveys in Slums

In this section, an attempt has been made to capture the political and social exclusion in slums through a micro level study in select slums in two cities (Delhi and Bengaluru). In the two selected cities, an attempt has been made to examine the role of

**Table 1** Housing conditions and status of basic civic amenities in sample households (in %)

| Housing and basic amenities | | Bengaluru | Delhi |
|---|---|---|---|
| Ownership status of house | | | |
| Owned | Owned/purchased | 36 | 53 |
| | Allotted by government on lease | 35 | 36 |
| Rented | | 29 | 11 |
| Total | | 100 | 100 |
| *Housing structure* | | | |
| Katcha | | 4 | 0 |
| Semi-concrete (Pucca) | | 29 | 6 |
| Concrete (Pucca) | | 67 | 94 |
| Total | | 100 | 100 |
| *Availability of kitchen* | | | |
| Yes | | 53 | 22 |
| No, outside premises | | 7 | 2 |
| No, inside premises | | 40 | 76 |
| Total | | 100 | 100 |
| *Types of cooking fuels* | | | |
| Firewood | | 17 | 0 |
| Kerosene | | 2 | 0 |
| LPG | | 80 | 100 |
| Electricity | | 1 | 0 |
| Total | | 100 | 100 |
| *Availability of bathroom facility* | | | |
| Attached (inside premises) | | 59 | 33 |
| Detached (outside premises) | | 40 | 10 |
| No bathroom | | 1 | 57 |
| Total | | 100 | 100 |
| *Availability of toilet facility* | | | |
| Inside | | 53 | 12 |
| Outside | | 27 | 6 |
| No toilet | | 20 | 82 |
| Total | | 100 | 100 |
| *Drainage facility of households* | | | |
| Covered Pucca | | 79 | 53 |
| Open Pucca | | 21 | 44 |
| Open Katcha | | 0 | 2 |
| No drainage | | 0 | 1 |

*Source* Primary Survey, 2017

governance in the provision of basic services and to identify the gaps in the coverage of basic services with regard to their access to marginalized sections of the population residing in selected slums. In Delhi and Bengaluru, two different types of slums are covered in the primary survey, viz,—(1) slums (locally known as JJ-Clusters in Delhi) and (2) resettlement colonies. From each type of slum, 50 households were surveyed during May–June, 2017.

The analysis (Table 1) shows that the living condition in resettlement colonies was slightly better in comparison to slums. This is the case in both cities, which may be attributed to the high tenure security of the households living in resettlement colonies as compared to households living in slums/JJ-Clusters. When one compared the city level scenario, power supply was better in Delhi as compared to Bengaluru. The frequency of water supply was poor in Bengaluru as compared to Delhi, mainly because of acute water shortage in city. In most part of Bengaluru, households are dependent on privately run water tankers. The condition is slightly better in Delhi where the sample slums were covered through piped water. However, in terms of quality of water, the condition was poor in Delhi as compared to Bengaluru mainly because of poor maintenance of pipes, which led to water-contamination. In the absence of toilets within premises, a significant percentage of households in Delhi reported uses of alternative sources of sanitation among which public toilet was one option. However, the number of toilet complexes and seats were not sufficient to cover the entire population in Delhi, where unlike Bengaluru, individual toilets were not common. These facilities were in very filthy and unhygienic conditions. Because of poor conditions of public toilets, some households also resorted to open defecation. The lack of tenure security mainly among slum dwellers is one of the main reasons for the low percentage of households with toilets within premises in Delhi. The garbage disposal mechanism in Bengaluru is yet to be institutionalized, as it is dumped at community dumping sites near slums. Slum dwellers were prone to water and air borne diseases because of their poor working and living environment.

A very high percentage of households across sub-groups use liquid petroleum gas (LPG) as cooking fuel. The positive impacts of interventions of government schemes to avail clean fuel such as 'Kerosene Free Delhi' and 'Ujjwala Yojana' is reflected in the use of fuels in the slums of Delhi and Bengaluru. All households in Delhi reported that they use LPG as cooking fuel, whereas 20% of the households in Bengaluru used other sources. Among them, firewood is the second most important fuel as 17% of the households reported that they use it for cooking. Notably, no household in Delhi or Bengaluru reported use of cow-dung as cooking fuel, which was one of the main sources of cooking until the last decade of 20th century. As mentioned, this positive transformation is the result of implementation of policies to eradicate kerosene and traditional source of cooking fuel in low-income settlements.

The results from primary survey show that a high percentage of households in Delhi and Bengaluru possessed ration card, voter id, electricity bill with residential address, and bank account (Table 2). A significant percentage of households expressed their dissatisfaction with the quality of ration provided through public distribution system to households living in slums and resettlement colonies of Bengaluru and Delhi.

**Table 2** Possession of different documents by sample households (in %)

| Documents owned | Bengaluru | Delhi |
|---|---|---|
| Ration card | 81 | 95 |
| Passport | 1 | 7 |
| Electricity bill | 75 | 97 |
| Driving license | 19 | 32 |
| Voter card | 81 | 80 |
| Bank account | 85 | 80 |
| Caste certificate | 46 | 49 |
| Pension documents | 5 | 20 |
| Disability card | 2 | 4 |
| Health card | 4 | 14 |
| Aadhar card | 97 | 99 |

*Source* Primary Survey, 2017

In addition, the percentage share of households with caste certificate, pension documents, disability and health card was negligible. Television and mobile phones were the main modes of communication and entertainment in slum/resettlement colonies of Delhi and Bengaluru (Table 3). The percentage of households who never

**Table 3** Frequency of use of communication media by respondents

| Communication media | Frequency of use (in %) | | | | | |
|---|---|---|---|---|---|---|
| | Never | Rarely | Occasionally | Regularly | Very regularly | Total |
| *Bengaluru* | | | | | | |
| Newspaper | 84 | 12 | 3 | 0 | 1 | 100 |
| Television | 10 | 0 | 1 | 22 | 67 | 100 |
| Mobile phone | 7 | 1 | 5 | 26 | 61 | 100 |
| Radio | 89 | 2 | 7 | 2 | 0 | 100 |
| Internet | 97 | 0 | 2 | 0 | 1 | 100 |
| Public announcement | 41 | 16 | 33 | 10 | 0 | 100 |
| *Delhi* | | | | | | |
| Newspaper | 64 | 6 | 11 | 8 | 11 | 100 |
| Television | 7 | 5 | 8 | 43 | 37 | 100 |
| Mobile phone | 15 | 9 | 9 | 34 | 33 | 100 |
| Radio | 86 | 3 | 7 | 2 | 2 | 100 |
| Internet | 70 | 3 | 8 | 11 | 8 | 100 |
| Public announcement | 78 | 8 | 10 | 1 | 3 | 100 |

*Source* Primary Survey, 2017

**Table 4** Functioning of ward committee/mohalla sabha and peoples' participation (in %)

| | Bengaluru | Delhi |
|---|---|---|
| *Ward committee/Mohalla Sabha in slum* | | |
| No | 24.0 | 83.0 |
| Yes | 76.0 | 17.0 |
| Total | 100.0 | 100.0 |
| *Frequency of meetings of ward committee/Mohalla Sabha* | | |
| Weekly | 16.0 | 0 |
| Fortnightly | 0 | 1.0 |
| Monthly | 59.0 | 12.0 |
| Never | 0 | 26.0 |
| Not aware | 25.0 | 61.0 |
| Total | 100.0 | 100.0 |
| *Awareness about last meeting of ward committee/Mohalla Sabha* | | |
| No | 11.0 | 28.0 |
| Yes | 65.0 | 11.0 |
| Not aware | 24.0 | 61.0 |
| Total | 100.0 | 100.0 |
| *Respondents' membership to ward committee/Mohalla Sabha* | | |
| No | 48.0 | 32.0 |
| Yes | 28.0 | 7.0 |
| Not aware | 24.0 | 61.0 |
| Total | 100.0 | 100.0 |
| *Membership of other family members to ward committee/Mohalla Sabha* | | |
| No | 56.0 | 32.0 |
| Yes | 20.0 | 7.0 |
| Not aware | 24.0 | 61.0 |
| Total | 100.0 | 100.0 |
| *Attended last meeting of ward committee/Mohalla Sabha* | | |
| No | 68.0 | 96.0 |
| Yes | 32.0 | 4.0 |
| Total | 100.0 | 100.0 |

*Source* Primary Survey, 2017

read newspapers was highest among Muslims and OBCs. This is a reflection of a low level of literacy among these communities. Youth were more tech savvy using internet and mobile phones regularly.

In comparison to Delhi, the functioning of Mohalla Sabhas/Ward Committee was better in Bengaluru, as a significant percentage of respondents reported that committee meetings took place every month regularly (Table 4). It could be because of high level of literary and political awareness among the slum dwellers of Bengaluru as compared to those of Delhi which ensured regular meetings of Mohalla Sabhas/Ward Committees. In comparison to households from resettlement colonies, community participation and active citizen engagement were more dominant among slum dwellers. This is a positive indicator for local democracy. The survey also showed that the households that were in poor condition had more confidence in local institutions and realized the importance of getting their voices heard through such avenues/platforms. Women and youth respondents stood out as more dynamic communities in slums. In Delhi, engagement with poor through Mohalla Sabhas was minimal.

It was observed during the field survey that in slums, the participation of women in public matters was more as compared to resettlement colonies (Table 5). Women from slum areas were proactive in putting their problems in public fora. Self-help groups were active in both cities. As captured in the primary survey, a high percentage of Muslims and women were part of self-help groups. NGOs were also operating in selected slums/resettlement colonies of both cities. Youth and women benefitted from NGO interventions. Slum dwellers deeply value Indian democracy and its ideals.

**Table 5** Community participation of slum households (in %)

| | Bengaluru | Delhi |
|---|---|---|
| *Final word in community issues* | | |
| Local leadership | 42.0 | 36.0 |
| Representative of leadership | 0.0 | 9.0 |
| Community members | 47.0 | 41.0 |
| Influential locals who are not in leadership | 0.0 | 5.0 |
| Others | 0.0 | 8.0 |
| No comments | 11.0 | 1.0 |
| Total | 100.0 | 100.0 |
| *Gender equality in community issues* | | |
| Equal participation | 78.0 | 41.0 |
| Men play more active roles | 15.0 | 42.0 |
| Women play more active roles | 7.0 | 17.0 |
| Total | 100.0 | 100.0 |
| *Minorities participation in community affairs* | | |
| No | 7.0 | 35.0 |
| Yes | 93.0 | 65.0 |
| Total | 100.0 | 100.0 |

*Source* Primary Survey, 2017

**Table 6** Level of satisfaction among respondents about the performance of government/NGOs/Mohalla Sabha/ward committee (in %)

| Organization | Levels of satisfaction (in %) | | | | | |
|---|---|---|---|---|---|---|
| | Not satisfied at all | Somewhat satisfied | Neither Satisfied nor dissatisfied | Satisfied | Highly satisfied | Total |
| *Bengaluru* | | | | | | |
| Central government | 30 | 40 | 30 | 0 | 0 | 100 |
| State government | 33 | 28 | 38 | 1 | 0 | 100 |
| Municipal government | 0 | 29 | 38 | 33 | 0 | 100 |
| NGOs | 4 | 15 | 30 | 19 | 32 | 100 |
| Mohalla Sabha/ward committee | 0 | 27 | 33 | 2 | 38 | 100 |
| *Delhi* | | | | | | |
| Central government | 24 | 13 | 31 | 26 | 6 | 100 |
| State government | 40 | 28 | 21 | 9 | 2 | 100 |
| Municipal government | 50 | 30 | 13 | 5 | 2 | 100 |
| NGOs | 15 | 13 | 19 | 41 | 12 | 100 |
| Mohalla Sabha/ward committee | 18 | 3 | 14 | 4 | 61 | 100 |

*Source* Primary Survey, 2017

Political participation at all levels, viz, local, state and central government was very high in both the cities. The gender equality in participation in in community affairs were high among the slum dwellers of Bengaluru as compared to Delhi. This may be attributed to higher literary rate in the former leading to able to voice their concerns with assertion.

The respondents from Delhi reported fire and health epidemic (Dengue and Chikungunya) as major natural hazard during the last five years. However, respondents from Bengaluru reported flooding and overflow of drains during intense rainfall and health epidemic as major natural hazard. Slum dwellers from both cities reported that they did not receive any help from government (central/state/local) during these calamities. A small share of respondents reported that they got help from local NGOs. A very high percentage of households across sub-groups believe that the government is not prepared/poorly prepared to deliver basic services.

Slum dwellers from both cities perceived that providing support to elderly and disabled population, women and child safety, improving connectivity to slums and developing infrastructure and community based preparedness for disaster should be on high priority for the government (Table 6). Several respondents had very low to medium level of trust for measures taken by government related to women and child safety. The level of confidence in local government was very low among slum dwellers. In contrast, the level of confidence in NGOs was medium to very high. Except youth, the respondents from other sub-groups were not satisfied with the performance of the government. Respondents from Bengaluru were highly satisfied with Mohalla Sabhas as compared to Delhi. A high percentage of respondents in Delhi were not aware of the functioning of Mohalla Sabha in their areas. Further, as per their opinion, the engagement of service providers was better in Delhi as compared to Bengaluru.

The assessment level of slum residents of Bengaluru was low on various parameters of local governance like corruption, transparency in functioning; level of trust by community members and gender sensitivity mentioned in Table 7. Interestingly,

**Table 7** Assessment of the institutional feature of local government (in percentage)

| Institutional features | Low | | Medium | | High | |
|---|---|---|---|---|---|---|
| | Bengaluru | Delhi | Bengaluru | Delhi | Bengaluru | Delhi |
| Corruption | 58 | 22 | 20 | 18 | 22 | 60 |
| Quality of service delivery | 35 | 72 | 45 | 26 | 20 | 2 |
| Transparency of activities | 56 | 83 | 33 | 16 | 11 | 1 |
| Level of trust by community members | 51 | 87 | 30 | 7 | 19 | 6 |
| Gender sensitivity | 59 | 71 | 28 | 24 | 13 | 5 |
| Response to marginalised section | 27 | 69 | 46 | 25 | 27 | 6 |

the assessment level of slum residents of Delhi was also low in most of the institutional features except corruption. However, the share of respondents who indicated their perception as "low" was very high in Delhi. The share of respondents who had medium and high level of assessment was higher in Bengaluru. However, the share of respondents with high level of assessment was least in Delhi, indicating poor level of local governance in the national capital in low income areas as compared to the same in Bengaluru.

## 5 Factors Influencing Access to Urban Services

The above analysis corroborates the fact that urbanisation, which has been exclusionary in nature has not been able to adequately distribute the benefits and opportunities arising from urbanization to all sections of the society in an egalitarian manner. Rather, it has led to an "urban divide" within cities and within different types of settlements such as slums and resettlement colonies. The following section summarizes the factors responsible for inadequate and inequitable access to basic services.

### *5.1 Democratic Local Governance and NGO Participation*

India's democratic political system mandates elections at three levels, viz, central, state and local. At all three levels, the representatives are elected for five years. At the central and state levels, there is a cabinet form of governance. To elucidate, at the centre, there is the Prime Minister, who is the leader of the ruling party and has cabinet ministers. A similar pattern is observed at the state level, with the Chief Minister as the leader of the ruling party. At the city level, however, the Mayor, who is the administrative head, may or may not be directly elected. His tenure varies from one year in some states to five years in others.

The level of empowerment is the least in case of Mayors. The executive powers at the local level are with the Commissioner, who is a civil servant appointed for a fixed tenure. In states like West Bengal, Kerala and Madhya Pradesh, there exists a Mayor-in-Council system of governance, which is akin to the cabinet system. In most of the other states, the local governments function as extended arm of the state governments with very little local autonomy. The 74th Constitutional Amendment Act (1992) sought to decentralize governance at the local level based on the principle of subsidiarity through devolution of funds, functions and functionaries. However, this Act is yet to be implemented in totality. Since urban development is a state subject, there are variations in the level of engagement of the local governments in urban governance. Under the 74th CAA, participation of people in the local democratic process was sought through ward committees (WCs). This was strengthened under JNNURM through mandatory formation of mohalla/area sabhas (neighbourhood committees). However, well-structured ward committees ensuring greater accountability of local governments and service providers did not materialize in most states. In the absence of WCs, non-governmental organisations (NGOs) have filled in the vacuum. In the surveyed cities and slums, several NGOs are working to voice the collective concerns of the urban poor to relevant agencies through appropriate channels and platforms. In the surveyed slums, a strong presence of the NGO, World Vision was found. Other than imparting education, the NGO was very active in providing skill training, employment generation activities, youth engagement in community mobilization and credit and thrift society formation through self-help groups etc.

It may be noted that the minimum age for voting in India is 18 years. The recent elections in India have shown a high turnout of voters. However, this voting behavior is not uniform across all types of settlements. A study conducted by Harris (2005) on the voting behavior of citizens of Delhi indicated that those living in unplanned and unregulated jhuggi-jhopris are much more likely to be politically active than those living in upper income planned colonies. This trend is common in most of the cities where the participation of poor in the election process is more as compared to higher income groups. Likewise, the participation of the elites is highly restricted in routine local level meetings of ward committees/mohalla sabhas.

## 5.2 *Lack of Coordinated Governance*

India is a federal system, with powers constitutionally divided between the central and state governments. India also has seven union territories among which five are under the direct control and administration of central government and thus prominently display the unitary features. However, Delhi and Puducherry have elected representatives and administrative work is divided between central and state government. Under this structure, housing and urban development is a state subject.[8] State governments define state-specific housing and urban development policies, establish institutions including urban local governments for advancing the policy agenda, design and implement housing and urban development programs and projects. The role of the central government is to define an overall approach to urban development, compatible with macroeconomic policy. It also provides funds for central schemes and centrally sponsored schemes. In 1992, the Constitution was amended for a major local government reform. This 74th Constitutional Amendment gave constitutional recognition to urban local bodies. A new Twelfth Schedule to the Constitution provided recommended a list of functions to be performed by Urban Local Bodies (ULBs). States were mandated to transfer various responsibilities to municipalities and to strengthen urban local governance. However, even till date the devolution of administrative and financial resources has not managed to keep pace with the devolution of responsibilities. Nevertheless, the 74th Constitutional Amendment Act reflects the intent of the government to recognize the importance of local governments in the national economy. Consequently, the central government has, in partnership with the state and local governments introduced a number of initiatives to enable the participation of local governments in the urban agenda, making urban development a shared responsibility.

A typical slum household suffers from several deprivations including lack of access to improved water and sanitation, insecure land tenure, unreliable power supply and intermittent water availability, insufficient treatment of wastewater, poor drainage and flooding, and uncollected garbage. These deficiencies in distribution are the manifestation of lack of coordination between parastatal and local agencies coupled by inadequate investments in infrastructure. For example, roads within slums are at times under Delhi Urban Shelter Improvement Board as well as the public works department (state level agency) and Municipal Corporation of Delhi. Due to the common responsibility of various stakeholders, most of the times it takes months to repair the potholes in slum roads. In addition, roads and pavements are often dug up multiple times to lay telephone or metro cables. Lack of coordination leads to

[8] Seventh Schedule of the Constitution.

the digging and remaking of roads several times which results in wastage of public resources.

## 5.3 Low Capacity and Resources of Local Governments

The Eleventh Plan launched an inclusive agenda and emphasized the need to bring about major changes in urban governance in order to boost investment in infrastructure development in urban areas (Kundu and Samanta 2011). Under Jawaharlal Nehru National Urban Renewable Mission (JNNURM) attempts were made to allocate substantial additional central assistance (ACA) to cities for infrastructure, housing and capacity building.

The mission succeeded in getting the state and the city governments to commit themselves to structural reforms which the central government failed to achieve despite adopting several measures and incentive schemes through other programs and legislations (Kundu 2007). It was also effective in renewing focus on the urban sector across the country. Yet, many states and cities lagged behind in program utilization due to lack of enabling capacity and capacity to generate matching funds (Planning Commission 2011) which explains the deficits.

Most of the ULBs in India are weak both in terms of capacity to function as independent entities, raise resources and in financial autonomy. Their precarious state of finances as well as their complex institutional and fiscal framework poses a serious challenge. An important step was taken to empower them to function as the third tier of government by providing them democratic status through the 74th Constitutional Amendment Act (CAA) in 1992. Despite the empowerment and delegation of powers envisioned in the 74th CAA more than two decades ago, most ULBs in India are still not in a position to carry out their routine functions efficiently. Several other initiatives such as the Financial Reforms Expansion (Debt) component of the USAID, Mega Cities Scheme and Urban Reforms Incentive Fund, launched in the nineties also tried to improve local governance in the country by bringing about financial and administrative discipline among the ULBs. However, these piecemeal efforts failed to bring about the desired change in the urban governance structure.

## 5.4 Elite Capture

There has been a sea change in urban governance in the country during the past few decades. The economic liberalisation initiated in the country followed by decentralisation measures adopted by all tiers of the government as an aftermath of the 74th Constitutional Amendment Act (CAA) has resulted in gradual withdrawal of the state and increasing private sector participation in capital investment and operation and maintenance of urban services. Further, the inability of the wards committees, institutionalised through the 74th CAA to usher in decentralised governance has led to the growth of middle class activism through the resident welfare associations (RWAs). The municipal responsibility of provision of services is being increasingly passed on to the RWAs. Their involvement has been broadly in areas of operation and management of civic services, capital investment in infrastructural projects, planning and participatory budgeting, and maintenance of neighborhood security. Importantly, their functioning has been restricted largely in better-off colonies. Correspondingly, the informal settlements, which house the urban poor, are unable to exercise their voice through the same, which has resulted in an accentuation of disparities. The responsibility of municipalities to provide crucial services is being increasingly passed on to the resident welfare associations. Similar tools of intervention are absent in the slums and low-income neighbourhoods and even the local ward committees fail to represent their needs and aspirations. The very mechanism of the functioning of RWAs is likely to accentuate and institutionalise disparity in urban areas.

## 5.5 Poor Implementation and Targeting of Schemes

The Eleventh and Twelfth Plan documents envisaged the government strategy to "establish the macroeconomic preconditions for rapid growth and support key drivers of this growth". The Eleventh and Twelfth Plan document further added that the strategy must also include sector-specific policies to ensure that the structure of growth and the institutional environment in which it occurs, achieves "the objective of inclusiveness in all its many dimensions".

Unfortunately, most of the ULBs do not have the mechanisms and the requisite skills to achieve the inclusive mandate mentioned in Plan documents. They lack the capacity and human resources to carry out project preparation. Capacity building of ULB officials is, perhaps the single most important activity required in today's urban sector. Thus, most of the smaller ULBs could not avail the JNNURM grants as they were unable to prepare detailed project reports (DPRs) and generate matching resources. It is a fact that JNNURM has provided substantial central assistance to cities for infrastructure development, and has indeed been effective in renewing the country's focus on the urban sector. However, the Mission brought about a move towards polarised development with an inbuilt big-city bias. The small towns could not benefit from this mission because of their lack of technical capacity, financial resources and adequate skilled manpower. This resulted in poor targeting of the highly acclaimed mission.

The present government replaced the former Mission with another mission, Atal Mission for Rejuvenation and Urban Transformation (AMRUT), increasing its coverage to all the 500 cities above 100,000 population. However, disbursements under this program are also linked to reform measures undertaken by the city and much is left to be desired from the small cities which do not have the wherewithal to adhere to reforms and prepare requisite DPRs for their projects. These cities also suffer from weak municipal finances as the coverage of property tax, which is the most important source of revenue, is very low. The Fourteenth Central Finance Commission (CFC) has also highlighted the decline in the share of own revenues of the local governments. Enforcement of local taxation is weak along with an absence of database on inventories in almost all local bodies. In addition to this, cities covered under the Smart Cities Mission are all large cities. These are likely to become more vulnerable as their most important and lucrative revenue streams are to be tied to financing the projects under the Mission.[9] In addition, given the recent inclusion of the non-poor in the PMAY or Housing for All Mission, the focus of this program is also likely to be diluted.

Several reforms have been introduced in recent years and many states are demonstrating innovations in moving towards effective delivery of programs. However, the overall returns of spending in terms of poverty reduction have not reached its potential. Poor administrative structures for delivery of services and lack of capacities of institution have hindered the effective implementation of programs. Further, there is multiplicity of programs and policies since they are administered by various ministries and departments all working in silos. The current Missions are all based on convergence of schemes. This notion should be taken to the grassroot level where the Missions are being implemented. This highlights the need for integrated urban and social policy at the national level. The target based social security programs have limited outreach as implementing agencies faces the problem in identifying the right beneficiary. Most of these schemes have not worked efficiently due to inclusion of non-beneficiaries. People working in the unorganised and informal sector are often excluded from such schemes due to income criteria. These people are very vulnerable as they face the risks associated with sickness, accident, unemployment, disability, maternity and old age but are not covered under any of the social security schemes. Many of the evaluation studies of the programs have reported high leakages of resources especially in public distribution scheme (PDS) and mid-day meal programs.

## 6 Conclusion

Indian urbanization has been exclusionary in nature. The country, after urbanizing very rapidly in the post-independence period in the decade of the fifties, registered a decline in urban growth rates since the 80s. Urbanization in India has been 'messy and hidden', with huge sprawls coming up beyond city boundaries.[10] It has been accompanied by increase in poverty and inequity, deteriorating quality of the urban environment, and deficiencies in the coverage of basic urban services. Urban India currently faces the challenges of congestion, pollution and disparity in the access to adequate urban infrastructure, including education and health facilities.

[9]New Urban Agenda in Asia-Pacific: Governance for Sustainable and Inclusive Cities, 2018.

[10]https://www.thehindubusinessline.com/economy/indian-urbanisation-messy-reforms-needed-world-bank/article7685068.ece.

Till recently, India did not have a national urban policy to guide the process of urbanisation in a planned manner. The existing policies of the government are in silos and integration of the same is essential for a balanced development. The National Urban Policy Framework (NUDF), 2018 aims to have a fluid plan that can be tweaked at regular intervals instead of one rigid master plan and which is at the root of the urban mess.

Exclusionary forces have affected adversely the marginalized communities in various forms. Two prominent categories of exclusions are discernible: institutional inequality including housing, land and basic services and structural inequality including barriers to participation of women in economic, social and political activities. These manifest in various forms of inadequacies for different vulnerable groups. At the macro level, in comparison to male headed households, the conditions of housing and basic amenities were slightly poor in female headed households. The migrants, living mostly in rented accommodation, have lower coverage to basic amenities as compared to the non-migrants. The condition of housing and other basic amenities among SC and ST households in urban India was poor as compared to OBCs and other categories. The condition of basic services was much better in the households from higher income groups.

At the micro level, the results show a better housing condition among slum dwellers in Delhi as compared to Bengaluru. However, the coverage of basic services was better in Bengaluru (except piped water) which could be attributed to the better functioning of Mohalla Sabhas/Ward Committees, and high participation level of slum dwellers in meetings of Mohalla Sabha/Ward Committees. The gender equality and participation of minorities in Mohalla Sabha/Ward Committees was also high in Bengaluru, mainly because of better literacy levels, and high level of political awareness among slum dwellers as compared to Delhi. The assessment of slum dwellers of the institutional features related to local government shows that as compared Bengaluru, the slum households of Delhi were not satisfied with the quality of service delivery, transparency, and sensitivity of official towards gender and minorities.

The slum dwellers from both cities perceived that providing support to elderly and disabled population, women and child safety, improving connectivity to slums and developing infrastructure and community based preparedness for disaster should be on high priority for the government (Table 6). Several respondents had very low to medium level of trust for the measures taken by government related to women and child safety. The level of confidence in local government was very low among slum dwellers. In contrast, the level of confidence in NGOs was medium to very high.

The 74th Constitutional Amendment Act (1992) sought to decentralize governance at the local level based on the principle of subsidiarity through devolution of funds, functions and functionaries. However, this Act is yet to be implemented in totality. Since urban development is a state subject, there are variations in the level of engagement of the local governments in urban governance. Under the 74th CAA, participation of people in the local democratic process was sought through ward committees (WCs). This was strengthened under JNNURM through mandatory formation of mohalla/area sabhas. However, well-structured ward committees ensuring greater accountability of local governments and service providers did not materialize in most states. In the absence of WCs, non-governmental organisations (NGOs) have filled in the vacuum.

Lack of coordinated governance at various levels, low capacity and resources of local governments, elite capture of urban governance and poor implementation and targeting of schemes are some of the challenges in this sector. Also, big city bias of the urban missions has accentuated the disparities in the coverage of basic services. An integrated approach to urban planning and management with proper targeting of beneficiaries may ensure inclusion of the disadvantaged sections of the society. In addition, an honest attempt should be made to institutionalize inclusive urban policies and programs that fill gaps between urban planning and realities. These policies need to accommodate the concerns of the marginalized groups by promoting greater participation in urban decision-making and holistic management of city regions with integration across jurisdictions and sectors.

## References

Bhagat RB (2012) Migrants' (denied) right to the city. In: Workshop compendium (Vol. II) workshop papers, national workshop on internal migration and human development in India. UNESCO/UNICEF, New Delhi, pp 86–99. http://www.unesco.org/new/fileadmin/MULTIMEDIA/FIELD/New_Delhi/pdf/Internal_Migration_Workshop_-_Vol_2_07.pdf

Harriss J (2005) Political participation, representation and the urban poor: findings from research in Delhi. Eco Pol Weekly 40(11):1041–1054

Kundu N (2003) The case of Kolkata, India. In: UN-habitat, global report on human settlement 2003, The challenge of slums. Earthscan, London. https://www.ucl.ac.uk/dpu-projects/Global_Report/pdfs/Kolkata.pdf

Kundu A (2007) A Strategy paper on migration and urbanisation in the context of development dynamics, governmental programmes and evolving institutional structure in India. United Nations Population Fund (UNFPA), India

Kundu D (2009) Elite capture and marginalisation of the poor in participatory urban governance-a case of resident welfare associations in Metro Cities. In: India urban poverty report, MoHUPA and UNDP. Oxford University Press, New Delhi

Kundu D (2011) Elite capture in participatory urban governance. Eco Pol Weekly 46(10):23–25

Kundu D (2016) Emerging perspectives on urban-rural linkages in the context of Asian urbanisation. Reg Dev Dialogue 35:180–195

Kundu D, Samanta D (2011) Redefining the inclusive urban agenda in India. Econ Political Weekly 46(5):55–63

Kundu A, Saraswati LR (2012) Migration and exclusionary urbanisation in India. Eco Pol Weekly 47(26–27):219–227

Planning Commission of India (2011) Urban development. In: Mid-term appraisal eleventh five year plan 2007–2012. Oxford University Press, New Delhi. http://planningcommission.gov.in/plans/mta/11th_mta/chapterwise/Comp_mta11th.pdf

United Nations Department of Economic and Social Affairs Population Division (2018) World urbanization prospects: the 2018 revision, online edition. https://www.un.org/development/desa/publications/2018-revision-of-world-urbanization-prospects.html

**Debolina Kundu** is a Professor at the National Institute of Urban Affairs, India and has over 20 years of professional experience in the field of development studies. She was a doctoral fellow with the ICSSR and has a Ph.D. from Jawaharlal Nehru University on urban governance. She has worked as consultant with, ADB, LSE, IIDS, UNDP, UNFPA, UNICEF, UNESCAP, KfW, GIZ, Urban Institute, Washington and East-West Centre Honolulu on issues of urbanization, migration, urban development policies, municipal finance, governance and exclusion. She is currently the HUDCO Chair and editor of journals Environment and Urbanisation, Asia (SAGE) and Urban India (NIUA). She is the Country Investigator—India for the Global Challenges Research Fund (GCRF) Centre for Sustainable Healthy and Learning Cities and Neighbourhoods (SHLC), which is supported by UK Research and Innovation. SHLC is an international, interdisciplinary and collaborative project, which aims to increase understanding of sustainable neighbourhoods and address urban, health and education challenges in fast growing cities across Africa and Asia. A member of the Fifth Delhi Finance Commission and various other committees of the Government of India, she has several articles published in books and journals. Currently, she is editing books on internal migration and national urban policies.

# Access of Low-Income Residents to Urban Services for Inclusive Development: The Case of Chengdu, China

**Bo Qin and Jian Yang**

**Abstract** This chapter focuses on the access to public services for low-income residents in urban China, taking Chengdu as the study area. After providing a brief introduction to the overall urbanization process in China, we choose to focus on the low-income residents in Chengdu, a major city of western China, to understand the extent to which public services are accessible to the low-income residents, what are the barriers to political and social inclusion that the low-income residents experience, and what are the good practices and innovations that might be relevant to the cities in other developing countries. The study was conducted by both using first-hand data (e.g., interviews with residents, officials at different levels, and questionnaire surveys), and second-hand data (e.g., archival documents, government policy reports, statistical yearbooks, etc.). Two neighborhoods, located in the central and suburban areas of Chengdu city respectively, were selected for questionnaire survey. Follow-up in-depth interviews were conducted for understanding the institutional and structural barriers that may hinder access to equalized public services. It is found that the governments in Chengdu put much effort to improve the quality and access to the basic public services for the low-income residents. However institutional barriers such as hukou system are still there, which impede better public participation of the low-income residents and thus lead to unequal access to some public services such as education and healthcare. Good practices, innovations, and take-away messages are also discussed.

**Keywords** Public services · Low-income residents · Inclusive development · Urban governance · Chengdu · China

B. Qin (✉)
School of Public Administration and Policy, Renmin University of China, Qiushi Building 211, Beijing 100872, P.R. China
e-mail: qinbo@ruc.edu.cn

J. Yang
School of Public Affairs and Law, Southwest Jiaotong University, Chengdu, Sichuan 610031, P.R. China



S. Cheema (ed.), *Governance for Urban Services*, Advances in 21st Century Human Settlements,
https://doi.org/10.1007/978-981-15-2973-3_9

This chapter focuses on low-income residents' access to public services in Chengdu, China. It is argued that promoting the equalized access to public services is the starting point to ensure inclusive development in Chinese cities, while institutional reforms are in desperate need. This study attempts to understand the extent to which public services are accessible to low-income residents; barriers to political and social inclusion that the marginalized groups experience; and policy options, best practices, and innovations that can meet the needs of low income residents, including women, minorities, migrants, the elderly, and youth.

This investigation is conducted using first-hand data (e.g., interviews with residents, officials at different levels, and two questionnaire surveys at different neighborhoods), as well as second-hand data (e.g., archival documents, internal government reports, statistical yearbooks, etc.). Two neighborhoods, located in the central and suburban areas of Chengdu city respectively, were selected for questionnaire survey. To align with the local context of the study, the standard questionnaire recommended by the Research Taskforce was slightly modified. A total of 30 questionnaires from randomly selected households in the two neighborhoods were collected and analyzed. Furthermore, follow-up in-depth interviews were conducted for understanding the institutional and structural barriers that may hinder access to quality public services.

# 1 National Context

## *1.1 Urbanization in China*

China is in the middle of dramatic processes of transition. It is a transition from a traditional to a modern social order, from an agricultural to an industrialized and digitalized society, from a planned to market-oriented economy, and from a rural to an urbanized country.

As a result of the transition in the past decades since the initiation of economic reforms in 1978, China has undergone rapid urbanization process, which may be considered one of the fastest and largest in the world history (World Bank 2002). The latest population census taken in 2010 reported that, among the total population of 1339.72 million in China, 665.58 million were urban residents (National Bureau of Statistics in China 2011a, b). It suggests that the urban population share in China rose significantly from 17.92% in 1978 to 49.68% in 2010, rising by 493.13 million. In other word, in the 32 years nearly 500 million people had moved from rural area to urban area in China. As a point of comparison, as of December 2010 the entire US population numbered 310.86 million.

The process of transition invariably brought new problems and challenges for cities. Notably amongst these is the increasing number of poor and marginalized urban residents crowding the expanding cities of China. These urban poor are at the bottom of the society, left behind in the accelerating achievements since the late 1970s. They are to this day facing the risk of being further marginalization due to the institutional and structural mechanisms that leave them at disadvantage.

Clearly, understanding the problem of urban low-income groups is of great importance in maintaining social equality while ensuring the rapid, healthy, and sustainable development of China's economy in the years to come (Xue and Weng 2017).

Rapid pace of urbanization in China has brought about a flood of rural migration into urban areas. Rural migrant workers, the main body of the floating urban population, often choose to live in areas with relatively abundant job opportunities as well as low housing and living costs. They form urban villages in city centers or heterogeneously mixed communities in peri-urban areas, where the economic sectors are largely informal, the public facilities are in short supply, the urban planning is disordered, and the level of education for the children of migrant workers is low (Yang 2011).

### *1.2 Definition of Low-Income Residents*

There is no standardized definition for low-income residents. The main features of low-income residents are that: (1) their per capita disposable income is limited, and (2) the Engel coefficient for living consumption is relatively high. Most existing literature on the subject adopts the definitions used by the United Nations and World Bank, which define low-income residents as those whose annual incomes are less than half of the per capita annual income of a given country.

Chinese scholars in the field generally conclude that, according to local living standards which vary across different cities and regions, low-income residents must meet the following conditions to be classified as such: (1) The per capita income of the low-income household is lower, or only slightly higher, than local living costs; (2) The per capita expenditure of the low-income household is higher, or only slightly less, than their household income; (3) The per capita housing area of the low-income resident is lower than for the local average housing area; (4) The per capita asset value of the low-income household is lower than that of the local average per capita asset value (Chen 2011). Taking Chengdu as an example, household is considered low-income if its monthly per capita income is below 550 Chinese Yuan or Renminbi (RMB) for a household in the central urban area, is below 500 RMB in the suburban area, or is below 450 RMB in the outer suburban area of Chengdu.

### *1.3 Scale and Composition of the Low-Income Residents*

In urban China, low-income residents and their families are mainly composed of two groups: low-income floating migrants lacking local urban household registration (commonly referred to as urban hukou) and local low-income families. The former makes up the majority, accounting for more than 80% of all low-income residents in urban China, according to the national statistical report in 2016. The latter group is more specifically made up of the local families with a proper urban household registration but is low-income and lives in poverty. Most of this latter group is made up of unemployed workers and the so-called "three no's people" (those with no source of income, no ability to work, and no legal dependents, such as the disabled as well as elderly single residents), and they account for about 20% of all low-income residents.

According to the most recent national report, there were about 13.71 million low-income residents in China that applied for urban subsistence and social security assistance in 2016.[1] According to the latest report for Chengdu, approximately 34,000 urban residents in the city received assistance in the form of minimum living allowances in 2016.

China's floating migrant populations do not enjoy access to urban minimum living allowances as a result of their household registration (*hukou*) status. Thus, their situation is direr than is the case for a city's local low-income residents. According to the data from the National Bureau of Statistics of China, migrant workers in China totaled 280 million in 2016, of which 110 million were migrant workers moving within the province where they were staying, while 170 million of them were migrant workers moving from other provinces. The average monthly income of migrant workers for 2016 was 3275 RMB, which is equivalent to about $5900 a year. It was reported that Chengdu's floating migrant population was 6.31 million.[2]

### *1.4 Institutional System of Public Service Delivery*

The institutional system of public service delivery in a country is no doubt closely related to its political and administrative system. China's administrative levels are fivefold. The five divisions, and their administrative mechanisms, are organized along the lines of the central, provincial, prefectural, county, and township levels. Among the range of public services offered amongst them, national defense and diplomacy are the exclusive purview of the central government of China. Other public services (e.g., education, health, culture, sports, and safety) are considered shared responsibilities between the nation's central government and local governments at or above the county level. Figure 1 shows the administrative hierarchy in China.

According to the existing literature and reports on the topic, there are three notable problems for the institutional system in providing public services in China (He 2005; Ye and Jiao 2013; Ye and LeGates 2013; Li 2017).

*Economic-growth driven* versus *public service-oriented government*—A notable problem in China is that the local governments put too much emphasis on economic growth, rather than public service delivery. One possible reason is that the central government of China has economy-oriented criteria for performance evaluation, which weakens the positive impact of national and local polices in providing public services. Therefore, in the organizational systems of county-level governments in

[1]From the website of Ministry of Civil Affairs of the People's Republic of China. http://www.mca.gov.cn/article/sj/tjyb/qgsj/201707/201708241429.html. Accessed 24 Aug 2017.

[2]From the website of Chengdu Bureau of Statistics. http://www.cdstats.chengdu.gov.cn/htm/detail_51267.html. Accessed 10 Apr 2017.

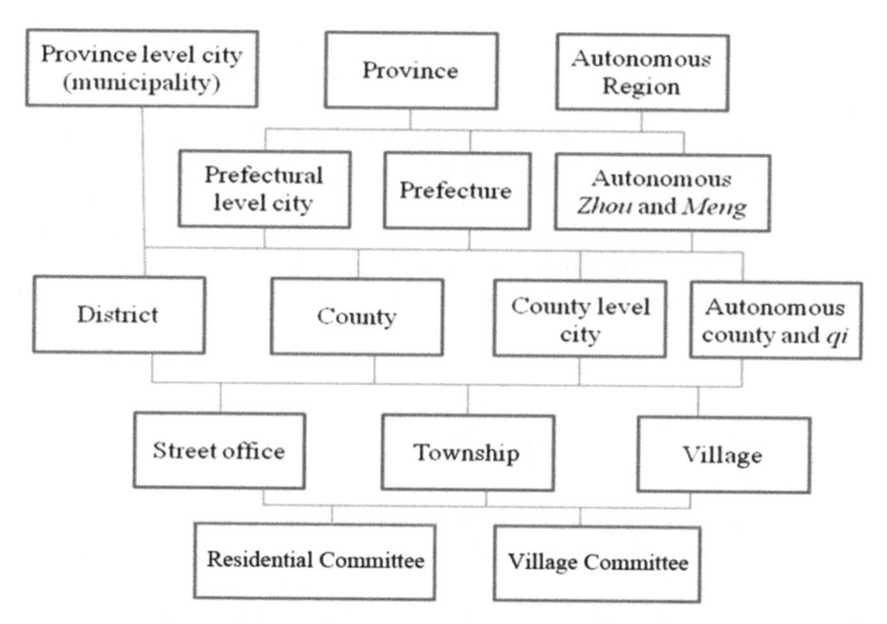


**Fig. 1** The administrative divisions of China

China, especially in western China, different departments and bureaus that are tasked with providing public services are generally not very well established. For example, when visiting the government website of Pidu District, one will find that only the online service platform for labor security (e.g., employment policy, security policy, etc.) and education services (e.g., pre-school education, compulsory education, high school education, vocational education, etc.) have been set up. The online services for seemingly lower priority public services, such as healthcare and social welfare, are neglected. The Pidu case is not rare in China.

*Horizontal* versus *vertical coordination problem*—The responsibilities for public service delivery among governments at different levels and the bureaus within the governments are unclear. This is caused by the fact that Chinese laws and regulations on the topic typically do not clearly define the responsibilities between county-level (or district) governments and provincial or municipal governments in providing public services. For instance, both the local government of Pidu District as well as the municipal government of Chengdu do not vary markedly in their public service provisions. This often leads to situations where the left hand does not know what the right hand is doing, and governments will turn a blind eye to focus on simpler, easier, and more affordable tasks in providing public services. Thus, the organizational structures of public service institutions need to be improved.

*Market oriented* versus *state dominated approach*—The provision of public services in China lacks diverse forces. The governments in China dominate the supply side of public service market, which foments a wealth of inherent problems (e.g., cases of moral hazard, soft budget constraints, performance assessment difficulties, etc.). Other types of enterprises that are capable of providing public services, such as non-governmental organizations, are to some extent unnecessarily excluded from participation due to government policies.

Keeping the problems in mind, the Central Government of China proposes further institutional reforms for providing better public services. The approaches include reducing red tape and bringing market role into play, increasing government spending on public service delivery, consolidating grass-roots democracy to engage the public. Above all, focusing more on public service delivery rather than economic growth in evaluating government performance is the most significant change in the institutional reforms.

## 1.5 Fiscal Decentralization in China

One challenge for local governments in China and other developing countries is the mismatch between administrative duties in providing quality public services and fiscal resources available. Fiscal decentralization is critical to understand Chinese governments' behavior and policy instruments, including in public service delivery.

The fiscal decentralization with centralized political power is identified as a unique and successful approach to promote China's economic growth by motivating local governments to pursue economic prosperity. As a unitary state, China is no doubt highly centralized politically. However, the intergovernmental fiscal system has become increasingly decentralized since the 1980s. Abundant empirical studies intend to test the association between fiscal decentralization and economic growth, service delivery, anti-corruption, including Zhang and Zou (1998), Lin and Liu (2000), Blanchard et al. (2001), and Niu (2013).

Measuring fiscal decentralization is a challenging task due to the complexities of fiscal instruments. In China, the sub-national government acts as the agent of the central government and is accountable to the latter. The central government decides the tax base and tax rates and determines policy overall for all levels of government. The sub-national government does not have any discretion. That is to say, revenue sharing and expenditure responsibilities are assigned from the top down to all five tiers of government.

It is generally regarded that the 1994 taxation system reform was the starting point of the fiscal decentralization process in China. On the one hand, the 1994 reform was intended to increase two ratios: the ratio of central government revenue to the total government revenue and the ratio of the government revenue to GDP. Thus, the central government assigned the major tax sources to itself. For example, 75% of the value-added tax (VAT), the largest source in China's taxation, was collected by the central government. Meanwhile, the sub-national governments were assigned substantive responsibilities for public services. On the other hand, in order to deal with the increased pressure on local government to deliver public services, the central government of China created a formal intergovernmental transfer system, allowing the local governments to receive fairly large transfers from the central government.

## 1.6 The Hukou System and Residence Permit Reform

A unique institutional barrier for the low-income residents to enjoy equalized public services in the cities in China is the household registration system (Hukou). On January 9, 1958, *The Regulations on Hukou Registration of the People's Republic of China* was promulgated. It has constituted a public service delivery system in favor of the urban population. The rural migrants have made significant contribution to economic development. But they still have difficulty in enjoying the public service as well as the local urban residents, such as education, healthcare, pension and housing security.

In recent years, the central government of China has made a series of major policy decisions to reform the Hukou system. For example, *The National New Urbanization Planning* (2014–2020) issued in March 2014, *The State Council's Opinion on Further Promoting the Reform of Hukou System* issued in July 2014, and *Provisional Regulations of Residence Permit (Order No. 663)* issued in January 1, 2016. These policies are employed to promote quality public services for the rural migrants. The key measures include:

(1) Citizens who leave their previous residence register in *Hukou* and live in other places for more than six months shall be eligible to apply for residence permits, if possessing legal and stable job, legal and stable residence, or continuous study status.
(2) Residence permit holders enjoy the same accessibility to public services as the local residents. The public services and facilities include social insurance, public housing fund, education, employment services, public health services and family planning services, public cultural and sports facilities, and other legal services, such as applying for passport, vehicle registration, driver license, signing up for the professional qualification examination etc.
(3) A gradual implementation mechanism has been established. *Provisional Regulations of Residence Permit* requires the relevant departments of the state council and local governments at various levels reform the Hukou System and implement the residence permit system in a gradual way. Local governments have the discretion to expand public service delivery for the residence permit holders and to improve service quality according to their development stage and fiscal power.
(4) Household Registration Application System is reformed as well. Different groups of the cities have different household registration requirements. Among them, towns and small cities tend to fully relax the requirements of applying for household registration, as long as the applicants have legal stable residence, whether leasing or purchased. The medium-size city and large cities in western China (e.g., Chengdu) tend to relax the requirements as well. For instance, any resident in Chengdu is eligible to apply for urban Hukou status after paying social security pension for five consecutive years.

The more stringent group is the cities with population more than 5 million (e.g., Shenzhen, Shanghai). A point-collecting approach is proposed for the applicants. A point index is created, which suggests that, for instance, having stable jobs and fixed residence equals to 2 points, obtaining Ph.D. degree equals to 3 points, paying for social security pension fund equals to 1 point. An applicant can apply for the Hukou status of the cities once enough points have been collected. The most strictly controlled city is Beijing, because the central government wants to control its population size.

The implementation of the residence permit system aims to replace the Hukou system and thus to abolish the identity difference between urban and rural residents. It is of great importance for providing equalized public services for both urban and rural citizens, and for both local and migrant population.

## 2 Urban Context Analysis of Chengdu

Chengdu is the capital city of Sichuan Province, located in southwestern China. It is a historically important and culturally rich city. Sichuan Province covers more land area and has a larger population than Germany. It consists of nineteen separate administrative units: nine districts, four county-level cities, and six counties. There are 257 townships and 2745 villages in Chengdu. The city's nine districts are located in its urban core, and are very dense and heavily urbanized. Predictably, Chengdu's surrounding counties are comparatively less dense. Chengdu is a key center in China's national Great Western Development Strategy, an economic initiative launched in 2000 to bring the level of development in western China closer to that of the country's more developed cities on the eastern coast (Goodman 2004). The city serves as an industrial, financial, communication, logistics, and technology hub for the southwest China.

### *2.1 Public Services for Low-Income Residents in Chengdu*

In recent years, Chengdu's municipal government has taken various initiatives to provide more and higher quality public services for the city's low-income residents. These initiatives have been coordinated with an impressive collection of departments and authorities, such as the Civil Affairs Bureau, Development and Reform Commission, Financial Bureau, Bureau of Human Resources and Social Security, Bureau of Housing Management, Health Bureau, Public Security Bureau, Local Taxation Bureau, and Bureau of Statistics (Song et al. 2014; Liu 2017). The goals of these initiatives have been set clearly, which include promoting the equalization of access to public services, improving the quality of public services for residents, and addressing other problems in urban and rural neighborhoods. More specifically, the policies include:

(1) Focusing more on residents' basic livelihood and public service needs, such as education, public health, employment, and social security. Since 2008, local authorities have initiated reforms aimed at village-level public services, and have been exploring innovative policies on the supply mechanisms for public service distribution between rural and urban populations (Li 2017; Zhou 2017).
(2) Formulating quality standards for urban and rural public services, and subsequently establishing and implementing quality check according to the standards. Specifically, these standards have been applied to urban-rural public infrastructure facilities, including primary and secondary schools, kindergartens, as well as grassroots health service providers in rural and suburban areas (Lu 2011; Zhang 2008).

After years of efforts, at the end of China's Twelfth Five-Year Plan (2010–2015), Chengdu completed a preliminarily system for providing basic public services for all the residents in the city. This system covered social security, employment, education, healthcare, and housing. The fruits of these efforts can be best observed in the following areas:

(1) *Education and training*: The city of Chengdu has successfully accommodated approximate 365,000 children of its migrant workers in local primary and middle schools. The average years of schooling for the city's incoming laborers has increased from 13.1 years in 2010 to 14.4 years in 2015. There are 56 colleges and universities in Chengdu, with about 800,000 students in attendance and 50,000 full-time teachers.
(2) *Employment and job training*: 1.7 million workers have found employment in Chengdu, and the city's registered unemployment rate in its urban area have stabilized around 4%. More than 1.6 million people take employment and entrepreneurship training each year.
(3) *Health care*: There are about 9800 medical institutions of all kinds in Chengdu, with 128,000 bed space and 148,000 health technicians available. The average life expectancy of Chengdu's residents has risen from 77 to 79 years of age. The maternal mortality rate has dropped from 12.2 to 9.0 per 100,000 and the infant mortality rate has dropped from 6.0 to 3.6‰ as well.
(4) *Social services*: City authorities have built a temporary assistance and supplementary medical assistance system for the needy. A mechanism for ensuring that minimum living standards and subsidy standards are adjusted with economic development in a timely manner has been established as well. Specifically, the average annual increase in the subsidy standards for the low-income residents of Chengdu's urban areas has been 16.99%, while average for those in Chengdu's rural areas has been 44.78%. Lastly, the implementation of integrated disaster prevention projects has led to the formation of a three-leveled (city-county-township) command system for disaster preparedness and emergency response.
(5) *Housing security*: Approximately 139,000 affordable housing apartments have been constructed and provided for the city's low-income residents. In the same vein, 59,900 apartments have been renovated in what was once Chengdu's urban shantytown, while 20,000 more have been brought up to safety code standards in Chengdu's rural areas.

(6) *Cultural and sports activities*: The coverage rates of radio and television have both reached 99%. The Plan of Promoting Residents' Physical Fitness has been implemented as well.
(7) *Welfare for the disabled*: Chengdu has also successful been officially designated as a role model for caring for the disabled in China by the nation's central government. This title was awarded for the city's extensive work in improving its accessibility for the physically disabled.

In spite of such progress however, there are still many shortcomings in practice. For example, the provision of public services in Chengdu's urban and rural areas is still not well-balanced. For instance, the schools and hospitals in urban areas are equipped much better than in rural areas. Basic public services for vulnerable social groups, such as migrant workers and the disabled, have yet to be improved substantially. Long-term mechanisms, such as institutional design, financial resources, as well as evaluation and supervision schemes, need to be improved further as well.

## 2.2 Satisfaction Evaluation of Public Services in 2016

In February 2017, Chengdu's municipal party committee released its "Report on the Satisfactory Assessment of Public Services in Chengdu for 2016". The report covers the areas of public safety, transportation, food and drug safety, ecology and the environment, health care, compulsory education, grassroots organization services, recreational activities, legal services, social security, and agricultural security.

The report was based on a large survey with a total sample size of 23,047 respondents, of which 9526 are urban and 13,521 are rural samples, accounting for 41.33% and 58.67% of total respondents, respectively. The survey covered 374 towns (or street offices) in the city's 20 districts (or urban) counties, as well as the Chengdu High-tech Zone and the Tianfu New District.

The results show that the total rate of satisfaction amongst the residents of Chengdu regarding public services is over 80% for 2016. The average of overall satisfaction was 81.18%. The average overall satisfaction for urban residents was 82.78%, while the average for rural residents was 79.47%. Urban residents were generally more satisfied with grassroots organization services, recreational activities, and legal services. Rural residents, on the other hand, generally had higher levels of satisfaction regarding recreational activities, health care, and public safety.

Judging from their rankings in satisfaction levels for the different services covered in the report, public service satisfaction for recreational activities and agricultural security rose highest, with their rankings rising six places. However, public safety, food and drug safety, ecology and the environment, health care, grassroots organization services, and legal services generally remained unchanged in rank from the year before. The results of the assessment are useful for helping the city's municipal party committee, municipal government, local governments, and relevant departments understand the satisfaction level that residents feel toward the public services provided by the city. They can thus help guide future policymaking for the effective and improved provision of public services to the people of China.

## 2.3 *Equalization of Basic Services in Chengdu*

Chengdu has made much effort to achieve the equalization of basic public services for both the local people and the migrants, for both the urban and rural residents, and for both the rich and the poor. Though 100% equalization is not yet realized, the effort has been recognized as outstanding among the Chinese cities. The journalists John and Naisbitt proposed in his book the concept of Chengdu Triangle, which is composed of the property rights reform, the equalization of basic public services, and the grassroots democracy (Naisbitt and Naisbitt 2012). From the perspective of public service equalization, Chengdu practices can be summarized as follows:

(1) Standardized Financial Transfer Payment System

The standardized fiscal transfer payment system is an important measure to promote social equity. Since 2003, the outer suburbs accounted for more than 70% of the financial transfer payments from Chengdu municipal level, and the suburban district accounted for more than 20%, with less than 10% in the central urban districts. Taking 2008 as the base, the expenditure on public services and public facilities construction of governmental at all levels has been required to be mainly used for rural public services, and to ensure equal access to basic public services among urban and rural residents. From 2003 to 2008, 12.32 billion yuan was invested to the public facility projects in rural Chengdu, in which education and social security are the major.

In 2009, Chengdu began to provide each village with 200,000 yuan of public fund, which could be used in any way so long as there is collective democratic decision-making process. The amount had been raised to 300,000 yuan in 2011.

(2) A Comprehensive Public Service System

Chengdu has tried to establish a comprehensive public service system, which contains almost all-important dimensions in public service. According to the interviews with local officials in Chengdu, it is suggested that a systematic and comprehensive institutional design at the municipal level is the key to coordinate various actions by different governments and bureaus.

(3) Standardized Public Facilities

Chengdu has comprehensively promoted the equalization of public services by constructing standardized public facilities. In the city public facilities are divided into 8 categories, including education, medical and health care, culture center, sports facilities, community services, security services, agricultural services, and business services. The facilities cover a wide range from middle school, elementary school, nursery school, community health service centers, health centers, farmer's market, cultural activity center, sports ground, gym, police station, community service centers, community center, community service facilities, aging people day care center, waste logistical station, renewable resources recycling station, to public toilets.

# 3 Analysis of Survey Data from Low Income Neighborhoods

Chengdu Municipality contains 22 districts and counties, forming a concentric ring pattern in urban spatial structure. For this study, two neighborhoods were selected for the extensive interview survey. The two neighborhoods are located in the central and suburban areas of Chengdu city, respectively. They are the neighborhood of Jiuli and the neighborhood of Linwan.

Jiuli neighborhood locates in Jinniu district, northwest of the central city of Chengdu. In 2017, the GDP of Jinniu district was 106.1 billion yuan, ranking the 3rd among the 22 districts and counties in Chengdu. Historically Jinniu district has been the center of Chengdu, with the largest population and the most prosperous business environment. The annual per capita income of Jiuli community is about 40,000 yuan.

Linwan neighborhood is located in the Pidu district, a suburban district in Chengdu. In 2017, the GDP of Pidu district was 52.5 billion yuan, ranking the 10th among 22 district and counties in Chengdu. Linwan has flower market, bird market, stone market, grain and oil market, second-hand furniture market and farmers market. Its annual per capita income is about 20,000 yuan, around half of that of Jiuli neighborhood.

## *3.1 The Socio-Economic Status of Surveyed Households*

This study's survey respondents, 40% are males and 60% are females. Those classified as household heads numbered 87% men and 13% women. Of the heads of the surveyed households, 50% were found to be in good physical condition. Respondents were aged between 16 and 65, and all who were surveyed considered themselves of the Chinese Han ethnicity. In terms of education, 40% respondents in Jiuli were found to have degrees that are high school or above, while 27% people in Linwan were found to have similar educational backgrounds. Of the respondents, 87% were found to be working in non-agricultural sectors.

The share of respondents who had ever migrated was 53%. Among them, 33% of respondents were in Jiuli neighborhood and 60% of respondents were in Linwan neighborhood who migrated from rural to urban areas. It indicates that suburbs such as Linwan neighborhood are often the first choice for farmers to move to cities.

## *3.2 Community Organization and Participation*

As to the mediums of information and communication, the survey covers the mediums of newspapers, radio, television, internet access, mobile phones and public announcements (see Table 1). By comparing access to these various means of com-

**Table 1** Penetration of mass communication mediums (in %)

| Frequency of use | Newspaper | | Television | | Mobile phone | |
|---|---|---|---|---|---|---|
| | Jiuli | Linwan | Jiuli | Linwan | Jiuli | Linwan |
| Low | 80 | 93 | 20 | 14 | 0 | 6 |
| Medium | 13 | 7 | 20 | 13 | 7 | 7 |
| High | 7 | 0 | 60 | 73 | 93 | 87 |
| Total | 100 | 100 | 100 | 100 | 100 | 100 |
| Frequency of use | Radio | | Internet | | Public announcement | |
| | Jiuli | Linwan | Jiuli | Linwan | Jiuli | Linwan |
| Low | 87 | 87 | 26 | 20 | 66 | 34 |
| Medium | 13 | 13 | 0 | 7 | 20 | 33 |
| High | 0 | 0 | 74 | 73 | 14 | 33 |
| Total | 100 | 100 | 100 | 100 | 100 | 100 |

*Source* Field Survey, 2017

munication, it was found that most residents have good access to televisions and mobile phones, with relatively good access to the Internet as well. On the usage of communication tools, the survey found a notable divergence among respondents. Mobile phones were found to be primary means of communication in most cases. What's more, respondents younger than 50 years of age were found to be highly dependent on the Internet, while respondents older than 50 years of age were found to be highly dependent on television. The results indicate that radio and newspapers are no longer primary carriers for information transmission in Chengdu.

On the question of their attitudes toward local elections, 67% respondents reported that "they and their family members vote in an active and voluntary way". The urban neighborhood Jiuli residents voted more actively, about 13% ahead of Linwan. 60% respondents reported that their neighborhood leadership had engaged them during the elections.

Overall, gender has an impact on neighborhood public affairs' participation. Of the survey, 60% of respondents noted that women were "seldom" involved in the public affairs of the neighborhood, and 20% of respondents claimed that women "almost never" took part in. Together, these responses accounted for 80% of all samples. Only 20% of respondents reported that women "often" participated in the public affairs of the neighborhood, and all of them were from the suburban neighborhood Linwan. One possible reason is that a large number of male adults in Linwan leave for jobs in the central city of Chengdu, thus leaving more opportunities for women to participate in the public affairs.

As to the role of gender in participating in and steering community affairs, 87% of respondents expressed the belief that men and women should be equal in their participation in community affairs. It suggests that the idea of equality between men and women in China is rooted in the hearts of people. However, there were

13% respondents who supported the idea that a "man's role is more important" in community affairs. There is still room for the promotion of gender equality.

As to willingness of participate in public affairs, of the two neighborhoods, more than 80% reported that they would like to participate more in community public affairs. A majority therein expressed a desire to participate in community public affairs as "volunteers", with others preferring to participate in community hearings. About 20% indicated no desire to participate in the public affairs of the community. This finding suggests that most residents are willing, in fact, to participate in the public affairs of their communities, but would like to do so in ways that are more flexible.

In terms of social inclusion, more than 80% of respondents stated that migrants and ethnic minorities had the right to participate fully in community affairs, with about 20% in disagreement. All of respondents agreed that it is important to communicate with the residents of other communities. If they need help, 93% people had confidence in their ability to borrow something from their neighbors. Only 7% respondents thought that they would not be able to borrow from their neighbors in times of need. This finding indicates that the social inclusiveness and social capital of the two communities in Chengdu are high.

## *3.3 Risk and Vulnerability Assessment*

As to the rate the preparedness of the government departments to provide basic urban services, 20% respondents reported that they were very satisfied, 26.7% respondents rated poorly prepared in this regard and 13.3% respondents expressed no opinion on the matter. Lastly, 36.7% respondents agreed that the local government had supplied public services to some extent. It suggests that the governments should put more effort to improve the quality of public services.

In terms of the basic public services provided by the government, 40% of the suburban Linwan residents thought that it was satisfactory. The satisfaction level of the urban Jiuli neighborhood (73.4%) was notably higher than that of the suburban Linwan neighborhood. As shown in Fig. 2, 56.7% respondents expressed that they were satisfied with the public services provided by the local government, while 43.3% respondents reported not. Residents who expressed their dissatisfaction believed that the government should increase the supply and enhance the quality of public services in the following aspects: health care, social security, environmental hygiene, food safety and so on.

When asked if your community had suffered disasters, 10% respondents noted that they had suffered from flood. 6.7% respondents had suffered from fires. And 46.7% respondents had suffered from "a lack of drainage after heavy rain". 30% respondents reported that they had suffered from other disasters. More than half of all residents reported that they were satisfied with the government's handling of natural disasters, with slight differences between the responses from the urban and suburban communities (Fig. 3).

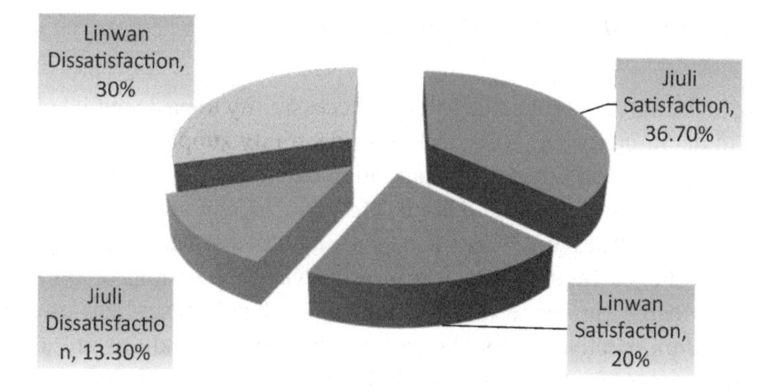


**Fig. 2** Whether the public services provided by local governments could meet local demand

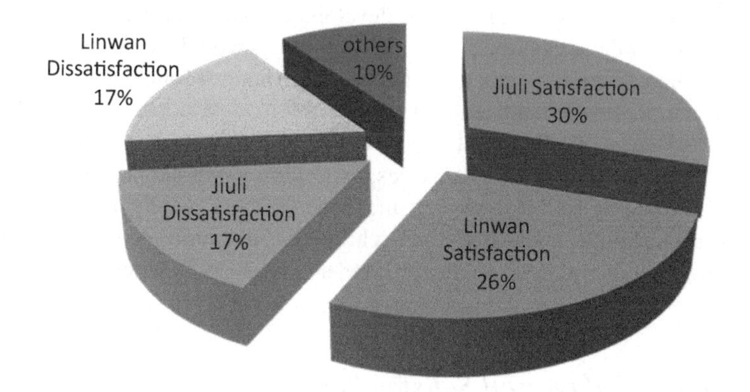


**Fig. 3** Respondents' satisfaction with the government's handling of disasters

When asked the awareness of development plans being undertaken by the local government focusing on public services delivery and management, a staggering 86.7% of respondents reported that they did not know of any such development plans at all. The situation in both communities was similar. Only 13.3% respondents had heard a little bit about this. No one respondent could offer a reply on knowledge of the local government's development plan. One notable reason is that there is no sufficient, sometimes not at all, public participation when the governments in Chengdu prepare the development plans.

When asked the awareness of local government or non-governmental organization provision for public services aiming at youth, women, immigrants, or minorities, 43.3% respondents reported that they were not aware of any such measures. On the other hand, 56.7% respondents reported that they were aware of such measures When asked if had participated in any open budget sessions of the local government, 93.3% of respondents reported that they had never attended such a meeting. Only 6.7% of resident from the Jiuli neighborhood reported any attendance.

### 3.4 Housing and Basic Public Services

According to the survey results, the levels of accessibility to basic public services for residents, in declining order, are as follows: electricity supply (100%), safe drinking water (86.7%), good community sanitation and waste disposal systems (80%), adequate housing conditions (70%), food safety (66.7%), public toilets (66.7%), and basic health care facilities (63.3%). When facing natural disasters, the residents would have access to shelter, food, and medicine, as provided by various organizations.

According to the survey results on what type of public services governments of different administrative levels choose to focus on, the findings are as follows: (1) the central government of China would focus more on agricultural related public services; (2) the provincial and municipal governments would focus more on raising funds; (3) the district and township governments would focus more on creating job opportunities, raising funds, urban development and construction, as well as improving the healthcare; (4) the community authorities would focus more on training for jobs and mental health services, and lastly that (5) nongovernmental organizations would focus more on mental health services.

Residents were generally found to believe that the local governments should play a major role in providing public services, especially when helping disadvantaged groups and providing for basic service infrastructure. Specifically, residents ranked priorities, in declining order, as follows: children's education, people's livelihood, local government credit, housing, health care, and recovery policies.

### 3.5 The Level of Overall Satisfaction

The overall levels of satisfaction for residents on specific public services are reported in Fig. 4, the main distribution is between 70 and 80%. Respondents generally demonstrated positive expectation for the future, specifically in the case of the future of resident's family. Faith in the policies of the local government and in society overall were high as well, but to a lower extent. Residents were generally satisfied with the urban planning and their local environments. The lowest levels of satisfaction were centered on shortcomings in the community and local nongovernmental organizations.

The overall levels of satisfaction for residents on specific public service organizations are reported in Fig. 5. The central government had the highest support of 86%. The satisfaction level with the social/village leadership was 65%. The results suggest that the higher the level of government, the higher the residents' satisfaction level. Surprisingly the satisfaction level of Nongovernmental organizations was the worst, with only 62.7%. It is probably due to that governments still dominate the supply of public services, leaving limited room for NGO in China.

The overall satisfaction levels of local governments were concentrated in the range of 65–74% (see Fig. 6). Among them, it is worth noting that the lowest level of satisfaction was government's efforts in anti-corruption, at only 58%. It was far

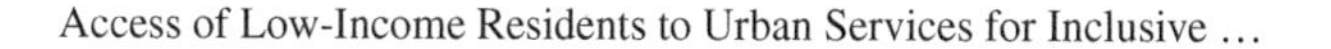


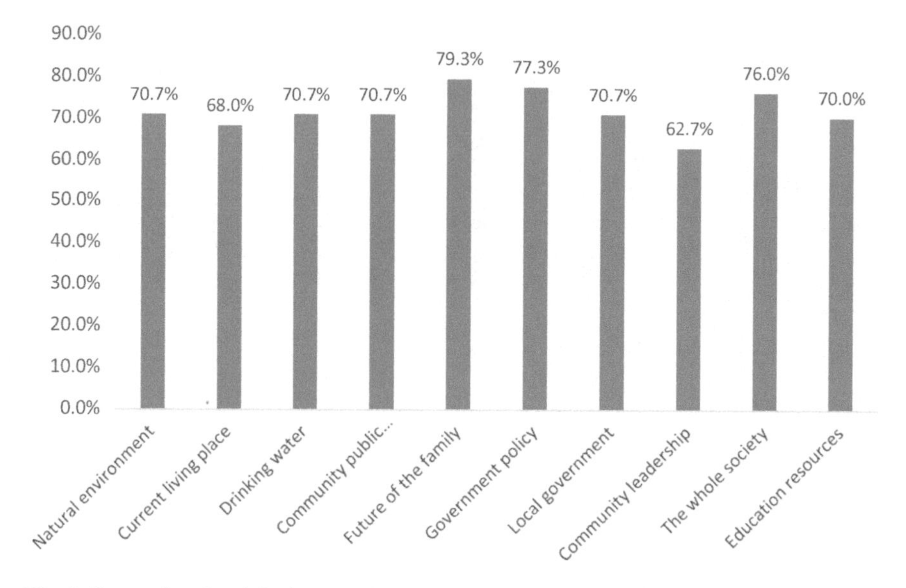


**Fig. 4** Respondents' satisfaction regarding specific areas of public services provision

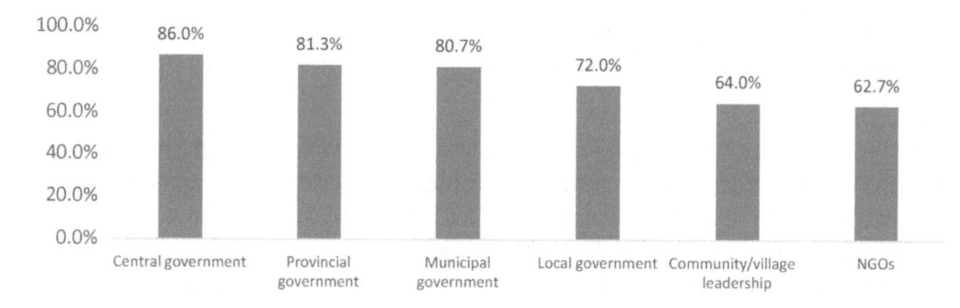


**Fig. 5** Respondents' overall satisfaction regarding the performance of specific public service organizations

below the average satisfaction level. This implies that, after years of efforts, anti-corruption initiatives are still one of the most widespread and concerned issues for the public in China. Corruption is still the dimension of governance in which the Chinese people are least satisfied.

The analyses of the first-hand surveyed data were consistent with the analyses of the secondary data from the government report and the existing literature. They suggest that an equalized public service system has been gradually established in Chengdu. The overall satisfaction levels of residents on public service delivery in the two neighborhoods were high. Even the low-income residents expressed fairly high level of satisfaction on the local governments' efforts, according to the survey results and our on-site interviews. The possible reasons are in the below.

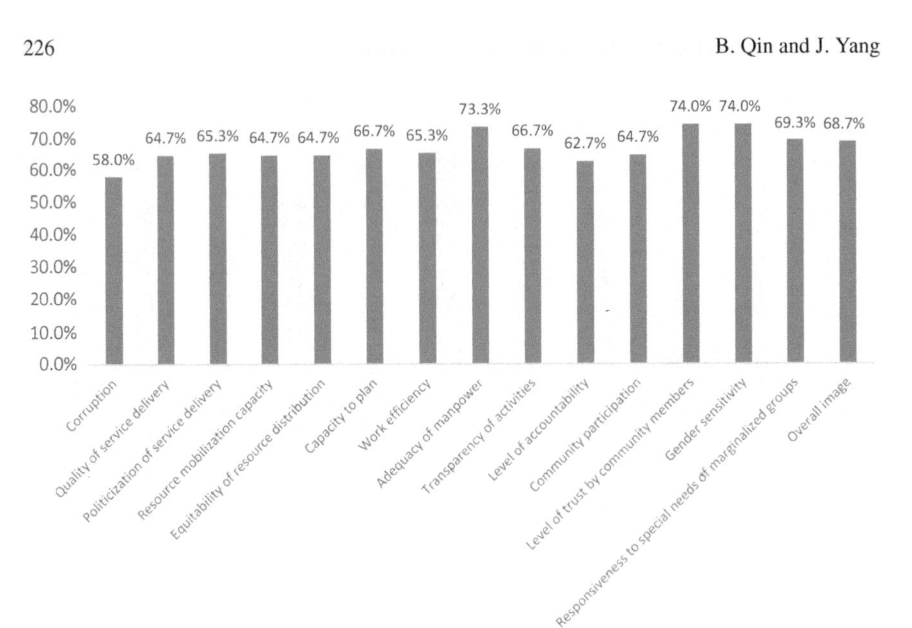


**Fig. 6** Respondents' overall assessment of specific aspects of local government

(1) *It is the result of continuous efforts of Chengdu municipal government in providing equalized public services for both urban and rural, for both local and migrant.*

In the past ten years, the government has been committed to meeting the needs of various groups of people, including the rural residents who have been lagged behind of development, the migrant workers who do not possess local hukou, and the laid-off, disabled, and elderly urban residents. The efforts of the Chengdu municipal government have been reported in this book chapter and also been well documented in John Naisbitt's book, *The Chengdu Triangle*, and Richard LeGates and YE Yumin's book, *Coordinating Urban and Rural Development in China: Learning from Chengdu.*

(2) *Economic growth and urban development are the material basis to improve public service satisfaction.*

On the one hand, the satisfaction level of the residents on public services would be influenced by economic conditions of each household. One could not give high remarks to the government's effort if he or she does not have enough food to eat or basic shelter. On the other hand, when the city has undergone rapid development and robust economic growth, the municipal government and lower level governments would have more fiscal power at disposal. With fiscal resource available quality public services could be provided by the government and by the partnership between the public and the private.

(3) *Public service assessments have been employed to facilitate government institutional reform.*

Since 2008, Chengdu municipal government has carried out the public satisfaction assessment survey on government's performance every year. The content covers all the dimensions of public services, such as social public security, transportation, ecological environment, basic health care, education, etc. The survey results have been employed to find the real and most urgent problems regarding to public services. The surveys have put great pressures on the local governments and various public agencies, forcing them to constantly strengthen and improve the public services respectively.

For example, if the public service assessment survey reflects the difficulty of accessing to primary education, this problem will be the subject for the next year's evaluation. The local governments and public agencies associated with public education would be motivated to take measures to tackle down the problem. Otherwise they would lose the opportunity for promotion and other awards.

# 4 Good Practices and Innovation

Institutional innovations in Chengdu might help explain the relatively high satisfaction level as well. For providing better public services, the authorities of Chengdu take various measures to improve public participation of the society and to enhance accountability and transparency of the government. Based on our field surveys and official archives, selected good practices and innovation are documented and analyzed in the following.

## *4.1 The Caojiaxiang Renewal Committee and Its Work in Urban Redevelopment*

The leadership of Chengdu intends to balance the lagged development of the city's northern urban neighborhoods through its "North Urban Redevelopment" program. The neighborhood of Caojiaxiang has served as a pilot project for this program. The area has many families with members who work in older, state-owned enterprises. The redevelopment project faces problems regarding how to best convert public housing to commercial housing, how to protect the interests of the community's retired workers, as well as many other challenges. With the support from the Municipal Government, the Caojiaxiang Resident Renewal Committee was established to play an effective bridging role between different administrative divisions. The self-governing committee collected the opinions and suggestions of its residents regarding the renovation of dilapidated buildings, organized all local households to encourage self-education, self-management, and self-restraint in order to promote the safeguarding of the legitimate rights and interests of the tenants concerned (He 2005).

At the same time, the committee relied on the input of the majority of the community's locals to prevent a minority of influential residents from asking for unjustifiably high compensation or obstructing the work of relocation. The committee was able to persuade the obstinate minority to support and cooperate with these urban redevelopment projects. It also published opinions and suggestions on the design and implementation schemes for the resettlement program as well as arguments for the approval of the programs many parts.

The committee also supervised the implementation of resettlement as well as the new housing construction. After the resettlement buildings were completed, the committee coordinated the relocation of the residents to their new homes. The Caojiaxiang Renewal Committee has played a decisive role in the urban rejuvenation project of the neighborhood overall.

## *4.2 The Dereliction of Duty Committed by the Staff at the Water Affairs Station of the Xindu District*

This case is one of ten representative cases of "micro-corruption affairs" as published by the Chengdu municipal party commission recently. On the afternoon of June 5, 2017, Lin and Bai were invited to come to the teashop of a local farmer to play mahjong with two local villagers.

As the managing personnel of the local waterworks, Lin and Bai should have been performing their duty to take care of the watering and irrigating for the township. However, they disregarded their obligations to the community. Not only did they fail to regulate water flows, but also played mahjong with others during their working hours, wreaking havoc on the local economy as a result of shameful neglect.

In June 2017, the two were severely chastised within the municipal party commission. Yong Wang, the director of the Juntun unit at the Water Affairs Station, as well as other party members implicated in the scandal, were held responsible for the incident. They were punished accordingly for dereliction of duty. Some were given warnings while others were dismissed from their posts.

## *4.3 Identifying "Low-Income Families" in Public*

According to the "Detailed Implementation Rules for Checking the Economic Conditions of Low-income Families in the Urban and Rural Areas of Chengdu", there is a prescribed procedure for defining and identifying low-income families, and the public authorities are subject to supervision throughout the entire process. The defined workflow is as follows: (1) A district or county civil affairs department will check for a family's declarations on social security, housing, vehicles, provident funds, business, taxation, household deposits, securities, and other information, in order to determine whether the household can be classified as low-income. (2) The community committees will publicize the results. In total, the publicity period should be five working days. Families or individuals who disagree with the contents of the public notice are empowered to report to the committee within this publicity period. (3) The committee visit pertinent households thereafter to conduct surveys and receive comments (Li 2007). For the last step of this first phase, there is firstly a dissenting family double check. Secondly the families that have a declared income less than 1.5 times of the minimum living standards of Chengdu are checked. This second step must be completed within seven working days, and reports issued to the community office of the township thereafter. (4) The authorities will audit and report to the district or city's County Civil Affairs Bureau. (5) The County Civil Affairs Bureau will issue a statement of opinion in return. The batch procedure will start on the 10th and 25th of each month. Within 45 working days from the start of the batch, the check shall be completed in full, and a statement shall be issued upon completion.

## *4.4 Instant Disclosure of Corruption Information for More Transparent Government*

As noted earlier, when asked the overall satisfaction with various aspects of local government, a low number of residents, only 58%, were found to be satisfied with local anti-corruption efforts. This is a significant finding of concern.

According to the website of the news agency Xinhua, Chinese people's satisfaction with the Communist Party of China's efforts and reduce corruption in government in 2016 reached a post-revaluation high of 86.42%. This evidences that, since the Eighteenth National Congress of the Communist Party of China launched its anti-corruption initiatives, the move has been welcomed by the people of China to a great extent.

In recent years, Chengdu has likewise stepped up its outreach and education on the topic through the launching of its "Online Anti-corruption Education Base". The concept of honesty and transparence in governance has been gradually strengthened over time. To understand the progress of anti-corruption programs in Chengdu, one can browse the "Morality in Chengdu" website and note how the city authority's social media outreach platform has become a source of considerable public debate and interaction on the topic. At present, the daily average page views of the website has exceeded 53,000, the popular subscription volume of the WeChat group has exceeded 80,000, and the city authority Weibo page's subscribers have exceeded 400,000.

## 4.5 Helping the Resident Population Through Vocational Training in Jinniu District

In early 2016, the district of Jinniu released "Document No. 1", which is an initiative aiming at helping low-income groups in both the urban and rural areas of the district. In determining the standards to apply to the initiative, Jinniu District delineated the minimum living allowance plan at a monthly income of 880 RMB. This value is more than twice as high as the defined provincial minimum for living standards. In other words, the residents of Jinniu District with an annual income of less than 10,560 RMB are empowered to benefit from the district's subsidies. So far, the government of Jinniu District has identified a total of 2184 households and 3906 low-income residents that are applicable for the public service. Among them, there are 1813 low-income families.

Based on the interviews conducted for this study, Mr. Zheng (aged 42) of the township of Dongzikou, went to work in the central city of Chengdu two years ago. Because of his lack of technical skills, his average income was less than 800 RMB a month. His life was extremely difficult during this period. In March of 2017, a large-scale job fair was held in Jinniu District, aimed at helping those in need find a job as soon as possible. That day, 76 employers recruited a total of 542 people. Some 112 people were also enrolled in vocational training. Mr. Zheng attended the multiple cooking classes and now commands a substantially increased income of about 3000 RMB a month.

## 4.6 Establishing a Health Archive for All Residents

The Xinping Public Health Center is an ordinary township hospital in Chengdu. Local farmers will usually see a doctor here if they get sick. After the doctor enters the name of the patient in the hospital's computers, the patient's health record, including any history of disease, immunization or chronic illness management, will be quickly retrieved and shown to the doctor. In recent years, Xinjin County conducted a nation-wide health checkup on more than 300,000 urban and rural residents to establish the comprehensive health file.

Zhou, a resident of the Wujin township interviewed by for this study, stated, "The condition of my health has been recorded and archived. At any hospital in the county, doctors can read my information through the database to avoid unnecessary duplication of checks, and as a result less resources are spent." Establishing and developing such information systems can also help township hospitals to enhance their service capacity.

### 4.7 *Residence Permit System*

Residence permit system is an important institutional innovation to make sure equalized public service accessibility for the migrants. Since January 1 of 2011, the municipal government of Chengdu formally abolished the temporary residence permit system, which had been enacted for 20 years, and implemented rather a comprehensive residence permit system. This was a historic step forward.

The floating migrant population in Chengdu that hold the new residence card can enjoy the same benefits and access to public services (e.g., medical and health care, motor vehicle driver's license applications, public transport, legal aid, employment, participation in community management, etc.) as the permanently registered people of Chengdu. In addition, and in accordance with the provisions for the professional and technical qualifications standards in Chengdu, these newly empowered migrant workers can apply for a permanent residency in Chengdu via the traditional urban household registration system once they pass certain criteria. At the same time, the children of these residence permit holders are permitted to receive a public primary and middle school education in the urban areas as well.

### 4.8 *Pioneering Grassroots Democracy*

Since 2008, Chengdu has explored new mechanisms for grassroots governance. Village councils have taken a leading role in the process of urban-rural integration. They have resolved various contradictions in the reform of the city's property rights system, and further advanced public outreach. Local villagers have been empowered to run their own affairs greatly as a result. For example, locals will determine for themselves the solutions to local issues such as whether urban planning should focus on developing high-rises, whether land should be used for circulation, or how village-level public service funds are to be utilized.

All meetings and elections are voluntary and independent, run by the villagers themselves. Through the village council system, democratic awareness has been strengthened in the village, and the relationship between party cadres and the common folk has been improved considerably. According to relevant surveys, satisfaction in Chengdu's farmers with the work of their village councils has reached a soaring 95%.

## 5 Structural and Institutional Barriers

Chengdu, like many other cities in China, faces certain institutional obstacles as well.

## *5.1 Market-Oriented Reforms Lead to Social Reorganization*

The deepening of market-oriented reforms in China has brought about a severe problem when it comes to poverty (e.g., laid off SOE workers, rising housing prices, increasing live costs) Since the 1990s, unemployed workers have fallen into vulnerable and low-income groups. During the urbanization processes occurring throughout all of China during this period, the reforms on pensions, medical care, housing, and education had an especially aggravating effect on the burdens already felt by China's low-income groups.

## *5.2 Lack of Comprehensive Social Security Coverage for Low-Income Groups*

Social security and welfare policies are important measures for regulating the gap between nations' rich and poor. However, social security reform in China has been slower than expected, characterized by narrow coverage and low levels of reliability. A relatively large number of low-income residents are not entitled to a corresponding minimum living allowance. There are blind spots and loopholes in basic living security policies, and many public services are not sufficiently in place.

## *5.3 The Hukou System Leads to the Discriminatory Provision of Public Services*

The reforms from the hukou system to the residence permit system, which is beneficial to the provision of basic public services, but still has a distance from the equalization of high-quality public services. Short-term migrant workers are typically unable to enjoy equal pay for work with the locals who hold the right urban household registration papers. Because of their lack of these registration papers, migrant workers also cannot enjoy public benefits where they live and work. They not only have to bear the relatively high cost of urban living, but also can hardly avoid being reduced to low-income groups in China's urban sprawl (Han and Tang 2013).

## 6 Main Findings and Conclusions

### *6.1 Improving Public Services is the Focus of Governments in Transitional China*

With market-based reforms deepening in China, the supply mechanisms of public services have undergone significant changes and also great stresses. The ability of different groups to access public services has shifted considerably over the years as a result. In particular, and especially due to their lack of effective platforms of political expression to have their demands heard, China's most vulnerable residents struggle to access the nation's public services in a timely and effective manner.

In October 2017, President Jinping Xi pointed out that as "socialism with Chinese characteristics enters a new era, the main social contradiction in China has become the one between the people's growing needs for good life and unbalanced, inadequate development". The implication is that all levels of governments will need focus on serving the people's livelihood with quality public services. Such steps will go a long way in guaranteeing the basic rights and promoting the interests of socially vulnerable groups as well as society as a whole.

### *6.2 The Quality of Basic Public Services in Chengdu is Rapidly Improving*

In 2007, Chengdu was assigned to serve as a pilot for urban and rural development programs by the Chinese government. Over the past ten years, the level of basic public services provided by the city of Chengdu has risen rapidly. This improvement is reflected primarily in seven specific aspects: an overall improvement in educational quality and equality, the realization of basic public services for equal employment and social security in both urban and rural areas, markedly improved accessibility in medical and health services, the substantial improvement of the social service system, a continued development in housing security, active construction of public facilities, as well as substantial improvements in public services accessibility for the disabled.

### *6.3 Low-Income Groups Still Lack Channels for Influencing Public Policy-Making*

The survey found that the overall level of citizens' political participation in Chengdu is not very high. The village council system is an innovative attempt to foster citizen participation in governance. However, the implementation of the council system is notably limited to local governance below the jurisdiction of a Chinese township. Likewise, the Residents' Representative Commission was a temporary arrangement brought about to promote the revitalization of specific areas. Unfortunately, this second policy can only serve to solve the difficulties of very few, privileged families, and is not universally applicable.

Although the citizens of Chengdu exhibited a certain willingness to participate in politics, the overall level of actual participation was found to be very low, especially amongst urban women. For the urban low-income groups represented by migrant workers, unemployed workers, as well as the disabled, a lack of smooth channels for political expression was found to remain a fundamental and unresolved issue. At present, feelings of satisfaction amongst low-income groups in regards to public services were reflected mostly in an awareness of local government planning.

## *6.4 The Take-Away*

In reviewing the efforts and innovative practices spearheaded by the local government in Chengdu, the following experience might be reference for other regions and nations as well. The first takeaway is the value of good leadership. The success in developing a comfortable and harmonious society in Chengdu is inseparable from the close attention of the city's authorities paid to Chengdu's most vulnerable residents.

The second takeaway is the importance of increasing fiscal capacity along with rapid economic development. Chengdu is one of the fastest growing cities in China. Therefore, the rapid and steady growth of its economy could provide a solid foundation for the financial support required to establish initiatives for the equalization of public services amongst the city's residents. If the economy is stagnant, it would be difficult to provide better public services for the poor and the marginalized. Facilitating economic growth is one of the central tasks of governments, along with providing public services (Deng and Zhao 2013).

Thirdly, the power of innovation is undeniable. It was through its innovative reforms, such as the reform of the household registration system, that Chengdu has been empowered to break down the legal barriers that once stood between its urban and rural areas. Such innovative thinking is the key to the inclusive growth in the years to come (Zhang and Wu 2017).

The last, but not least, is the value of grassroots democracy and equalization in governance. The fiscal decentralization of governance stimulated enthusiasm in local citizen participation, fostering pluralistic governance in Chengdu (Huang and Wang 2014). The townships and rural villages have gained larger discretion on how to use their fiscal income and land resources, and thus built up the capacity to provide better public services gradually. However, it is still worth exploring further. It is not a choice, but rather a necessity, for local governments to strengthen grassroots democracy, to streamline governance, and to promote government transparency as well public engagement.

## References

Blanchard O, Shleifer A (2001) Federalism with and without political centralization: China versus Russia. IMF Staff Pap 48(1):171–179
Chen Y (2011) Construction and application of the index system of urban low-income population's planning acceptability. MS thesis, Huazhong University of Science and Technology (in Chinese)
Deng Z, Zhao JW (2013) Public service capability and improvement strategy of local government—take Pixian County of Chengdu City as an example. J Wuhan Univ Technol Soc Sci 2:91–95 (in Chinese)
Goodman D (2004) The campaign to "Open up the West": national, provincial-level and local perspectives. China Q 178:317–334
Han Z, Wen T (2013) Research on the living environment of low and middle income people under the background of urbanization—taking Chengdu City as an example. Sichuan Build Sci 39(3):316–320 (in Chinese)
He C (2005) Exploration of progressive renewal mode of low-income communities in Chengdu old city. MS thesis, Southwest Jiaotong University (in Chinese)
Huang R, Jun W (2014) A study on public satisfaction with urban–rural integration by structural equation modeling (SEM). In: Proceedings of the seventh international conference on management science and engineering management. Springer, Heidelberg
Li J (2007) Chengdu: build a housing security system for low and middle-income families through pragmatic innovation. Urban Rural Constr 2:51–52 (in Chinese)
Li Z (2017) Increasing income of low-income groups will be the focus of income distribution reform. Econ Inf Dly 2017-10-10(003) (in Chinese)
Lin JY, Liu Z (2000) Fiscal decentralization and economic growth in China. Econ Dev Cult Chang 49(1):1–21
Liu Y (2017) Public goods acquisition and social stratification—an empirical study of chengdu's characteristic population based on ArcGIS. Manager 23:001 (in Chinese)
Lu M (2011) Remodeling of rural public service supply mechanism under the background of new village construction. Asian Agric Res 37:50–53
Naisbitt J, Naisbitt D (2012) Innovation in China: the Chengdu triangle. Argo-Navis
Niu M (2013) Fiscal decentralization in China revisited. Aust J Public Adm 72(3):251–263
National Bureau of Statistics in China (2011a) China statistical yearbook in 2001. National Statistical Press, Beijing
National Bureau of Statistics in China (2011b) The communiqué on the sixth population census
Song XQ, Wei D, Liu Y (2014) Spatial spillover and the factors influencing public service supply in Sichuan province, China. Coord Urban Rural Dev China: Learn Chengdu J Mt Sci 11(5):1356–1371
World Bank (2002) Urbanization of China: patterns and policies: draft summary. In: Wang Y (ed) Income status and existing problems of low-income groups in China. Commercial Economy 4:28–29 (in Chinese)
Xue L, Weng L (2017) China's policy opportunities and challenges to achieve the UN 2030 sustainable development goals. China Soft Sci 01:1–12 (in Chinese)
Yang F (2011) Migrant workers' integration into cities from a lifestyle perspective: Chengdu evidence. Chongqing Soc Sci 4:70–75 (in Chinese)
Ye Y, Jiao Y (2013) The system structure and implementation path of China's balancing urban and rural development. Urban Plan Forum 01:1–9 (in Chinese)
Ye Y, LeGates R (2013) Coordinating urban and rural development in China: learning from Chengdu. Edward Elgar Publishing
Zhang T, Zou HF (1998) Fiscal decentralization, public spending, and economic growth in China. J Public Econ 67(2):221–240
Zhang QF, Wu J (2017) Providing rural public services through land commodification: policy innovations and rural–urban integration in Chengdu. In: Public service innovations in China, Palgrave, Singapore pp 67–91

Zhang Z (2008) The new housing guarantee policy enables low-income families in Chengdu to live and work in peace and contentment. Hous Ind 10:78–80 (in Chinese)
Zhou M (2017) Study on the current situation and problems in the implementation of housing security policy for low-income groups in Chengdu. Econ Trade Pract 01:237–243 (in Chinese)

**Dr. Bo Qin** is a professor and former Head of the Department of Urban Planning and Management, School of Public Administration and Policy, Renmin University of China. Professor Bo Qin holds a Bachelor of Engineering in urban planning from the Department of Architecture at Wuhan University in 2000. He also obtained a Master of Science in Human Geography from the Department of Urban and Environment Studies at Peking University in 2003, and a Ph.D. from the School of Design and Environment at the National University of Singapore in 2007. His research interests include urban internal spatial structure, urban sustainable development and cities in transitional economy. Recently his research focuses on spatial planning for healthy city and urban governance in transitional economy. He has published more than 70 articles in international and Chinese academic journals in urban planning and urban studies, including the journals such as the *Journal of American Planning Association, Landscape and Urban Planning, Urban Studies*, etc. He has also completed seven projects funded by the National Natural Science Foundation of China, authored and co-authored four books, and co-edited five books.

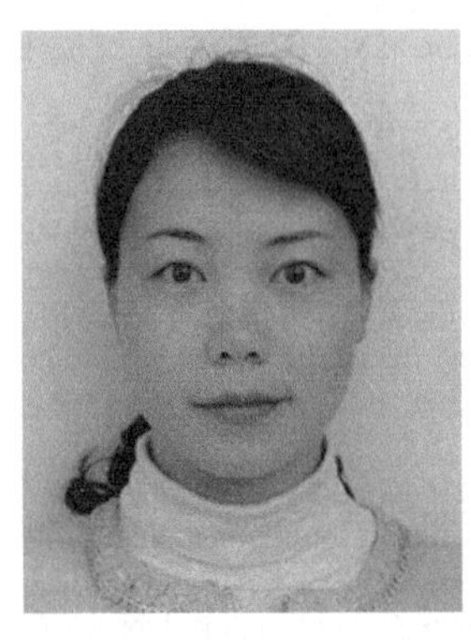

**Jian Yang** is an associate professor at the School of Public Affairs and Law, Southwest Jiaotong University, China. Currently she is also a Ph.D. candidate at the School of Public Administration and Policy, Renmin University of China. Professor Jian Yang obtained her Bachelor of Economics and Master of Economics from the College of Humanitie in Southwest Jiaotong University, China. Her research interests include regional economics and urban management, with emphases on urban spatial expansion and socio-economic development in the cities of western China. She has published more than 10 articles in peer-review journals and a book by Science Press of China. Professor Jian Yang has also provided consultant services for Chengdu Municipal Government and Sichuan Provincial Government, including industrial development strategies and coordinated urban-rural development policy suggestions.

# Access to Urban Services for Political and Social Inclusion in Pakistan

**Nasir Javed and Kiran Farhan**

**Abstract** Pakistan, the fastest urbanizing country in South-Asia, is facing the challenge of emerging pockets of marginalization and poverty in the large cities. The issues of social and political exclusion being faced by urban residents is rooted in institutional, political and structural factors including legislative provisions for municipal autonomy at provincial level. Present system of services provision by the service providers especially in the public sector, is inadequate in responding to the requirements of fast expanding urban economy. The main challenge is to introduce a well-equipped system of governance with resourceful communication technologies, environmental economics, urban finance, water and sanitation systems, alternative transport systems, traffic management and skills in conflict resolution. Another challenge is to develop an effective legislation which clarifies the balance of power and functions between the provincial and local governments. This chapter discusses some of the structural and institutional barriers to political and social inclusion of the marginalized groups including women, youth, migrants and ethnic minorities in the urban areas and some innovations and good practices in a decentralized governance. Primary data was collected through a survey in low income areas of two cities of Punjab and KPK. Further, statistical analysis of access to basic services and satisfaction level was comprehensively gauged to link it with the prevailing governance structure. Additionally, five case studies are quoted for their originality and innovation which can serve as good practice models for replication and scaling up.

**Keywords** Governance · Urban services · Community · Inclusive cities · Local governments

It is widely recognized that different forms of exclusion and barriers to political and social equity need to be addressed to ensure full engagement of the marginalized communities including women, youth, migrants and ethnic minorities in the processes of local democracy in cities and towns. This can be accomplished through inclusive urban policies and programs that fill gaps between urban planning and urban realities.

---

N. Javed (✉) · K. Farhan
The Urban Unit, Shaheen Complex, 503 Egerton Rd, Garhi Shahu, Lahore, Punjab 54000, Pakistan



S. Cheema (ed.), *Governance for Urban Services*,
Advances in 21st Century Human Settlements,
https://doi.org/10.1007/978-981-15-2973-3_10

These policies must accommodate marginalized groups in urban governance by promoting greater transparency, accountability and community participation in urban decision-making and holistic management of city regions that is integrated across jurisdictions and sectors. In developing countries, economic disparities among different groups and regions lead to political and social exclusion. In a properly functioning democratic system, citizens need to devote some time and energy to the democratic practice (Bhalla and Lapeyre 1997). When people are excluded from the main sources of income, however, their first priority becomes survival and a basic livelihood rather than political participation.

Like many other developing countries, Pakistani masses are also suffering from various forms of exclusion. For instance, Castillejo (2012) narrates four main forms of exclusion in Pakistan which includes; (i) the political and economic exclusion of some regions within the socio-political system; (ii) the exclusion from access to land mostly in rural areas; (iii) the exclusion and violence faced by religious minorities; and (iv) the exclusion of many young people and women, which contributes to Pakistan's demographic instability.

The high rate of urbanization in Pakistan is fundamental to understanding the issues of social and political inclusiveness being faced by urban residents. According to the recent census, Pakistan's annual rate of urbanization stands at 2.7%-estimated to be the highest in South Asia. This spike in urbanization is largely linked to rural inhabitants migrating to cities in search of socio-economic opportunities and access to services. Pakistani cities are manifestly under- prepared for the myriad challenges being ushered in by rapid urbanization. Cities have grown exponentially in both size and density during the last decades. The majority of this increase in urban areas has been in the province of Punjab. This growth was initially concentrated in existing cities; however areas that were once small towns have now been transformed into major urban centers. The role of transport and connectivity patterns has determined the pattern of urbanization for centuries and this is evident still in the number of cities dotted along the main road infrastructure of Punjab.

This chapter discusses some of the structural and institutional barriers to political and social inclusion of the marginalized groups including women, youth, migrants and ethnic minorities in the urban areas of Pakistan and innovations and good practices in decentralized governance. It describes the national and city level context, including the decentralization policy, institutional mapping and service deficits (Sect. 1); findings of household surveys in selected slums and squatter settlements in Lahore and Peshawar (Sect. 2); good practices and innovations (Sect. 3); and barriers to political and social inclusion (Sect. 4). The final section presents conclusions.

## 1 National and City Context

With a population of 207 million (Pakistan Bureau of Statistics 2017), Pakistan is undergoing significant socio-economic transformations. A large proportion of the population is young and the majority is part of the national labor force. With an advantageous and beneficial amount of human capital, the government in Pakistan has a stronger responsibility to strengthen its institutions and social policies to ensure inclusion, productivity and capacity within the population. A negligent institutional arena can risk wastage of precious human resources and can exacerbate costly socio-economic challenges. One such challenge is of provision of basic services as mentioned in the national constitution.

As a consequence of rapid urbanization, pockets of marginalization and poverty are becoming increasingly apparent in urban Pakistan. One of the most obvious areas of exclusion and marginalization is urban housing. The lack of housing is due to higher numbers of migrants into major cities. As cities struggle to provide affordable, secure, and tenured living conditions to all citizens, spatially marginalized settlements crop up, a natural phenomenon given the income levels and institutional gaps. These settlements end up as unregulated areas, often referred to as slums. Their residents are sidelined in society with poor and inadequate access to basic amenities such as clean drinking water, electricity, gas, sanitation, health facilities, and education.

### *1.1 Decentralization Policy*

Despite notable gains in democratization, effective implementation of decentralization has been constrained due to lack of political support for decentralization at the national level in the 1960s. The 1973 Constitution of Pakistan, however, recognized a third tier of local governance, calling on the State to establish '*local government institutions composed of elected representatives of the areas concerned and in such institutions special representation will be given to peasants, workers and women*' (Jamil 2016). This was a significant institutional provision for the inclusion of marginalized voices at the local level. However, with the military-led government in the late 1970 and 1980s, decentralization efforts were derailed and more powers and resources were concentrated at the central government level.

In 2010, however, the 18th Constitutional Amendment was passed by the National Assembly of Pakistan and the federating units of the country. It was a breakthrough legislation because it decentralized numerous important Ministries previously held by the Federal Government. The Amendment provides a legal framework for the structural reshaping of the state into a decentralized federation, along with considerable legislative and policy making power in key areas such as health, education, and social welfare firmly concentrated in the hands of Pakistan's second administrative tier of provinces. The key features of the 18th Amendment were that it:

- reasserted the role of two institutions of the federation—the Council of Common Interests (CCI) and the National Economic Council—to harmonize central-provincial relations;

- deleted the list of federal/provincial concurrent responsibilities and reassigned selective functions to the federation under the guidance of the CCI and devolved others to the provinces;
- resulted in the abolition of 17 ministries at the federal level;
- ended the extension of Federal regulatory authority to provincially owned entities or private entities operating in a single province; mandated the Federal government to consult the provinces prior to initiating any hydro-electric projects;
- gave a free hand to provinces in all public services delivered within their territory, and control over all local government institutions;
- extended the Residual functions not enumerated in the constitution to the provinces' domain; and
- designed to cede the responsibility of Federal government for taxes on immovable property, estate and inheritance taxes, value-added tax (VAT) on services, and zakat and usher (religious taxes) to the provinces; expanded the Provincial borrowing privileges to include domestic and foreign loans, subject to limits and conditions imposed by the National Economic Council (NEC).

Despite lapse of almost 9 years, there are a number of constraints being faced in the implementation of the 18th Amendment to the Constitution. The existing capacity of Provincial governments is weak vis-à-vis their new responsibilities. The Amendment ignored building local governments as the third tier and to reduce Provinces/States and local government capacity gaps. Also, political support for the devolution plan was not adequately mobilized. Consequently, the subject still remains as a political issue at various levels.

In Pakistan, decentralization has been an arena for a power tussle between the country's military and political parties. Military governments have historically introduced local governments at the district and sub-district level to bypass the role of national political parties. Thus, it is not surprising that the three most determined attempts at devolution to local governments have all been under military rule: The Basic Democracy Order of 1958 under the regime of General Ayub Khan, the Local Government Ordinance of 1979 under the regime of General Zia-ul-Haq, and the Local Government Ordinance of 2001 under the regime of General Pervez Musharraf. During the last ten years of democratic rule, much of the decentralization of 2001 ordinance related to powers and resources of local governments have been rolled back.

Despite being marked by power struggles between the military and elected governments, and infighting between major political parties and regional coalitions, Pakistan has completed its first uninterrupted decade of democracy (2008–18). It is undoubtedly clear that the country is undergoing a gradual deepening of institutionalized democratic politics. The general elections of July 2018 resulted in a new political government by the Pakistan Tehreek-e-Insaf, at the federal level as well as at the two provinces of Punjab and Khyber Pakhtunkhwa. Tehreek-e-Insaf is a relatively new party in the country, and a firm believer in the robust and meaningful devolution of political, administrative and financial powers to the local level, even at further lowest of Union Council or village/neighborhood level. The new government in Punjab has already started work on empowering the local governments and the Punjab Local Government Act 2013 is being amended to enable implementation of this policy. However, the implementation on ground might take a few years, since the current local governments would have to complete their tenure and law could only be amended at the turn of an election cycle.

## *1.2 Functional Gaps, Overlaps and Coordination*

There are visible differences amongst the provinces in terms of service delivery. The districts under Punjab Local Government Ordinance, 2001 (PLGO) has responsibility for delivering elementary and secondary education, primary and secondary health and dispensaries, agriculture and intra-district roads, street services (e.g. street lights, water supply systems, sewers and sanitation); markets and cultural events. Metropolitan/Municipal Corporations under the PLGO 2013, (previously City District Governments) delivered integrated services, supervising these on a city-wide basis. Councils at the union level are responsible for libraries, local streets and street services. Union councils work closely with village and neighborhood councils in promoting and coordinating development activities, and presenting proposals through annual plans to the district and tehsil levels to help appraise the development budget allocation of the requirements of local residents.

Despite the legislative provision for municipal autonomy and local representation, municipalities across Pakistan, remain toothless in the face of the Province due to institutional, political and structural factors. The Local Government Act is a provincial subject and a provincial law. The local Councils have very little financial resources of their own and are largely dependent upon the province, especially for all development expenditure, which is very limited anyway. Moreover, all recruitment at local level has to be sanctioned by the province. During the last couple of years, services like waste management, water supply & Sewerage, parking, among others have been transferred to companies, established and controlled largely by the province. These factors make the devolution granted under the law, more or less a theoretical concept. In addition, since 1985, the National and Provincial Legislatures have been provided with funds for local development, which are largely utilized to develop infrastructure and services, primarily falling in the domain of local councils, thus creating a kind of competition, obviously to the disadvantage of local politicians and local councils.

## *1.3 Institutional Mapping of Punjab and KPK*

Currently, the cities examined in this chapter, Lahore and Peshawar, have differing approaches to local government and representation. On ground these differences translate into a variation in forms of barriers faced by marginalized groups in both cities.

An overview of institutional mapping of both provinces shows that electricity and gas supply are federal subjects, while all other urban services are either provincial or local-including land ownership, parks and horticulture, building permits, water supply, waste management and building regulations. Since Local Government itself is a provincial subject, much of what is 'local' also remains within the ambit of the province. Local Government Ordinance 2001 had brought in some sort of uniformity amongst the systems in all the four provinces. However, since 2010, every province has its own Local Government Act. Whereas the Khyber Pakhtunkhwa (KPK) has retained more or less the same framework of District Governments and Tehsil Municipal Administrations, Punjab has significantly amended the system, reverting many functions to the province and creating a divide between urban and rural councils. Since the latest national elections in 2018, the government has already amended the Local Government Act (2019) that further lowers the level of local councils to neighborhood and village levels. However, since almost a year, the previous elected councils have been dissolved and bureaucratic officers are using powers of the councils, until elections are held.

Present system of services provision by service provider agencies especially in public sector, is inadequate in responding to the requirements of fast expanding urban economy. The main challenge is to introduce a well-equipped system of governance with supporting communication technologies, environmental economics, urban finance, geographic information systems, water and sanitation systems, alternative transport systems, traffic management and skills in conflict resolution. Another challenge is to develop an effective legislation which clarify the balance of power and functions between the provincial and LGs (Ministry of Climate Change, National Report of Pakistan for HABITAT III).

## *1.4 Service Deficit in Lahore and Peshawar*

Lahore, the provincial capital of Punjab and the country's second largest city, with a population of 11 Million indicates annual growth rate of 3% over the last twenty years and an average household size of 6.3 (Pakistan Bureau of Statistics 2017). Lahore, scores below average 61.1 (mean value of 64.3) in the UN-HABITAT (2010) Global Urban Indicator Database of 162 countries and is a 'low developed city' in the Asian Development Bank Cities Data Book for Asia and the Pacific. The Integrated Master Plan for Lahore 2021 has acknowledged that there is still no control on peripheral growth and land sub-divisions that consume prime agricultural land. The annual housing production shortfall has resulted in illegal sub-division of peri-urban agricultural land and squatter settlements (Yuen and Choi 2012). Four variables including; income, education, health and housing were taken for calculation of socio economic opportunity index. The result shows that the basic income deprivation is the highest among all other socio-economic variables, followed by housing, education and health. Results show that about 65% of inhabitants of squatter settlements have no access to socio-economic opportunities (Kalim and Bhatty 2006).

Peshawar, the capital of *Khyber Pakhtunkhwa* province, has a population of almost two million. It has a growth rate of 3.73% since 1998 (Pakistan Bureau of Statistics 2017) and has become one of the largest recipient cities for refugees and Internally Displaced People (IDPs) in South Asia (Mosel and Jackson 2013). Afghan refugees and IDPs are considered to be the major factors of rapid urbanization in Peshawar in addition to the rural–urban migration for economic reasons. All these factors contributing towards the relatively fast growth rate of almost 4%, over the last 30 years when Afghan war started. Afghan refugees and IDPs who arrived with limited means often chose to live in low-income urban settlements, thereby constituting a major proportion of the slums. Due to high population densities and absence of proper hygiene and sanitation mechanisms, the slums are prone to high incidence of diseases and health related problems since more than 95% of toilets are non-flush type. A community sewage system in the surveyed slums is mostly nonexistent (Cynosure Consultants 2013).

## 2 Analysis of Access to Urban Services Based on the Household Survey

The survey in low income areas of two cities of Punjab and KPK was conducted by the Urban Unit. The low income area named as Kotli Ghazi in Lahore was selected for survey. The reason behind selection of this area was that, after the construction of Lahore Ring Road, people started to move in this area which was once the periphery of Lahore. Access to Canal Road is also one of the reasons that this agricultural land has rapidly transformed into residential area, albeit mostly through informal land subdivisions and unapproved housing constructions. In Peshawar, Ganj Bazar located in the center of the city was selected. This area turned into low-income settlement due largely to the migration of Afghan refugees.

### *2.1 Socio-demographic Profile of the Respondents*

The objective of this part of the survey was to assess the overall socio-economic and demographic status of the sample and to see how that compares with the national and provincial statistics. The results show that majority of the respondents were males, and so were the heads of the households. With an average age of 48, the majority is married more than half are illiterate and at least 50% have an average health. The collected information shows that people living in the areas selected for the survey have below average income as compared to the provincial as well as national averages. When this information is compared with the income source, it is evident that one third of the respondents are involved in their own small-scale businesses and agriculture, which might be one reason why their average income is lesser than other comparative figures.

## 2.2 *Community Organization and Participation*

Two thirds of the respondents were aware of community leadership. But when asked if they would like to participate in community, majority of them expressed their willingness to be a part of their community organizations. As per the frequency of the respondents to use the communication media, the survey results illustrate that 62% and 40% respondents regularly use Cell Phones and TV, respectively. Meanwhile 57% respondents stated that they have never used internet followed by 52% respondents have never used public announcements. The results showed that 33% of respondents had voted every time while 33% had voted at least once for their community leadership. More than 75% of the respondents informed that women generally do not participate in community affairs. Moreover, 40% respondents pointed out that whole members' conference of villagers has the final word in important community issues, while 33% thought that community leadership has the final word in community matters.

Regarding the relationship of the surveyed community with their adjoining communities, the vast majority of the respondents not only realize it is important to communicate with the neighboring communities, they actually are at good terms with their neighboring communities in general and their neighbors in particular. They not only trust their neighbors, they look out to them when they need their help. They also look forward to participating in the community affairs of neighboring areas, when requested. These responses show that the communal bond is very much in place in these low-income settlements, where people look up to their neighbors for support. The questions related to community bonding and satisfaction living in the present community indicate very positive attitude of the respondents. They are well settled in their neighborhood, and in case of Lahore most of the respondents don't feel they would shift to another locality. However, in Peshawar, they may shift to some other area because many of them are Afghan refugees who are more likely to move to other locations in search of employment opportunities.

## 2.3 *Risk and Vulnerability Assessment*

With reference to the preparation of government departments in providing basic urban services such as clean water, electricity, healthcare and education, only 6% of respondents have evaluated that they are satisfactorily prepared, while 28% said that government departments are poorly prepared to provide basic urban services. Almost 26% respondents were not aware of any preparedness by the government departments. Concerning the awareness of the residents about development plans of local government, 72% of respondents were aware of development plans of local government which focus on urban service delivery and management. Up to 60% respondents said that they have faced the issue of drainage after heavy rain. This rate in Lahore is 70 and 50% in KPK. Only 25% respondents of Peshawar said that they have faced flood in past. All respondents from both cities stated that during disasters they have not received any kind of support from local government. Almost 55% of respondents said that no government project is focused on development of marginalized groups in their area while 45% were of the view that they do know that Government is focusing on such projects for the marginalized groups. Almost 70% respondents communicated that they have never participated in the budget sessions. This highlights the fact that people expect more from local governments and they hear about the development plans and somehow don't see their direct benefit for their own communities.

## 2.4 *Access to Basic Services*

The household survey shows that 52% of respondents have access to safe drinking water, 42% to community level sanitation and waste disposal, 58% to primary health care facilities, and 76% to education, 47% to food, 65% to electricity and 42% in terms of restrooms. Most notable point is that the proportions of respondents who have access to sanitation, food and restrooms are the lowest of all services. The survey shows that the availability of drinking water and sanitation facilities is low in both the surveyed areas as compared to national and provincial levels. Therefore, it can be said with confidence that these people are living a below average life, when it comes to basic necessities of life. Both the areas represent low income communities in provincial capitals and both lack these services. The surveyed area of Peshawar is below national, provincial and the city coverage.

One of the important feedback received from surveyed areas is that what actually people expect from government to prioritize while designing the projects. In terms of expected role of government in dealing with service provision and access, the issues that respondents want the local government to put on high priority and very high priority are door to door awareness campaign and support for the elders respectively. Respondents ranked, mobilizing community members and helping them to build community level shelters, as priority.

## 2.5 *Overall Satisfaction Level*

Concerning the overall satisfaction about performance of organizations, 27% of respondents were satisfied with Provincial Government, 68% of respondents were somewhat satisfied by Municipal Government, Local Government and Community Leadership. Meanwhile, nearly a half of respondents were not satisfied at all by Central Government (Table 1).

**Table 1** Overall satisfaction about the performance of the following organizations

| Both Provinces | Not satisfied at all | | Somewhat satisfied | | Satisfied | |
|---|---|---|---|---|---|---|
| | No. | % | No. | % | No. | % |
| Central Govt. | 28 | 35.0 | 40 | 50.0 | 12 | 15.0 |
| Provincial Govt. | 8 | 10.0 | 50 | 62.5 | 22 | 27.5 |
| Municipal Govt. | 15 | 18.8 | 55 | 68.8 | 10 | 12.5 |
| Local Govt. | 18 | 22.5 | 56 | 70.0 | 6 | 7.5 |
| Community Leadership | 15 | 18.8 | 55 | 68.8 | 10 | 12.5 |
| NGOs | 15 | 20.0 | 59 | 73.8 | 6 | 7.5 |

As per critical task of service providers and the overall assessment of respondents about institutional features of local governance, 57% of the respondents think that politicization of service delivery is high; and 41% respondents said that corruption is high. Nearly half of respondents thought that work efficiency of local government departments is low and same number of respondents believed that equability of resource distribution is very low. The respondents' overall assessment of the other features of local governance that were rated low included quality of service (40%), resource mobilization capacity (43%), capacity to plan (48%), adequacy of manpower (68%), transparency of activities (81%), and responsiveness to special needs of marginalized groups (83%) (Table 2).

**Table 2** Overall assessment of service providers at the local government level

| Both province | Low | | Medium | | High | |
|---|---|---|---|---|---|---|
| | No. | % | No. | % | No. | % |
| Corruption | 22 | 27.5 | 25 | 30.0 | 33 | 41.3 |
| Quality of service | 32 | 40.0 | 38 | 47.5 | 10 | 12.5 |
| Politicization of service delivery | 18 | 22.5 | 18 | 21.3 | 46 | 57.5 |
| Resource mobilization capacity | 35 | 43.8 | 35 | 43.8 | 10 | 12.5 |
| Resource management capacity | 34 | 42.5 | 36 | 46.3 | 10 | 12.5 |
| Equitability of Resource distribution | 41 | 51.3 | 33 | 41.3 | 6 | 7.5 |
| Capacity to plan | 39 | 48.8 | 35 | 43.8 | 6 | 7.5 |
| Work efficiency | 40 | 50.0 | 32 | 40.0 | 8 | 10.0 |
| Adequacy of manpower | 55 | 68.8 | 19 | 23.8 | 6 | 7.5 |
| Transparency of activities | 65 | 81.3 | 11 | 13.8 | 4 | 5.0 |
| Level of accountability | 67 | 83.8 | 5 | 6.3 | 8 | 10.0 |
| Community participation | 51 | 63.8 | 19 | 23.8 | 10 | 12.5 |
| Level of trust by community members | 48 | 60.0 | 15 | 18.8 | 17 | 21.3 |
| Gender sensitivity | 53 | 66.3 | 19 | 23.8 | 8 | 10.0 |
| Responsiveness to special needs of marginalized groups | 67 | 83.8 | 11 | 13.8 | 2 | 2.5 |

# 3 Good Practices and Innovations

## *3.1 Badar Colony Water Supply and Sewerage Initiative*

There has been significant awareness in the underprivileged and low-income communities regarding their rights to basic services and their attempts at improving their participation in development and maintenance of these facilities. In a locality, not very far from the selected low-income study area in Lahore, the residents of Badar colony embarked upon a project to provide water supply and sewerage to their locality of around 3000 households. It was done through the NGO, Anjuman Samaji Behbood (Social Welfare Society). The city-wide Water & Sanitation Agency (WASA) had limited funds and hence there were no immediate plans for service extension to this suburban area.

In 2008, the residents formed the NGO, took ownership of the project, and offered to contribute 38% of the capital investment, through which the system was laid. The same NGO entered into an agreement with WASA, purchasing bulk water and setting up their own distribution, billing and maintenance system. Now almost eight years down the line, the community is getting regular water supply, with 98% billing compliance and self-maintenance of routine faults. The responsibility for major breakdowns of tube well or sewerage disposal station remains with WASA. This is an excellent example of community participation in solving their local problems. The same NGO later developed a system of waste collection and disposal, on self-contribution basis and even a small dispensary, showing the confidence of citizens.

## *3.2 Lahore Waste Management Company*

The Public Private Partnership (PPP) in solid waste management has been recognized all over the world. Different types of waste management strategies have been applied to Lahore in different era to provide municipal services with the collaboration of private sector. But some of these strategies failed. Based on lessons learned, in 2010, Lahore Waste Management Company (LWMC) was established under section 42 of companies' ordinance 1984. Till 2009–2010 the solid waste management department of City District Government Lahore (CDGL) was responsible for the solid waste management services of the Lahore city under the Local Government Ordinance 2001. Solid Waste Management Department (SWMD) of City District Government Lahore (CDLG) signed a Services and Asset Management Agreement (SAMA) with M/S Lahore Waste Management Company. The objective of this agreement was to plan, implement and manage different Public-Private Partnership programs. Under this agreement, CDGL had transferred possession, management, use, maintenance and control of machinery, equipment, tools and plants, vehicles, lands, buildings, structures and all other moveable and immoveable assets that were owned, managed or controlled by the CDGL for solid waste management. After this agreement, LWMC is responsible for solid waste services of the Lahore city.

In 2011, LWMC outsourced the solid waste collection and transportation services through International Bidding. After proper assessments, contract was awarded to the two Turkish companies i.e. Albayrak and Ozpak against a total amount of US$320 million for seven years. The new Solid Waste Management Operations by Turkish companies were launched in 2012 and this contract is running successfully. This PPP model has increased the technical capacity of LWMC to monitor the activities of private sector. Due to this success story of LWMC, this model has been replicated and six more Waste Management Companies in other cities (Rawalpindi, Sialkot, Gujranwala, Multan, Faisalabad and Bahawalpur) of Punjab has been established in recent years.

### *3.3 The Aga Khan Development Network*

In Pakistan's cities there is an ever growing need for developed urban land, housing and the amenities. Yet, there are constraints of public sector in provision of these to an adequate level. Therefore, the private sector has been a big player in this field. This offers a big relief to the growing population needs but there is a downside to this as it leads to relative exclusion of the low income housing. However the opportunity lies in this market. Keeping this in view, two institutions of the Aga Khan Development Network i.e. the Micro Finance Bank Ltd Pakistan (FMFB) and the Aga Khan Planning and Building Service, Pakistan (AKPBS) launched a program designed to help poor and low-income families make their homes safer, healthier and more energy efficient.

FMFB provides loans ranging from PKR 10,000–500,000 to groups of poor people at a relatively low interest rate, from short to medium-terms for both structural and non-structural improvements to the houses. AKPBS provides technical expertise and training to people of rural areas. With over 157 locations in all over Pakistan and access to 68 Pakistan post outlets, FMFB has disbursed over PKR 11.8 billion (104 m $) through 677,000 loans while AKPBS has been working to improve the built environment in Pakistan since 1980. Its areas of focus include housing design and construction, village planning, natural hazard mitigation, water supply and sanitation and improved indoor living conditions for the most disadvantaged members of society.

### *3.4 Citizens Feedback Monitoring Program*

Basic urban services are often provided by the departments and agencies of government, with limited public accountability to ensure adequate access to services. The rigidity of bureaucratic procedures and a lack of customer orientation has led to limited impact in terms of access to services. In a bid to hold the service providers accountable to the customers, the Government of Punjab has developed an ICT based system of "Citizens Feedback Monitoring Program CFMP" through Punjab Information Technology Board, commonly known as the Jhang Model, as it was developed by a local District officer in Jhang, a rural district in Punjab.

The CFMP works on a simple mechanism of reaching proactively to all the users of public services. At every point, where a citizen comes in contact with a public service provider, his/her phone number is recorded and registered in the system. Then through a mechanized system, the citizen receives a robot call in the voice of the Chief Minister Punjab, asking for an honest and frank feedback on the quality of service provided and any issues or suggestions. The data so collected is analyzed and presented to the chief Minister and shared with the departments on a monthly basis. Over the years, this database has accumulated millions of records relating to dozens of departments and is proving extremely useful in disciplinary actions against officers and systemic improvements in service delivery.

### 3.5 *Dengue Activity Tracking System*

In 2011, when Lahore and its surrounding areas had been hit by the vector borne disease i.e. Dengue, all departments under Government of Punjab joined hands to combat dengue and to proactively gear up against its spread in coming season. They took precautionary measures to stop dengue carrier larvae and mosquitoes through anti-dengue activities. To track department's efforts for dengue surveillance, Government of Punjab has devised monitoring mechanism named as "Dengue Activity Tracking System" to record real-time, all field activities allied to prevention and eradication of the dengue. It was launched in March 2012 and is successfully operational till date.

The monitoring mechanism of "Dengue Activity Tracking System" is a real time reporting portal through GPS enabled android based mobile application. The android mobile application covers multiple activities ranging from larviciding, dengue patient tagging, dewatering, debris removal, and surveillance of graveyards, junkyards, workshops, schools, abandoned buildings, nurseries, and pools. Each official in the field takes geo-tagged photographs of performing their designated tasks, identifies larvae breeding hotspots and removes them accordingly; two separate photographs are submitted highlighting "before" and "after" actions. Data stream as submitted via android based mobile application gets plotted on Google maps in real time as the mobile application captures latitude and longitude along with photographs. Along with this surveillance, a massive media campaign was also launched to keep every citizen aware about precautionary measure. Owners were also arrested by Police if their premises are not clean, dry and were source of breeding place for the dengue Larvae.

This surveillance is successfully operational throughout Punjab and also at ICT (Islamabad Capital Territory). Recently Government of KPK has also sought help from Punjab regarding combat of dengue in Peshawar. There are several help desks and dedicated helpline to provide all patients with guidelines and tips for preventing and fighting dengue. More than 6.0 Million anti-denguesurveillanceactivitiesareregisteredviaandroidmobilessincethelaunch.Refresher trainings are also conducted at all divisional headquarters to combat the dengue menace. This efficient management has controlled the widespread of dengue in Punjab.

In all above mentioned innovations, the two key factors for success have been (1) finding need specific institutional design and (2) a project champion, either individual or an institution. The nature of success stories varies from housing, to water and sanitation, cleanliness, preventive health and citizens' empowerment, thus covering almost all facets of urban life. At an institutional design level, we see that whereas Aga Khan Housing is a purely NGO program, the citizens' feedback model on the other extreme is a hardcore government departmental initiative sponsored and monitored by none less than the provincial chief minister. The Badar colony water and sanitation program is an excellent example of a partnership between the public sector service provider WASA and an NGO; while LWMC is an innovation in delivering high quality waste management services using public sector corporate entity, thus by-passing red tape and bureaucratic delays. Last but not the least, the dengue monitoring program is an innovative use of ICT in public health, creating model of cross learning and use of state of the art technologies.

The good practices and innovations described above provide a significant ray of hope for the cities of Pakistan. The classical model of public sector service delivery, based on the 'one size fits all' kind of solution has not been successful. On the other hand, this bouquet of innovative solutions has the inherent flexibility to adapt the institutional solution to address the specific issue or bottleneck. This approach could be used as the guiding platform for re-designing the whole urban paradigm of service delivery towards a sustainable and equitable outcome.

## 4 Understanding Barriers to Political and Social Inclusion

Despite provision for wider representation and democratic inclusion in governance laws in Pakistani cities, there remain visible and palpable instances of marginalization especially in low income areas.

In the context of decentralization, barriers to inclusion can be seen via conflicts of interest at two levels: among decision makers located at the local level; and between the local, sub-national and national governments. In practice, Pakistani devolution has empowered the Provincial governments (as described above) but restricts local government discretion. Local governments, for example, cannot exercise hiring and firing autonomy for service providers in their jurisdiction; they are restricted in their ability to raise taxes (a critical element in restricting autonomy and thereby restricting effective local representation); and cannot effectively bid for an increase in the allocation of funds coming from higher tiers of government. This simply means that despite institutional provisions for decentralization in the country through a constitutional amendment, cities in Pakistan are less likely to enjoy the intended political and social inclusivity and democratic ideals. The main reason is that some Provincial governments are not willing to decentralize powers and resources to local governments. In practice, political leaders and government officials at the Provincial level are reluctant to cede powers and resources to plan and implement local development projects. Top-down decision making process is the norm. In Punjab, for example, the elected government of PTI abolished the previously elected local governments and has not organized local elections over the past one year.

The extent and determinants of exclusion in Pakistan are summed up by Jamil (2012): "*some estimates suggest that around 40% of Pakistan's population experience significant political, social and economic exclusion based on their identity (e.g. religion, kinship group, and language), their location or their gender*". To varying degrees, the same is true in the context of Pakistani cities and towns. For instance, the role of women in society varies traditionally from region to region. Peshawar, one of our case studies, is a city with a traditionally conservative social environment for women. Women are normally excluded from community level decision- making and are relegated to other areas of communal life. Specifically, slum-dwellers in Pakistan, are normally regarded as under-privileged and those that do attain a degree of education find extremely difficult to find adequate employment opportunities. This creates further barriers for an already marginalized urban class.

Despite constraints in streamlining effective local democracy, Pakistan has still made significant strides in recognition of its marginalized groups. The Punjab government has launched various programs for the systematic inclusion of women in all tiers of governance and economic life. For this purpose, in 2014 the Government of Punjab created the Punjab Commission for the Status of Women (PCSW), a statutory, autonomous body established for the promotion of women's rights. After the devolution of women's development to the provinces under the 18th Amendment to the Constitution of Pakistan, PCSW was conceived as an oversight body to ensure policies and programs of the government to promote gender equity in the province of Punjab" (Punjab Commission on the Status of Women 2015). Its major objectives are stated to be the elimination of discrimination against women in all forms and the empowerment of women. As part of the important work being carried out by the PCSW, the creation of a Gender Management Information System (GMIS) has allowed for resources to build data and maintain and sustain a database of women and gender issues that have implications on the formulation of provincial policy and strategic action for women empowerment.

As far as spatial marginalization is concerned, Pakistan has legislation in place known as the ***Katchi Abadsi Act of 1992***—*Katchi Abadi* is the local term for a squatter or unregulated urban settlement—this was most recently amended in 2012. According to the amended Katchi Abadis Act, those residents that have been residing in dwellings since prior to December of 2011 would be given proprietary rights. Moreover, the Katchi Abadi Act recognizes these dwellings as legitimate urban settlements and gives provision for mechanisms to be put in place for the regulated supply of basic services and provisions.

Pakistan, like many developing nations, has a complex history of political instability, poverty, shortage of resources and socio-economic under-development. While, much advancement has been made in the legislative and institutional arena in Pakistan, marginalization still occurs and barriers to political and social inclusion can still be observed. Their existence can be attributed to an arguably non-serious implementation process and a lack of consistency and monitoring of laws and inclusive programs and policies. In Lahore, for example, the satellite imagery and property tax data analysis shows that the rich quintile (measured in term of size of housing units) occupies almost 60% of land area, while the poorest quintile is crammed in just 5.8%. Yet, effective measures are not taken to implement efficient land use policies and programs that can make it more equitable.

## 5 Conclusion

Despite being marked by power struggles between the military and elected governments and infighting between major political parties and regional coalitions, Pakistan has completed its first uninterrupted decade of democracy. It is clear that the country is undergoing a gradual deepening of institutionalized democratic politics. Despite its federal system of government, Pakistan has been a centralized state in practice. The 18th Amendment to the Constitution changed that by giving more powers and resources to provincial governments. Local government is a Provincial subject. However, the Provinces have not extended decentralization from provincial level to local governments.

Institutional mapping of both provinces shows that public services are usually provided either by the government or the state-owned (non-business) organization/companies. However, in some cases, private sector is also playing its role, especially in big cities. But service deficit exists in both cities. Lahore with a population of 11 million indicates annual growth rate of 3% over the last twenty years but scores a below average 61.1 (mean value of 64.3) in the UN-HABITAT (2010) Global Urban Indicator Database of 162 countries and is a 'low developed city'. In Peshawar because of migration of Afghan refugees and Internally Displaced People (IDPs) caused rapid increase in urban population. Due to high population densities and absence of proper hygiene and sanitation mechanisms, the slums in Peshawar are prone to high health related problems.

Regarding different forms of political and social exclusion of marginalized urban communities, the survey in selected slums shows that service delivery of basic urban services are far better in Lahore (Punjab) where access to different urban services ranges from 62.5 to 87.5% as compared to Peshawar (KPK) where it ranges from 22.5 to 65.0%. In terms of expected role of government in dealing with service provision and access, the issues that respondents want the local government to put on high priority and very high priority are door to door awareness campaign and support for the elders respectively. Respondents ranked mobilizing community members and helping them to build community level shelters, as priority. Regarding the overall satisfaction about performance of organizations, 27% of respondents were satisfied with Provincial Government; 68% were somewhat satisfied by Municipal Government and nearly a half of respondents were not satisfied with the Central Government.

Rapid urbanization in Pakistan, coupled with weak public sector institutions and reluctance by the federal and provincial political elite to share power with the local leaderships, has resulted in the poor levels of services and evidence is shown by the survey results, highlighting the deficiencies, inequities and elite capture. However, there is another side of the story as well. A number of success stories unfolded by a range of institutions from pure public sector, government owned companies, public private partnerships to NGOs highlight the potential that innovations and institutional creativity carry towards addressing the urban challenges. This lesson could be the way forward toward an innovative urban governance framework. A fuller documentation of these success stories and a mechanism for replication and institutionalization of these models with flexibility for local adaptation are needed to cope with urban service deficits in poor urban settlements. With the new Local Government Act 2019 having been promulgated in both Provinces, we can hope for a change.

## References

Anjuman Samaji Behbood (ASB) (2007) Chnaga Pani Program. Anjuman Samaji Behbood, Faisalabad. http://www.asb.org.pk/index.html

Agha Khan Development Network (AKDN) (2010) Microfinance helps poor families build safer housing in Pakistan. http://www.akdn.org

Bhalla A, Lapeyre F (1997) Social exclusion: towards an analytical and operational framework. Dev Change 28(3):413–433

Castillejo C (2012) Exclusion: a hidden driver of Pakistan's fragility. NOREF Policy Brief. Norwegian Peace Building Resource Centre

Cynosure Consultants (2013) Study on slums of Peshawar, KPK. Cynosure Consultants Pvt. Ltd., Islamabad

Government of Pakistan (1973) The constitution of Islamic Republic of Pakistan. Government of Pakistan, Islamabad

Jamil R (2012) Why caste and social exclusion need to be addressed in policy making in Pakistan. End Poverty in South Asia [Blog]. The World Bank Group

Jamil R (2016) Making the state local: the politics of decentralized governance in Pakistan. Agenda for International Development (A-id) [website]

Kalim R, Bhatty SA (2006) Quantification of socio economic deprivations of squatter settlement's inhabitants: a case study of Lahore. In: Sixth global conference on business & economics

Local Government, Elections and Rural Development Department of Khyber Pakhtunkhwa (2019) Tehsil and Town Municipal Administration. Local Government, Elections and Rural Development Department, Government of Khyber Pakhtunkhwa, Peshawar [website]

Mosel I, Jackson A (2013) Sanctuary in the city? Urban displacement and vulnerability in Peshawar, Pakistan. Humanitarian Policy Group, London

Ministry of Planning Development and Reforms, Pakistan Vision 2025 (2014). http://pc.gov.pk/web/vision

Pakistan Bureau of Statistics (2017) Pakistan Population and Housing Census 2017: provisional summary results. Pakistan Bureau of Statistics, Government of Pakistan, Islamabad

Punjab Commission on the Status of Women (2015) Introduction (of PCSW). Punjab Commission on the Status of Women (PCSW), Lahore [website]

Punjab Information Technology Board, Citizen Feedback and Monitoring Program (2017) Punjab Information Technology Board, Government of Punjab, Lahore. https://www.pitb.gov.pk/

UN-Habitat (2010) The state of Asian cities 2010/11. State of Cities-Regional Reports. UN Habitat, Fukuoka

Yuen B, Choi S (2012) Making spatial change in Pakistan cities growth enhancing. World Bank Policy paper series on Pakistan

**Dr. Nasir Javed** is a senior civil servant with more than 30 years of experience, including twenty years in the urban and municipal sectors. He has served in the field as head of municipal services in Karachi and Lahore, the two largest cities in the country. He designed, set up and lead the Punjab Urban Unit, a policy and planning think tank in the urban sector. Dr. Javed holds degrees in Medicine, Psychiatry, Law and MBA. He also worked as a consultant with ADB, Unicef, UNDP and many others, mostly in the urban sector. He has authored a couple of papers and book chapters and is currently working as a free lance consultant.

**Dr. Kiran Farhan** A civil Engineer by profession, Dr. Farhan holds a masters in Environmental and Geotechnical Engineering from University of Grenoble (France) and a Doctorate in Environment and Earth Sciences from the same institution. She has nearly two decades of experience in national and international markets utilizing her sound knowledge of the region's economic, social and cultural characteristics and comprehension of development trends. Her prime areas of diligence are project designing and management regarding advancement of Water Supply, Sanitation and Solid Waste Management sector in close collaboration with national as well as international standards.

# Governance for Urban Services in Vietnam

**Nguyen Duc Thanh, Pham Van Long and Nguyen Khac Giang**

**Abstract** The chapter investigates the relationship between local democracy and different forms of barriers to political and social inclusion of marginalized urban communities, particularly migrants living in slums areas in Vietnam. As a one-party regime, Vietnam had been a highly centralized state until early 1990s, with the government exerting tight control over the society. The launch of market-oriented "Doi moi" policy in 1986 has contributed to the country's economic boom, lifting millions out of poverty, opening up the civic environment, and significantly improving the Vietnam's public services. The government has allowed the private sector to engage in the provision of certain services, in addition to the previous state-controlled public service companies and state non-business organizations. Access to basic services like electricity, water, health and education has been much improved. However, economic growth has brought numerous significant social and political implications, particularly for the new social class of migrant workers. The government's residence-based social policy created many barriers for marginalized groups, consisting mostly of migrants, to urban services including water and sanitation, health, education, and other socio—political rights that urban residents enjoy. The Vietnamese government has recently made significant efforts to remove the barriers, which were expected to make the accessibility of marginalized groups to urban services become easier, especially for immigrants. Nevertheless, the results are mixed and there is room for further policy improvement.

N. D. Thanh (✉) · P. Van Long
Vietnam Institute for Economic and Policy Research (VEPR), University of Economics and Business at Vietnam National University, Room 707, E4 Building, 144 Xuan Thuy Street, Cau Giay, Ha Noi 10000, Vietnam
e-mail: nguyen.ducthanh@vepr.org.vn

P. Van Long
e-mail: pham.vanlong@vepr.org.vn

*Present Address:*
N. K. Giang
6 Levy Street, Mount Victoria Wellington 6011, New Zealand
e-mail: nguyen.khacgiang@vepr.or.vn


S. Cheema (ed.), *Governance for Urban Services*, Advances in 21st Century Human Settlements,
https://doi.org/10.1007/978-981-15-2973-3_11

**Keywords** Social inclusion · Local democracy · Decentralization · Urbanization · Marginalized · Migration · Vietnam

# 1 Context

## *1.1 Introduction*

The notions of political inclusion and democratic governance are getting more significant in developing countries, particularly in places with high urbanization rates and rapid migration from rural to urban areas, which has led to a new generation of "second-class" city dwellers who have less access to public goods and political rights in their residence. It is, therefore, important to examine relationships between local governance and different forms of barriers to political and social inclusion of marginalized urban communities and groups such as women, youth, migrants and ethnic minorities.

Vietnam is a particular case for examination. As a transitional country with a booming economy, it has the social and economic characteristics of a typical developing nation. Since initiating market-oriented reforms in 1980s, Vietnam has been one of the fastest growing Asian economies, with the annual growth rate of around 6–7% for the past 30 years. As a communist one-party regime, the country also possesses a specific political and welfare system that aims to provide universal coverage for its citizens. Vietnam has been considered as a successful model for fulfilling Millennium Development Goals (MDGs), with particular regard to poverty reduction. Despite such achievements, Vietnam still faces many problems in terms of quality of growth, sustainable investment and social protection of vulnerable population groups, who have been left behind in Vietnam's successful stories.

One important consequence of the overall economic growth has been the increase in domestic migration. Nationwide, 13.6% of the population are migrants, of which the migration rate of the population aged 15–59 is 17.3% (UNFPA 2016). Increasing migration reflects not only economic growth but also important regional socioeconomic disparities, particularly between cities and the countryside, and the growing labour market in large cities and the expanding industrial zones. Given that the urbanization rate continues to accelerate, with the closest estimation or urban dwellers at 40% of the population in 2020, it is certain that domestic migration and its impacts will pose different challenges for policy makers.

This chapter aims to examine relationships between local governance and different forms of barriers to political and social inclusion of marginalized urban communities, particularly women, youth, migrants and ethnic minorities in Vietnam. The research focused on two mega cities of Vietnam, the capital Hanoi and the economic center Ho Chi Minh City (HCMC), with estimated population of 8–10 million for each city. These cities account for the largest number of migrants, who share the biggest portion of marginalized communities in urban areas. Migrants account for above 16 and 20% in Hanoi and HCMC respectively (UNFPA 2016).

In this study, "inclusion" is defined as the equality of opportunities for participation in the society (European Commission 2004; Hayes et al. 2008; Stewart 2000; World Bank 2011a; Boushey et al. 2007; Levitas 2003). This requires multiple factors such as a standard well-being with basic needs met (European Commission 2004; Cappo 2002; Boushey et al. 2007), access to markets and public services (European Commission 2004), political voice (European Commission 2004) and feelings of respect and recognition (European Commission 2004; Hayes et al. 2008; Cappo 2002; World Bank 2013; Levitas 2003). World Bank (2013), in its most general and up-to-date review of the concept of social inclusion, suggested the definition as "the process of improving the terms for individuals and groups to take part in society", with the term being elaborated as ability, opportunity and dignity. This definition, as such, generalizes most of the aforementioned factors.

The chapter consists of five parts: The first describes national and city context, including government decentralization policies, urban development programs, and government policies and programs for slum improvement. The second part presents institutional mapping and mechanisms for service delivery and access. The third part analyses the survey data from selected from low income settlements. The fourth part examines factors that influenced effectiveness of service delivery and access. The last part presents conclusions.

## 1.2 National and City Context Analysis

Vietnam had been a communist regime with strict control on the economy, domestic migration, and social policy in a Leninist style before it adopted market-oriented reform policies in 1986, following what had been done in China since their own "Open Door" policy in 1978. Consequently, Vietnam has gradually become one of the best economic performers in Asia. The country has been dubbed as the new "Asian tiger", implying its trajectory of development that resembles that of Asian newly-industrialized countries. The economic boom has led to many social and political implications, particularly for the new social class of migrant workers. From over 80% of population living in rural areas, drastic waves of domestic migration have pushed a large number of population to newly urban areas in city and industrial zones, reducing rural population to 65% in 2016 and is expected to fall under 60% in 2020 (GSO 2017). By 1990s, the number of cities and towns was about 500. As of December 2016, the country had 795 cities and towns.

Vietnam is experiencing a high rate of urbanization. In a critical review of urbanization in East Asia, World Bank assessed that urbanization in Vietnam has dramatically increased, both spatially and demographically, as during the 2000–10 decade the country overtook Thailand and the Republic of Korea in the amount of urban land. In terms of urban population, Vietnam is the sixth-largest in East Asia with 23 million people (World Bank 2015). The spatial growth of urban land in Vietnam stays third in East Asia, only behind China and Indonesia. World Bank also notes that the rapid rate of expansion in Vietnam's two biggest cities (3.8% in Hanoi and

4.0% in HCMC per year, respectively) is much faster than other East Asian countries, except China.

Nevertheless, the country is suffering from failure to steer urbanization for its development. The huge increase in urban population puts pressure on the cities' housing, infrastructure, services and social-welfare system. Infrastructure in Hanoi and HCMC is failing to meet demands of the population and new development projects. When it has been generally agreed that infrastructure stock needs to reach around 70% of GDP for an economy to sustain urban growth, Vietnam falls short of this benchmark with its share standing at 47% in 2013 (World Bank 2035 report 2016).

In terms of housing and accommodation, many migrants live under poor conditions. More than 40% of migrants live in places with areas less than 10 $m^2$, while this number in non-migrant population is just 16% (UNFPA 2016). Vietnam is expected to service, upgrade or rebuild an estimated 4.8 million housing units in two main categories: (i) non-permanent houses and (ii) houses that lack basic services (World Bank 2015), yet this can only be seen as a long-term promise.

The household registration system, or *ho khau* (*hukou* in China), which was used as a tool for social control in the pre-1986 period, is one of the major barriers that discriminates migrants from non-migrant population. Many migrants (49%) only register temporary residence and 13.5% of migrants were not even registered for temporary residence/temporary absence. This poses various difficulties for migrants without *ho khau*, from non-qualification for formal banking credits and difficulty for registration of vehicles such as motorbikes to education for their children. At the same time, qualitative interviews show that registration procedures for *ho khau* in many places are complex and even require informal costs, i.e. bribes (UNFPA 2016).

In addition to housing, other major challenges for Vietnam's rising urban areas are providing reliable services of education, health, water and waste management. In terms of education for migrant children, about 13.4% of migrants with school-aged children (5–18 years old) say that their children are not attending school, according to the report entitled *National Domestic Migration Surveillance 2015* by the United Nations Population Fund. Health care is generally better in urban areas than in the countryside; however, there is still a lot of room for improvement. For example, according to General Statistical Office (2018), the rate of under 5 years old children with malnutrition in urban areas (according to their weights) was 7.5% in 2016. In addition, the environmental concerns of migrants in big cities have been rising quickly, mainly due to high population density and pollution. Issues such as "rising temperatures", "air pollution", and "water pollution" are repeatedly mentioned in our in-depth interviews with migrants in the two cities. Only a half of new urban areas have centralized waste water treatment stations, the remaining half have no wastewater treatment stations, leading to serious environmental problems (Vietnam Environment Administration 2016).

## *1.3 Government Decentralization Policies*

Vietnam had been a highly centralized state with strict control over the social and economic life of its citizens. Yet after market-oriented reforms were introduced in late 1980s, the centrally-planned economic system was gradually demolished. The economic changes inevitably brought about policy changes in terms of political and social governance. In 1990s, the governing Communist Party of Vietnam (CPV) put in place the legal framework for the expansion of direct citizens' participation in local government, or the so-called "grass-roots democracy" in a series of Party guiding documents and then applying the mechanism in government decrees.[1] In 2007, the National Assembly issued the Order No. 34/2007/PL-UBTVQH11 on implementing democracy at local level, officially realizing their CPV's guidance into governing orders. A study by Vu and Zouikri (2011) states that decentralization has generally improved Vietnamese local government efficiency, but some indicators are more sensitive than the others, both in de jure and de facto terms. Anh (2016) believes that there are three co-existent and interrelating decentralization processes in Vietnam: fiscal, administrative and political decentralization.

**Fiscal decentralization**—The process of fiscal decentralization in Vietnam can be seen via three versions of the State Budget Law 1996, 2002 and 2015 (amended). State Budget Law 1996 marks a milestone for fiscal decentralization in Vietnam by, for the very first time, clearly specifying the division of rights and responsibilities between central and provincial governments as well as among different levels of local government with respect to revenue and expenditure. The amended State Budget Law 2002 states that tax revenue is shared between provincial and central governments and the rate is kept stable for intervals of five years. The most recent State Budget Law was approved in 2015 and became effective from the 2017 fiscal year. With regard to decentralization, despite enormous pressures from many provinces demanding a fair share of import tax collected at the provincial level, the 2015 State Budget Law rejects this demand. It nevertheless adds income tax, including from SOEs, private enterprises, and personal income, to the tax-sharing list (Anh 2016). With those changes, the government wants to strengthen fiscal discipline in the system, from central to provincial level. The 2015 State Budget Law specifies the maximum level of debt that provincial governments can mobilize. For Hanoi and HCMC specifically, the ratio between debt and decentralized revenue should not exceed 60%. For other provinces with decentralized revenue, maximum ratio is only 20%. It should be noted that with decentralizing policies on state budget, people at the local level, in theory, can influence budget distribution via their elected representatives (i.e. member of the People's Council).

At lower administrative level (districts and communes), the government issued the Circular No. 60/2003/TT-BTC in 2003 to instruct the operation of budget. The Circular then was upgraded into Circular No. 344/2016/TT-BTC in 2016 in accordance with the new amended State Budget Law (2015). Both circulars specify the

[1]Most notably Instruction No. 30-CT/TW (1998) by the Central Committee, Decree 29/1998/NĐ-CP by the Government.

responsibilities and rights of grassroots authorities to use budget that they mobilize, as well as set the shares that they can spend. This gives a considerable room for local governments to use state budget, including spending for social affairs such as supporting marginalized communities.

Regarding the decentralization of public investment, the government first issued a regulation regarding investment management and construction in 1999, under which provincial governments are entitled to public investment projects of Category B and C, while decisions concerning the most important projects (i.e., Category A) are retained at the central government. In 2005, the government issued a decree allowing provincial governments to decide on all public investment projects. However, the list of Category A projects is still decided by the Prime Minister, and the capital amount must be jointly decided by the central and local governments. Decentralization of public investment was extended in 2007 when provincial governments were allowed to ratify the list of and grant licenses for Build-Operate-Transfer (BOT), Build-Transfer-Operation (BTO) and Build-Transfer (BT) projects. One example of public investment decentralization was that provincial governments were given "block funding" for all National Targeted Programs (NTPs), which allowed them to allocate this funding among different NTPs. However, in 2004, the central government retook the rights to allocate NTP funding, and provinces can only allocate funds within each NTP.

**Administrative decentralization**—The administrative decentralization in Vietnam has gone through three phases of reforms, starting in early 1990s and continuing until today (Phase 3). The reform, generally, concentrates on two aspects: the provision of public services and socio-economic development plans (SEDP). In the first phase (1991–2001), the main focus was to restructure the state bureaucracy and specify responsibilities of different state branches. The second phase (2001–2010) concentrated on four main goals: institutional reform, administrative reform, improving state officials' capacity, and fiscal reform. The third phase (2011–2020) focuses on perfecting the institution of "market economy with socialist orientation" and building a competitive business environment, in addition to other goals set up in previous phases.

For the provision of public services, the process of decentralization in education and health care has been accelerated at the provincial and lower levels in the 1996, 2002 and 2015 State Budget Laws. However, it should be noted that provincial governments must follow guidance from the central government, specified in terms of quotas, standardization, and cost norms. For poor provinces, the share of predetermined expenditure in local budgets can reach 80–90% (Ninh and Anh 2008). Since 2003, local People's Councils (at provincial/district/and commune levels) have been given more authority to draft and verify SEDPs for their levels. Starting from 2004, Provincial People's Councils were able to issue documents in the areas of socio-economic development; budget allocation; defense and security; and people's livelihood. Accompanied by fiscal decentralization, this is a significant progress towards decentralization in Vietnam.

Vietnam consists of 63 provinces. Officially, provincial administrative units are classified into four groups. The first group (special administrative units) includes

Hanoi and HCMC which enjoy special status. Group I includes three other cities directly under the central government—Hai Phong, Da Nang, and Can Tho—which have higher level of central oversight, but have more space for local affairs and potentially have more access to central funding. The two other groups (II and III) are subjected to a common decentralization framework, depending much on their population sizes, economic development, and to some extent geographical position, according to the Law on Local Government Organizations (2015). Under the provincial/central city level, the local governments are organized under smaller units by districts and then communes. Officially, communes are the most grassroots level of local governance in Vietnam; however, under communes there are wards/villages/blocks level, which do the jobs of communicating between the people and the commune governance. Village/block chiefs do not have direct salaries from the state budget or full status as state employees, rather they act as a "freelance" communicator and receive a minuscule payment from the commune governance.

**Political decentralization**—Despite much progress in fiscal and administrative decentralizations, political decentralization remains a challenging topic, given Vietnam's strictly-controlled one-party regime. In principle, CPV monopolises the absolute political power, which is realized via their ability to nominate, allocate, and dismiss high-level officials in all main government structures from central to local levels. This can be done because all senior government posts require Party membership. As a result, even the elected provincial People's Councils and Committees are chosen via a mechanism known as "Party nominates, people vote" (*Đảng cử; dân b`âu*), as the people can only vote for the candidates that CPV nominates.

All key provincial officials are under direct management and supervision by CPV, who can be classified into three categories. The first category includes Party Secretary, Chairman of People's Council, and Chairman of People's Committee of Hanoi and HCMC, who are under the supervision of CPV Politburo. It is worth noting that Party secretaries of Hanoi and HCMC are members of the Politburo, which is the de facto group of supreme leaders in Vietnam. The second tier are the positions decided by the Central Party Committee and Central Party Secretariat, including chairman and party secretaries of the other provincial People's Council and People's Committee. In a normal process, the position of provincial party secretaries is elected by the provincial People's Committees, then the selection must be ratified by higher levels of CPV (Central Party Committee). The chairman of the People's Committee usually co-chairs the Deputy Secretary post of the CPV provincial branch. The third category includes those positions that need ratification by the Central Committee. Since 2007, the Vice Chairman of People's Council and People's Committee are no longer subjected to pre-evaluation by the Central Committee, but fully depends on the local provincial People's Committee. It is noted that these positions in Hanoi and HCMC still needs to be verified by the Central Committee, given the strategic importance of these two cities.

Under a lower level of districts and communes, the most important guiding document is the Party Directive No. 30/CT-TW, dated February 18, 1998. This Directive was issued right after the social unrest in Thai Binh province in 1997, which, according to the Politburo's perception, resulted from "undemocratic practices" of local governments. Later, this was issued as Decree 29 in 1998. These "grassroots" policies, along with other various guiding documents, provide a new mechanism for citizens to have more power over local government activities. These were summed up in the widely known motto of the CPV "people know, people discuss, people implement, and people supervise" (*Dân bi´êt, dân bàn, dân làm, dân kiêm tra*). However, the effectiveness of this principle is still limited, given the tight control of CPV in all administrative levels.

## 1.4 Urban Development Programs

Since early 2000s, with the appearance of the Law on Construction (2003) and its guiding documents (Decree No. 08/2005/ND-CP and Decision No. 03/2008/QD-BXD), city governance and urban planning have improved in Vietnam. This process is enhanced by the Law on Urban Planning, ratified in 2010. As a result, the legal framework for urban planning has been consolidated for local governance, in terms of both master planning for cities/economic areas and lower level of administrative governance. However, Vietnam's urban planning programs are still plagued with the low quality of master planning, lack of implementation capability, and lack of efficient coordination mechanism (Centre for Urban Forecast and Research (PADDI) 2014). Table 1 shows more details on the timeline of Vietnam's urban development policies, which is combined by the Vietnam Urbanization Review, a technical report carried out by World Bank (2011b).

## 1.5 Overview of Slum Policies and Other Government Programs

Officially, Vietnam does not have a slum policy, as the country's policymakers are reluctant to recognize that "slums" even exists. In 1998, Vietnamese government authorized a master plan on urban policy, yet did not have practical strategies for efficient urban planning and management. For revision, the government consulted with international donors to address its urgent need for urban policy, especially focusing on marginalized urban communities. As the result, from 2001 to 2003, World Bank-managed consultants carried out four in-depth studies which aimed to assess the housing and infrastructure constraints faced by the urban poor, review ongoing national and international urban upgrading programs, develop a detailed action plan for the city of Can Tho, and prepare a national strategy for expanding supports for better housing and services to the urban poor (World Bank 2003). Following the consultation with World Bank and other partners, the government launched the Vietnam Urban Upgrading Project (VUUP), a 10-year, USD 417 million project which is partly funded by international donors to improve living conditions for the poor and make planning processes more inclusive and pro-poor in urban areas. It was

**Table 1** Government policies to control and guide urban development in Vietnam

| Urban Development Policies | Consequences |
|---|---|
| Central control of administrative boundary shifts | From 1954 to present, administrative boundary changes require approval from the central government. This has historically been viewed as an effective tool to control city size and encroachment of urban areas onto agricultural lands. However, with rapid urbanization since late 1980s, the loss of agricultural land to urban use is increasing conflicts at the urban fringe of many cities |
| Controlling population movements | Demographic transition has been largely controlled by the urban residency permission system. Since 1990 this policy has been relaxed leading to the rise of urban population rose from 19.5% in 1990 to roughly 30% in 2009 |
| Urban service provision and the welfare transition | Since the 1990s and the 2000s, reforms in service provision have been made to allow for cost recovery in tariffs and an orientation to commercial practices. This has led to increased access to basic services across all urban classifications. Quality of services remains a problem |
| Urban finance and the economic transition | Urban construction finance has been largely controlled by the state. This has had a positive impact on equity among regions. Yet many cities struggle to make essential infrastructure investments |
| Land markets and the physical transition | The 1993 Land Law was a step forward to release land for housing market. Conversion of farmland to urban use accelerated rapidly, though it was chaotic due to low levels of legally recognized land use rights and many informal transactions |
| Transition towards pro-urban policies? | The Government Decree No. 72 (2001) and Decree No. 42 (2009) set up city and town classification requirements in an attempt to distinguish between the roles of different cities. The classification system has implications for administrative functions, tax collection and state funding allocations. The 2011–2020 Socio Economic Development Strategy states that urbanization will be necessary to promote the country's goals of industrialization and modernization |

*Source* World Bank (2011b)

piloted in four main cities (Hai Phong, Nam Dinh, HCMC, and Can Tho), serving as a test case for the national urban upgrading program.

With the success from Phase One, the government has recently authorized Phase Two of the program that extends to seven cities in the Mekong Delta (Vinh Long, Ben Tre, Long an, Hau Giang, SocTrang, An Giang, and Bac Lieu) (Government Office 2016). Given the success of the pilot project, in 2009 the government announced the Master plan on Vietnam's urban development to 2025 with the prospect to 2050, specified in Decision No. 445/QĐ-TTg. In this amended master plan, the government raised significant goals for urban development, such as basic areas per person, the rates of land used for traffic infrastructure, and so on.

As Hanoi and HCMC are the most important urban areas in Vietnam and have special status, these two cities also have their own programs in urban development and urban poverty reduction. For example, Hanoi built the Plan on Housing Development

**Table 2** Access to basic services in urban areas in Vietnam

| Urban services | Accessibility |
|---|---|
| Electricity | Vietnam has achieved high coverage of electricity from just 14% of total households in 1993 to above 96% in 1993–2009 period. In urban areas, 100% households have access to electricity |
| Clean water and sanitation | Data from 65 utility companies show that more than 67% of households in urban areas had access to a toilet |
| Education | Vietnam has achieved high primary education enrolments in both its urban and rural regions (almost 90%). However, national averages mask inequalities among regions and minorities |
| Urban transport and land use | 80% of all trips in Vietnam's major cities are made by private vehicles, creating huge problems for Vietnam's main urban areas as the car ownership has been rising rapidly in recent years. Hanoi and HCMC are building their first metros and trying to improve the bus system. However, these efforts are strained by the lack of financial resources and inexperience of urban planning |

*Source* World Bank (2011b)

2012–2020 towards 2030, approved by the Prime Minister Decision No. 996/QĐ-TTg, in which Hanoi set the goals to demolish makeshift and unsafe housing by 2020.

Overall, Vietnam has been doing relatively well in terms of basic services such as electricity, water, and education. However, the country is seeing more difficulty in pushing for more hi-quality services. Details can be seen below (World Bank 2011b) (Table 2).

## *1.6 Different Schemes and Programs of NGOs for Slum Improvements*

Activities of many international NGOs, particularly in the areas of urban poverty reduction and supporting marginalized urban communities, are focused on Hanoi and Ho Chi Minh City (HCMC). The most notable organizations include Oxfam, ActionAid, Plan International, Care International, among others. Oxfam and ActionAid, for example, did a five-year project on urban poverty (2008–2012) (Oxfam and ActionAid 2012).

Vietnamese NGOs have also been playing a more active part in slum improvements in Hanoi and HCMC. Different groups of NGOs, ranging from LGBTs to education, poverty reduction to rights groups, have made relentless efforts in helping urban poor and marginalized communities. This is possible because NGOs in Hanoi and HCMC usually enjoy more freedom than social organizations in other provinces. However, activities from local NGOs remain modest and do not have an effective cooperation mechanism among different stakeholders. In addition to their lack of capacity and financial resources, the fact that Vietnam's civil society is strictly controlled by the state and has limited space for activities also contributes to the NGOs' underperformance.

## 2 Institutional Mapping and Mechanisms

### 2.1 *Institutional Mapping and Functional Mandates and Boundaries of the Agencies*

According to the 2015 State Budget Law, the National Assembly approves the annual budget of the Government as well as allocating to each public sector. The ministries and agencies, including local departments, are responsible for overseeing the implementation of general regulations of their specialism. For example, the Ministry of Planning and Investment is responsible for issuing investment licenses and approving public investment projects and programs, the Ministry of Finance is responsible for setting up fees and taxes, and the Ministry of Home Affairs is responsible for regulations of personnel in the public sector. In addition, given Vietnam's political

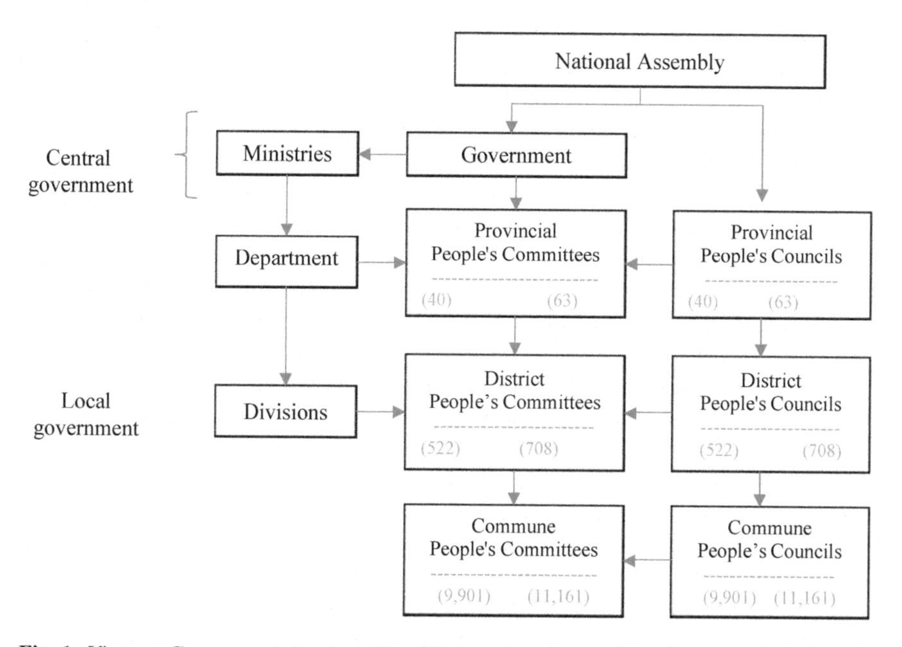


**Fig. 1** Vietnam Government structure. *Note* The numbers in round brackets are of 1986 and in square brackets are of 2015. *Source* Anh (2016)

system of a one-party regime, CPV frequently issues guidelines and directions in specific areas such as public management, culture, education and health. As such, state agencies are in fact under two governance system: the state and the party (Fig. 1).

Before Doi moi (1986), the government controlled almost all service activities, ranging from trade, finance, tourism and transportation to education, health, security and defense. After 1986, the government has allowed the private sector to engage in the provision of certain services such as security, aviation and notary services. However, in the current context of Vietnam, state agencies remain the primary public service providers. Public services are provided by "public service companies" and "state non-business organizations", which are established under state agencies such as the Government, Ministries, departments, People's Committees. According to (CIEM 2007), public services in Vietnam can be divided into three groups: (i) services are directly provided by state organizations such as public security, defense, public health and other service groups; (ii) services are provided by non-governmental organizations and private entities under government mandates such as the construction of public projects (public buildings, roads and other type of infrastructure); and (iii) services are provided via public-private partnership such as education, health, water and sanitation to the community with joint efforts between the Government, non-governmental organizations and community organizations.

Also regulated in the 2015 State Budget Law, at the local level, the government is responsible for providing and executing public expenditure tasks for public service activities. Specifically, these are: (i) education—training and vocational training; (ii) science and technology; (iii) national defense, security, social order and safety, and local management; (iv) health care, population and family; (v) culture and information; (vi) radio and television broadcasting; (vii) physical training and sports; (viii) environmental protection; (ix) economic activities; (x) activities of state management agencies, political organizations and socio-political organizations to support the activities of socio-political and professional organizations, social organizations, socio-professional organizations in accordance with the provisions of law; (xi) social security expenditures, including expenditures on implementation of social policies in accordance with law; (xii) investment in projects managed by localities in the areas specified in Clause 2 of this Article; (xiii) to invest and provide capital support to enterprises providing public goods and services ordered by the State, economic organizations and financial organizations in their respective localities in accordance with the provisions of law; and (xiv) other expenses as prescribed by law.

It is important to note, as aforementioned, that the current structure does not cover "spontaneous" migrants, as the government wants to discourage this form of migration. The government believes that spontaneous migration creates pressures on the overloaded urban social services, infrastructure and employment capacity, as well as social orders. As such, institutionally, there is no government agency to deal with the issues related to spontaneous migration. For example, the Ministry of Public Security deals with *ho khau* registration and management, and the Ministry of Labour-Invalids and Social Affairs (MOLISA) do not have a policy tailored to the particular risks posed to these migrants.

**Participation**—According to the Law on the Election of Deputies to the Parliament and the People's Council (Law No. 85/2015/QH13), the election of Parliament deputies and People's Council deputies shall be conducted on the principle of popularity and equality, by direct and secret ballot. However, in reality, the election of deputies to the Parliament as well as the People's Councils from the provincial to the commune level are based on the *Party nominates—people vote* mechanism. The election of these vacancies is not democratic, competitive, and still heavily planned to ensure representation (World Bank 2016). Ordinary people do not vote directly for the leaders of these agencies, except chiefs at the village/street level. This is defined in the Joint Circular No. 01/2005/TTLT/BTTUBTWMTTQVN-BNV which instructs the process of electing, dismissing village/block chiefs. Accordingly, the Circular states that the representative of the household or all eligible voters in accordance with the regulations will participate in the election in the form of a show of hands or secret ballot. Article 13 of the Ordinance No. 34/2007/PL-UBTVQH11 on the implementation of democracy at the commune, ward and township levels specifies that the village chief is elected by the people in that village. However, there are no regulations guaranteeing that the migrants, who do not have permanent residence registration, will be allowed to participate in the election process of the village chief where they live. According to our survey results, the percentage of voters in the residential quarters was rather low, only 32.5% of the respondents participated in the election for at least once.

**Accountability**—The Ordinance on the Implementation of Democracy at the grassroots level has focused on communal governments and stipulates in the motto "people know, people discuss, people implement, and people supervise." It is applied in various activities at the commune level, such as the use of public budgets and contributions, socio-economics development plans projects and plans for public investment, land use management, legal documents, administrative procedures and many other activities. In order to ensure the accountability of local budget, Provincial People's Councils must approve the allocation of local budget revenues to lower levels, determine criteria and mechanisms for allocation and collection of fees and contributions of local citizens. Local government financial supervision has also been strengthened by the establishment of the State Audit (in 1994) with the function of an independent audit of all state budget revenues and expenditures, including local budgets. Local governments are accountable to the superiors for budget use. The central authority regulates certain criteria that provincial governments must apply when allocating funds to achieve development goals.

**Transparency**—Since 1998, a number of regulations have been enacted to implement grassroots democracy, aiming to encourage people's participation in public meetings. However, these regulations have not yet made any evident changes in the structure of commune governments (World Bank 2016). Article 5 of the Ordinance 34/2007/PL-UBTVQH11 requires a set of information to be publicized under this ordinance, including regulations on public finance transparency, revenues and expenditures of commune governments. Article 6 stipulates that the commune authorities should apply the following forms of publicity: (i) listed publicly at the headquarters of the People's Council, the People's Committee; (ii) publicity on the commune radio system; (iii) publicity through the village head, the head of the residential quarter to inform people.

In 2004, the regulation on public finance transparency was promulgated under the Prime Minister's Decision No. 192/2004/QD-TTg, requiring local governments to publicize their estimation and actual use the annual state budget, supplements from the higher level budget, the collection and payment from the contribution of their citizens. In addition, other mechanisms may also enhance the transparency of local governments, such as the mandatory declaration of income and assets for high-level local officials and the obligation to respond citizen's inquiries within 10 days.

**Access**—Public services in developing countries are often subject to criticism for lack of adequate infrastructure, lack of reliable service providers in areas such as education and health care, with issues of low quality, corruption, non-transparency, bias and discrimination (MUTRAP 2009). In Vietnam, electricity and access to water used to be serious problems, negatively affecting the lives of people in the low-income residential areas. The situation has improved greatly over the past decade, particularly with electricity as nearly 100% urban population has access to electricity.

However, in marginalized communities of several low-income areas there are still problems with access to basic public services, as shown in our survey. Many houses there do not have direct access to water and the connection to the sewage system is limited. At the time of the survey, access to water and electricity services of households living in slums along canals and boarding houses in HCMC are easier but remains an issue (In-depth interviewed with Government officer in HCMC).

Related to sanitation issues, most surveyed households do not have septic tanks. The sanitary area is made in a diaphragm, covered by old iron sheets or plastic sheets and waste are discharged directly into the canal. Rented households do not have their own toilets, about 5–6 households share one. Local authorities have limited financial support to build toilets for disadvantaged households. However, the amount is usually not enough, and the disadvantaged households have to pay additional amount of money, which is quite considerable compared to their income.

The survey results show that 76.3% of respondents have access to educational services. Yet there is still a large percentage of people who do not have access to education because of economic difficulties. Although access to basic education is still available, there is a high probability of dropping out of school because of insufficient funds to cover tuition fees. According to the Vietnam 2035 Report (World Bank 2016), the poorer a student is, the less likely he/she graduates from high school or goes on to college or university compared to other students.

Access to health services is divided. According to the Vietnam 2035 Report, 20% of the poorest account for about one third (33%) of visits to medical examination and treatment at the commune level. Meanwhile, 20% of the richest make up nearly half (46%) of visits to central hospitals (World Bank 2016), where service quality is much better. The urban poor still face more health risks than the rural poor, with particular regards to the quality of the living environment, traffic safety, air quality, water supply, sanitation and solid waste management, while infectious diseases are easier to be widespread in urban areas such as dengue fever and pulmonary tuberculosis (World Bank 2016).

As temporary residents have no residence registration, migrants' access to local community institutions and activities are limited. Their general social exclusion and isolation is evident in several ways: difficulties in finding jobs, low and unstable income, poor living arrangements, home sickness, poor health care and labour exploitation. Half of the surveyed respondents said that they took no actions to address these problems, while the other half rely on the pre-existing social network of relatives and friends for support. Almost no migrants seek help from official sources, even regarding their safety, which is one of their key concerns.

## 3 Analysis Based on the Surveys Data from the Selected Low-Income Settlements

### *3.1 Demographic of Survey Households*

A survey of a selected slum settlements in Hanoi and HCMC was conducted to assess practical barriers to political and social inclusion of the marginalized groups. In each city, 40 households who are living in makeshift housing and slums were interviewed. The research sample is mainly focused on the poor working class who have limited living conditions. The average age of respondents is 44.9 and the gender of the respondents are relatively balanced between male and female (46.3% and 53.8% respectively). Most of them only finished secondary school and lower education level. Regarding marital status, 63.8% of them are married. The research team also asked households for other information such as housing status, toilets types and drinking water supplies. According to the survey, there are 86.3% of households with roofs made from metal sheets, 73.8% of floors are paved and 67.5% of houses are built from brick. In terms of house ownership, only 43.8% are privately owned and 46.3% are rental properties. Almost all respondents in Hanoi and HCMC use daily clean water sources such as tap water or bottled water. Only 56.3% of respondents said their families have their own toilets. Most of households' members living in slums in HCMC are unskilled labors, who receive rather low and unstable income.

### *3.2 Community Organization and Participation*

The survey finds a very low level of community participation among the migrants. For example, less than 5% of the interviewed migrants participated in community events, such as sports and cultural activities, or local meetings to learn about government policies and programs. By doing so, migrants exclude themselves from useful information and potential supports. When being asked why, the majority answered that they did not fit the official categories that would permit them to attend, and that they are generally not invited by the local authorities. Instead, they are often informed about the local news and activities via their landlords, who attend the local meetings.

At several surveyed communities, the landlords are requested by the local authorities to "keep an eye on tenants [migrants]". The security concern for the local community is often mentioned by local police in the regular meetings of residential clusters. Migrants are sometimes referred to as those making the community 'unsafe". This reinforces the negative view of the community towards them. Migrants consider themselves as "outsiders" and have minimum or no contact with local residents.

As regards the number of participation in local election, 32.5% of respondents take part in at least once, meanwhile 67.5% have never voted. The main reason for not participating in the community affairs of the migrants and ethnic minorities is that they are too busy to earn for their living and they are not local. The participation of women in the public affairs of the community/village was quite positive, as 56.3% of respondents frequently participate while only 20% said they rarely do. 77.5% of respondents said that the participation of women in the public affairs of the community/village is equal to that of men.

In terms of their willingness to participate in community services, 67.5% of respondents said "yes", and about 58.8% of respondents are willing to encourage other community residents to do so. These rates in Hanoi were 85% and 70% respectively, much higher than that of HCMC (only 50% and 47.5% respectively said "yes").

### 3.3 *Vulnerability Assessment*

The highest risks that the respondents in the surveyed areas have been confronted with are flood and fire. Up to 57.5% of respondents said that they have encountered flood. This rate in Hanoi is 32.5 and 82.5% in HCMC. Meanwhile, 12.5% of respondents have dealt with fire. This rate in Hanoi is 20% and only 5% in HCMC. In the question "Are the service provision projects of the government or NGOs targeted for youth, women, migrants or ethnic minorities?", 56.3% of respondents choose "Yes". However, up to 31.3% said that they "Did not know". These rates in Hanoi are 52.5 and 27.5%, while in HCMC the numbers are 60% and 35%, respectively. Only 13.8% of respondents ever participated in any open budget sessions of the local government. This rate in Hanoi is modest, at 5, and 22.5% in HCMC.

### 3.4 *Access to Urban Services*

92.5% of respondents have access to safe drinking water, 94.9% to community level sanitation and waste disposal, 83.8% to primary health care facilities, 76.3% to education facilities, 98.8% to food, 97.5% to electricity and 75% to toilets. It is worth noting that the proportion of respondents who have access to primary health care and education is the lowest of all services. Many households said that they do not have enough money to go to the clinic or send their children to school. In terms of the expected role of local government in dealing with service provision and access, the respondents want the local governments to prioritize on children's education, support for the elderly and people with disabilities, health care, livelihood, housing, door-to-door awareness campaign and preferential loans.

### 3.5 *Level of Overall Satisfaction*

In terms of overall assessment on safety/level of confidence, drinking water and community public safety are rated as "high" and "very high" by the respondents. Other issues, namely environment, living place, future of their family, government policies, local government, community leadership, and governance in general, are mostly rated at medium level. Regarding the overall satisfaction about the government performance, 75% of respondents are satisfied with the community/village leadership. 57.5% are satisfied and 22.5% are somewhat satisfied with local government. Meanwhile, nearly a half of respondents are neither satisfied nor dissatisfied with the government agencies at high level such as the central government, provincial and municipal governments (Table 3).

In term of institutional features of local government, most characteristics get medium or high rate of satisfaction from the respondents. Particularly, gender sensitivity is highly appreciated (47.5%). Local government also is also highly rated regarding providing basic urban services, including support for the marginalized groups. 23.8% of respondents evaluate corruption of local government at high level while 23.8% evaluate the issue at low level. However, corruption is still considered as a sensitive and complex matter. When being asked about this issue, respondents often avoid answering directly and said that they do not know (up to 33.8%). Admittedly, it is not easy for the respondents to make a fine assessment on this characteristic

**Table 3** Overall satisfaction about the performance of the organizations

| Both provinces | Somewhat satisfied | | Neither satisfied nor dissatisfied | | Satisfied | |
|---|---|---|---|---|---|---|
| | No. | % | No. | % | No. | % |
| Central government | 11 | 13.8 | 37 | 46.3 | 32 | 40.0 |
| Provincial government | 10 | 12.5 | 39 | 48.8 | 31 | 38.8 |
| Municipal government | 9 | 11.3 | 36 | 45.0 | 35 | 43.8 |
| Local government | 18 | 22.5 | 16 | 20.0 | 46 | 57.5 |
| Community/village leadership | 12 | 15.0 | 8 | 10.0 | 60 | 75.0 |
| NGOs | 2 | 2.5 | 70 | 87.5 | 8 | 10.0 |

*Source* Results of the survey

when they do not have regular contact with government officers, in addition to their fear of appearing hostile to the authority (Table 4).

**Table 4** Overall assessment of some selected institutional features of local government

| Both provinces | Low | | Medium | | High | | Don't know | |
|---|---|---|---|---|---|---|---|---|
| | No. | % | No. | % | No. | % | No. | % |
| Corruption | 19 | 23.8 | 15 | 18.8 | 19 | 23.8 | 27 | 33.8 |
| Quality of service delivery | 4 | 5.0 | 33 | 41.3 | 41 | 51.3 | 2 | 2.5 |
| Politicization of service delivery | 8 | 10.0 | 32 | 40.0 | 35 | 43.8 | 5 | 6.3 |
| Transparency of activities | 10 | 12.8 | 22 | 28.2 | 19 | 24.4 | 27 | 34.6 |
| Level of accountability | 8 | 10.0 | 26 | 32.5 | 29 | 36.3 | 17 | 21.3 |
| Community participation | 9 | 11.3 | 35 | 43.8 | 31 | 38.8 | 5 | 6.3 |
| Level of trust by community members | 4 | 5.0 | 35 | 43.8 | 35 | 43.8 | 6 | 7.5 |
| Gender sensitivity | 0 | 0.0 | 33 | 41.3 | 38 | 47.5 | 9 | 11.3 |
| Responsiveness to special needs of marginalized groups | 2 | 2.5 | 24 | 30.0 | 36 | 45.0 | 18 | 22.5 |

*Source* Results of the survey

## 4 Factors that Influenced Performance

A number of factors have contributed to effective performance of Vietnam in delivering services to the urban residents.

### *4.1 Human Resource Capacity*

Reform of public employment has been one of the priorities of Vietnamese government after 1986. The government approved the General Program on Administrative Reform, including three phases (1991–2000, 2001–2010, and 2011–2020), which partly aims to enhance the quality of public employment. There have been about 200 regulated standard titles for cadres and public workers, which play an important role in the management of public employees. Particularly, the Ordinance No. 11/2003/PL-UBTVQH11 prescribes the titles and criteria for people who are recruited and assigned professional titles at the commune level, which speeds up the standardization and training for public employees at grassroots level.

According to the data from the portal of public administrative reform of the Ministry of Home Affairs, during 2001–2005, the total number of trained cadres and public employees was about 2,510,000 (around half of public employment) of which 407,000 employees received training in political theory, 894,000 in state management and 1,076,000 in specialized knowledge. In terms of training in languages and information technology, the number were 37,000 and 96,000 respectively. The training materials for public employees are differentiated, depending on the level of seniority of public officers. Public data also showed that in 2004, there were about 292,000 delegates of People's Councils at different levels being trained to enhance their knowledge and skills. The government also initiates different policies—including state scholarships, cash payouts, and subsidized accommodation—to attract more high-quality labour into the state sector, particularly at local level.

## *4.2 Integrated Plans and Policy Framework*

After 30 years of Doi moi policy (since 1986), the capacity of the government has been greatly improved. The procedures of policymaking and lawmaking are gradually upgraded which depends more on evidence-based impact evaluations, consultation of targeted subjects and the general public. This procedure is defined in the Laws on Promulgation of Legislative Documents in 2008 and 2015.

Administrative reform has been implemented since 1990s, with three phases as mentioned in the previous section. The amended 2013 Constitution continued to create more space for legislative, executive and judicial reforms. So far, there have been legal documents on most of basic areas in economic, political, cultural, and social aspects and in terms of organization of the state apparatus (World Bank 2016). The number of laws and ordinances approved by the National Assembly and the Standing Committee of the National Assembly from January 1987 to June 2015 increased more than eightfold compared to those of 41 years before the Doi moi Policy. In more details, there were 63 laws and ordinances released from 2nd September 1945 to 30th December 1986 while this figure for the period from 1st January 1987 to 30th June 2015 was 524 (World Bank 2016).

## *4.3 Political Support and Coordination of the Political System*

As a one-party regime, the Communist Party of Vietnam (CPV) plays the pivotal role in leading and guiding the cooperation among state agencies. This is reflected in the forms of guidelines and policies, and the organization of the state apparatus (Kim 2006). Since the CPV 6th National Congress (1986), the Party has emphasized the mechanism of "the Party leads, the State manages, and the people own", which means the state is the enforcer of the Party's policy, while the people hold a largely vague "supervision" role. However, as the economic reform intensifies, there has been gradual perception of the Party leadership on enhancing public participation in the policy making process. The state increasingly encourages the community to participate in different forms, such as consultation meetings, Q&A with National Assembly and People's Council members, lax policy on associations, and so on.

## *4.4 Structural and Institutional Barriers to Full Engagement*

According to Oxfam (2015), one chronic barrier which prevents migrants and their families from public service access (health care, education, housing and clean water) is regulations in budget allocation. Budget allocation is currently based on the number of people who register for permanent residence in the area, which creates burdens for areas with high proportion of migrants, posing pressures on infrastructure. Therefore, local governments will prioritize permanently resided people, especially in terms of health care and education service access.

Administrative procedures remain complicated and are difficult to be implemented, which hinders the access to services of beneficiaries. There are many procedures which require people to register for permanent residence, making it difficult for migrants to benefit from social welfare policies. Examples include access to electricity, clean water as well as housing procedure, and policies for supporting poor households. According to the study entitled "Legal and Practice Barriers for Migrant Workers in their Access to Social Protection"[2] by Oxfam in 2015, about two third of migrant workers have to pay three times higher than the official rate for using clean water and about twice higher for using electricity compared to local residents.

The bright side is that in areas where local officials are more flexible and sympathetic in dealing with administrative procedures relating to migrants, they will have more chances to integrate into the community and have access to social welfare programs such as job introduction, vocational training, poverty reduction, health care and education services.

## *4.5 The Barrier of Household Registration System*

The government's residence-based social policy has created many barriers for marginalized groups, who mostly consist of migrants, from accessing urban services including water and sanitation, health, education, and other socio—political rights that permanent residents enjoy. The government, with expectation to have tight control over the population that they did under Vietnam's Soviet-style state management before 1986, discourages spontaneous migration as it would make it harder for population control. As a result, the government uses housing policy to prevent temporary migrants from purchasing and possessing residence. This residence-based policy depends on *ho khau,* a Vietnamese version of Chinese *Hokou* system.

[2]The study of Oxfam was implemented in 2015 four provinces including Hanoi, Bac Ninh, HCMC and Dong Nai with a total of 808 respondents who were migrant workers.

This strict policy has made it almost impossible for migrants to own a house, consequently, almost all the migrants have to rent a residence. Most of them seek accommodation in low-quality houses in poor neighborhoods with poor infrastructure. In addition, higher charges are normally applied to temporary residents for electricity and clean water, and other charges such as fees for rubbish collection, community sanitation, road/lane reparation, security, and so on. Migrants without *ho khau* also have restricted access to affordable health care services, as they are not allowed to buy health insurance at their temporary residence, but at their homeland. It is thus not surprising that migrants are more susceptible to illnesses and diseases. The situation is getting worse with the increasing privatization of Vietnamese health service, which inevitably leads to rising costs of meditation.

According to Vietnam 2035 Report, Vietnam witnesses the discrimination between urban residence and migrants (World Bank 2016). Residence registration through *ho khau* has shrunk the opportunities of urban migrants to access to public services, especially education and health care services. People living in the urban areas without permanent residence registration will confront challenges in health care, education, social welfare services, as well employment and social relationship (World Bank 2016). According to Marx and Fleischer (2010), *ho khau* is considered as a prerequisite of basic service access and of some administrative procedures. In Vietnam, schools are often overloaded, so it can only receive children with permanent residence registration in the area (Vietnam Academy of Social Sciences 2015). Children without *ho khau* or only having temporary residence registration will have to pay unofficial fees to enroll in public schools. Some cannot attend schools or have to pay a high amount of money to study in private schools (Oxfam and ActionAid 2012). Unofficial fee is a burden for migrant workers, especially when this group often has difficulties in finding jobs and ensuring stable income.

Access to health care and support for the poor are challenges for those without permanent residence registration. According to Haughton (2010), such people rarely seek professional health care services and only a minority of them register to health insurance. Children at the age of 0–6 years old without permanent residence registration also have difficulties in having free health care service which they are entitled to. According to the Residence Registration Survey in 2015, in HCMC, there were 2.7 million people without permanent residence registration while the figure for Hanoi was 1.2 million people. Such large figures indicate the fact that there are a lot of migrants from rural to urban areas despite barriers and high living costs (World Bank 2016).

### 4.6 *Efforts to Remove Barriers*

The Vietnamese government has taken its first steps to apply a new population management system as an alternative to the current household registration system. Since 2008, the government has assigned the Ministry of Public Security to construct the National database on population, which is expected to be accomplished by 2020.

This database will be used to build a system of personal identification number for Vietnamese citizens, incorporating various identification papers including citizen card and household registration. Accordingly, the government issued Resolution No. 112/NQ-CP in 2017, which removes the household registration book system as a population management instrument, replacing it with the personal identification number. Nonetheless, it will take a period of transition for this new method to come into effect. Additionally, it is important to note that while the household registration books might be abolished, the control mindset lives on. In the press conference announcing the plan in 2017, the Ministry of Public Security argued that they only change the method of population control "from paper to digital form", which means the residence-based welfare policy might not be abolished along with *ho khau*.

## 5 Conclusion

Despite the economic boom after adopting the market economy reforms, Vietnam has remained a highly centralized state, with strict control over the social, economic, and social life of its citizens. State agencies are still the primary public services provider, although the government has allowed private sector to engage in the provision of certain services. Rapid urbanization has produced an exodus of migrants from the countryside to urban areas, particularly to big economic centers such as Hanoi and HCMC. Internal migration is mainly one way: most migrants first come to cities for jobs and then bring their families with them to reside there permanently. They are the main components of the marginalized groups in urban areas.

On the one hand, the government, both at the central and local levels, has put a great deal of efforts in solving the issues of social inclusions for marginalized groups. Access to basic services like electricity, water, health and education is now much better compared to the pre-*Doi moi* era. Housing is a particular area where government policies have produced positive results. In both Hanoi and HCMC, the local governments have specific plans to remove makeshift housing and provide affordable housing for marginalized groups. As regards political participation, there are efforts, albeit small, in encouraging migrants to participate in the socio-political life at the grassroots level as shown in our surveys and in-depth interviews. The survey results show that 92.5% of respondents had access to safe drinking water, 94.9% to community level sanitation and waste disposal, 83.8% to primary health care facilities, 76.3% to education facilities, 98.8% to food, 97.5% to electricity and 75% in terms of restrooms. The survey, in line with other previous studies, shows that there is not much difference in public service deliveries between the two biggest economic centers of Vietnam. This confirms the limit of administrative decentralization, when urban governments cannot fully design their systems according to their own needs. However, for the past decade, both cities have been lobbying for "specialized mechanisms" which allow the city governments greater freedom to craft their welfare policies. The National Assembly has been discussing a potential "trial" of a new urban management model for both cities, which will definitely empower local authorities in terms of financial resources and policy independence. How this will help solve the problems mentioned in this chapter, however, remains to be seen.

On the other hand, the most serious policy challenge for Vietnamese government is to change the residence-based approach in social policy. This approach has created a few barriers for marginalized groups, most of whom are migrants, from accessing urban services including water and sanitation, health, education, and other socio—political rights that urban residents enjoy. For example, according to the survey results, only 32.5% of respondents had participated at least once in voting for their community/village leaders, meanwhile 67.5% had never voted. Despite recent efforts to remove the household registration system, there is little sign that the government will change the residence-based approach, given the worsening overpopulation problem and lack of resources in big cities.

## References

Anh VTT (2016) Vietnam decentralization amidst fragmentation. J Southeast Asian Econ 33(2):188–208

Boushey H, Fremstad S, Gragg R, Waller M (2007) Social inclusion for the United States. Center for Economic and Social Inclusion, London. http://inclusionist.org/files/socialinclusionusa.pdf. Accessed 10 Aug 2013

Cappo D (2002) Social inclusion initiative. Social inclusion, participation and empowerment. Address to Australian Council of Social Services National Congress, Hobart

Centre for Urban Forecast and Research (PADDI) (2014) Urban planning in Vietnam: comparison of method, tools, and implementation of urban planning in France and Vietnam. Labour-Society Publishing House, Hanoi

CIEM (2007) Public service reform in Vietnam

European Commission (2004) Joint report on social inclusion. Report 7101/04. Brussels: European commission. http://ec.europa.eu/employment_social/socprot/soc-incl/joint_rep_en.htm. Accessed 12 Jan 2013

General Statistics Office (GSO) (2017) Data on population and labour. Statistics Publishing House, Hanoi

General Statistics Office (GSO) (2018) Statistics on disposing solid waste and waste water by types of urban areas. http://www.gso.gov.vn/default_en.aspx?tabid=783. Accessed 12 Jan 2018

Government Office (2016) Mở rộng nâng c´âp đô thị Việt Nam [Upgrading Vietnamese urban areas]. Retrieved from Van Phong Chinh phu [Government Office]: http://vpcp.chinhphu.vn/Home/Mo-rong-nang-cap-do-thi-Viet-Nam/201610/20119.vgp

Haughton J (2010) Urban poverty assessment in Hanoi and Ho Chi Minh city. United Nations Development Programe, Hanoi. http://dl.is.vnu.edu.vn/handle/123456789/94

Hayes A, Gray M, Edwards B (2008) Social inclusion origins, concepts and key themes. Australian Institute of Family Studies, prepared for the Social Inclusion Unit, Department of the Prime Minister and Cabinet, Canberra

Kim VT (2006) Strengthening the organic relationship among the party, the state, the Fatherland front and mass organizations to create a good foundation for implementing the mastery of the people. People's Daily, May 20th, 2006

Levitas R (2003) The imaginary reconstitution of society: Utopia as method. Paper presented at the University of Limerick/University of Ireland, Galway, as part of the Utopia-Method-Vision project led by Tom Moylan

Marx V, Fleischer K (2010) Internal migration: opportunities and challenges for social economic development in Vietnam. United Nations Vietnam, Hanoi
MUTRAP (2009) The overall strategy for the development of the service sector in Vietnam by 2020 (CSSSD) and vision to 2025
Ninh KNB, Anh VTT (2008) Decentralization in Vietnam: challenges and policy implication for sustainable growth. Research paper prepared for Vietnam competitiveness institute (VNCI) of United State Agency for International Development (USAID)
Oxfam (2015) Summarized report: legal and practice barriers for migrant workers in their access to social protection. Program on work rights in Viet Nam. http://www.oxfamblogs.org/vietnam/2015/12/22/bao-cao-cua-oxfam-ve-rao-can-phap-luat-thuc-tien-doi-voi-nguoi-lao-dong-di-cu-trong-tiep-can-an-sinh-xa-hoi/
Oxfam and ActionAid (2012) Participatory monitoring of urban poverty in Vietnam: five years synthesis report (2008–2012). Hanoi
Stewart A (2000) Social inclusion: an introduction. In: Askonas P, Stewart A (eds) Social inclusion: possibilities and tensions. Macmillan, London, pp 1–16
UNFPA (2016) National domestic migration surveillance 2015: major findings. News Publishing House, Hanoi
Vietnam Academy of Social Sciences (VASS) (2015) Household registration system in Vietnam
Vietnam Environment Administration (2016) National report on environment
Vu TT, Zouikri M (2011) Decentralization and sub-national governance. Retrieved from Paris X University: http://www.vcharite.univ-mrs.fr/ocs/index.php/LAGV/LAGV10/paper/viewFile/562/34
World Bank (2003) Transforming urban policy-making in Vietnam. World Bank, Hanoi
World Bank (2011a) Social safety nets in Nepal. Draft report. World Bank, Washington, DC
World Bank (2011b) Vietnam urbanization review. World Bank, Hanoi
World Bank (2013) Inclusion matters: The foundation for shared prosperity. New Frontiers of Social Policy, Washington, DC
World Bank (2015) East Asia's changing urban landscape: measuring a decade of spatial growth. World Bank, Washington
World Bank (2016) Vietnam 2035. Toward prosperity, creativity, equity, and democracy

**Nguyen Duc Thanh** is President of Vietnam Institute for Economic and Policy Research (VEPR), a think tank he co-founded in 2008. VEPR is widely recognized in Vietnam as a strong advocate of the market economy reform, civil society empowerment and the implementation of the rule of law. He earned his Ph.D. in Development Economics from National Graduate Institute for Policy Studies (GRIPS), Tokyo, Japan in 2008. Nguyen Duc Thanh was a member of the Economic Advisory Group to the Vietnamese Prime Minister during 2011–2016. He has been publishing extensively in academic journals and been involved actively in the country's policy debates. He is the founder and chief editor of the influential *Vietnam Annual Economic Reports*, which has been annually published by VEPR since 2009.

**Pham Van Long** is a researcher on microeconomics and development issues at the Vietnam Institute for Economic and Policy Research (VEPR). He obtained his M.A. degree in Policy Analysis at Fulbright School of Public Policy and Management (HCMC, Vietnam) and a B.A. in Economics from Vietnam National Economics University, Hanoi. His research interests include public policy and industrial economics. He was a co-author of a book chapter on Vietnam's productivity enhancement through international labour market integration in *Vietnam Annual Economic Report 2018: Understanding Labor Market for Productivity Enhancement* published by Vietnam National University Press in Hanoi.

**Nguyen Khac Giang** is a Ph.D. candidate in political science at Victoria University of Wellington, New Zealand, and a senior research fellow at Vietnam Institute for Economic and Policy Research (VEPR), Vietnam. He has a B.A. in International Economics at Ha Noi Foreign Trade University, and M.A. in Media and Globalization at Aarhus University (Denmark) and City University London (UK). He writes regularly for major Vietnamese news media such as VnExpress, Vietnamnet, and Saigon Times. His work also appears on the Asia & Pacific Policy Studies and the East Asia Forum. His area of expertise is democratization and Vietnamese politics. His recent publication includes "Civil Society in Vietnam—an Institutional Approach" (Danang Publishing House 2018).

# Serving Africa's Citizens: Governance and Urban Service Delivery

**Camilla Rocca and Diego Fernández Fernández**

**Abstract** Between 1950 and 2050, Africa's urban population is expected to multiply by nearly 46 times, from 32.7 million to 1.5 billion. In 2040 African cities will host around a billion people, equivalent to Africa's total population in 2009. This urban growth is characterized by a massive youth surge, while economic growth is often taking place without job creation; inequalities are widening; and per capita incomes are lower than in other world regions at similar urbanization levels. Moreover, the new century brings unprecedented pressures linked to climate change, global pandemics, worsening security threats and growing migration flows. Growing urban populations put cities at the heart of governance, with the future of the continent thus placed into the hands of its cities. This poses a huge and immediate challenge, putting to test the capacity of local governments to step up public service provision for a population increasingly critical about the access and quality of these services. However, all these intertwined challenges could also constitute a transformative opportunity for Africa. Besides presenting an overview of the present context of urban public service delivery in Africa and an assessment of public service delivery in key sectors—from health to education to housing to waste management—this chapter will focus on urban policies and planning to foster and trigger sustainable and equitable development, improved governance and public service delivery, as well as a renewed sense of participation and citizenship.

**Keywords** Africa · Governance · Urban population · Cities · Youth · Public service delivery · Decentralisation · Urban policies and planning

C. Rocca (✉) · D. F. Fernández
Mo Ibrahim Foundation, 35 Portman Square, Marylebone, London W1H 6LR, UK
e-mail: rocca.c@moibrahimfoundation.org

D. F. Fernández
e-mail: fernandez.d@moibrahimfoundation.org


S. Cheema (ed.), *Governance for Urban Services*,
Advances in 21st Century Human Settlements,
https://doi.org/10.1007/978-981-15-2973-3_12

# 1 Introduction

Between 1950 and 2050, Africa's urban population is expected to multiply by nearly 46 times, from 32.7 million to 1.5 billion. Between now and 2050, the number of people living in urban areas in Nigeria is projected to increase by 184.3 million. In 2040 African cities will host around a billion people, equivalent to Africa's total population in 2009 (UNDESA 2018c, d). The future of the continent is thus in the hands of its cities.

This urban growth is characterized by a massive youth surge. Meanwhile, economic growth is often taking place without job creation; inequalities are widening; and per capita incomes are lower than in other world regions at similar urbanization levels. Moreover, the new century brings unprecedented pressures linked to climate change, global pandemics, worsening security threats and growing migration flows.

This poses a huge and immediate challenge. The already strained capacities of local governments will be put to test, as Africa's urban revolution puts additional pressure to effectively provide traditional and new public services to a population that is increasingly dissatisfied with how African governments are providing education and health services (MIF 2018c). However, properly managed, with sound governance and focused leadership, all these intertwined challenges could also constitute a transformative opportunity for Africa. Focusing on urban policies and planning has the potential to foster and trigger sustainable and equitable development, improved governance and public service delivery, as well as a renewed sense of participation and citizenship.

This chapter examines governance and urban service delivery in Africa. It presents Africa's urban landscape (Sect. 2), decentralization and urban structures for service delivery (Sect. 3), access to, and deficits of, urban services in Africa (Sect. 4), challenges for urban authorities (Sect. 5), and factors and good practices that influence urban governance and access to services (Sect. 6). The final section presents our main conclusions.

# 2 Africa's Urban Landscape

## *2.1 Demographic Trends*

Africa's demographic trends, including a growing youth bulge and urban population growth, are shaping the continent's urban landscape. African urban centers are critical hubs where fundamental public services are provided (e.g. energy, water and sanitation, housing, transport, health, education, waste management) to an ever-growing number of Africa's citizens, with some cities managing populations larger than those of countries.

**A growing African population**—From 1950 to 2015, the population in Africa, now the fastest growing in the world, expanded by more than +400%. Growing at an average rate of +1.6% per year, Africa's population is expected to continue rising until 2100. Between 2018 and 2050, Africa's population will more than double, from

1.2 billion to more than 2.5 billion. During that period, half of the world's population growth will be concentrated in nine countries, five of which are African: Democratic Republic of Congo (DRC), Ethiopia, Nigeria, Tanzania and Uganda.[1] By 2050, 26 African countries are expected to double their current population size. By 2100, six of them are projected to increase it by more than five times: Angola, Burundi, Niger, Somalia, Tanzania and Zambia (MIF 2018a).

**Urban population: exponential growth and specific demands**—In 2019, approximately 567.4 million people in Africa live in urban areas, equivalent to nearly 43.0% of Africa's population. Africa's urban population, as a share of the total population, is still considerably smaller than in Asia, Europe and Latin America & the Caribbean (43.0% compared to 50.5%, 74.7% and 80.9%, respectively) (UNDESA 2018c, f). In 2034, Africa will be the last continent to become on average 50% urban (UNDESA 2018f). This will be around 20 years later than Asia, and over 70 and 80 years later than Latin America & the Caribbean and Europe respectively. Despite this, Africa's share of the global urban population is growing faster than any other continent. In 2050, Africa is expected to host slightly more than ¼ of the global urban population (MIF 2015a).

By 2050, the continent's urban population will amount to nearly 1.5 billion, 58.9% of the total population (compared to nearly 43.0% today) (UNDESA 2018c, f). Africa's urban population growth rate in the period 2015–2020 is the fastest in the world (+3.58% compared to +2.16% in Asia, +1.30% in Latin America & the Caribbean and +0.35% in Europe). Of the ten countries with the fastest urban growth rates in the world for the period 2015–2020, eight are African: Burkina Faso, Burundi, DRC, Ethiopia, Madagascar, Mali, Uganda and Tanzania (UNDESA 2018a).

These growing urban populations will predictably put significant strain on African cities. The delivery of services such as the traditional government functions (security, justice, rule of law), as well as basic welfare needs (education, health), will have to meet a demand in expansion due to a growing population, which is requesting specific public delivery in health, education, transport, housing, safety and security, water and sanitation, waste management, cultural life and entertainment.

Besides having the world's fastest urban population growth rate, Africa is also a steadily urbanizing continent. For urbanization to occur, it is not enough that the urban population is growing, but its growth rate should exceed that of the overall population (MIF 2015a). Even though Africa's urban population growth is the fastest in the world, the rate at which it has urbanized in recent years is slower than that of Asia (1.11% for Africa and 1.39% for Asia in the period 2010–2015). However, by 2025–2030, Africa's urbanization rate should exceed Asia's for the first time since 1985–1990 (1.06% and 0.98%, respectively) (UNDESA 2018b).

**The challenge of youth**—Africa is already the youngest continent in the world. In 2015, more than 60.0% of Africa's population were below age 25, with 41.0% being under 15. The percentage of Africans under 25 will fall only slightly, to 57.1% in 2030 and to 50.4% in 2050, remaining a higher percentage than in other world regions (MIF 2018a). In 2015, 42.7% of Africa's youth (ages 15–24) lived in urban

[1]The nine countries in which half of the world's population growth will be concentrated between now and 2050 are DRC, Ethiopia, India, Indonesia, Nigeria, Pakistan, Tanzania, Uganda and the US.

areas. The number of youth living in urban areas has grown by +262% since 1980, from 26.7 million to 96.5 million in 2015. Between 2010 and 2015, Africa's urban youth population grew by +15.5%. This contrasts with the negative rate of growth seen in Asia for the same period (−0.9%) (UNDESA 2014).

A key challenge within African urban agglomerations will be providing jobs for the growing youth population. There is the potential for African economies to reap the so-called "demographic dividend", as this youth bulge joins the working-age population (ages 15–64). However, if Africa's growth patterns remain unchanged and not enough jobs are created, this dividend could turn into a threat. Consequently, young people may be unable to find a job, increasing unemployment and employment informality. Furthermore, they may be discouraged from actively seeking employment, lowering the labor market participation rate. Apart from economic losses, this could also lead to a brain drain, political and social unrest, instability and armed conflict (MIF 2018c).

## 2.2 *Urban Configurations*

The continent's fast urban growth can have positive or negative impacts: it can either lead to economic growth, structural transformation, and poverty reduction, or, alternatively, to increased inequality, urban poverty, and the proliferation of slums (MIF 2015a).

**City populations equivalent or larger than country populations**—The 20 biggest cities of the continent currently manage populations bigger than many countries. Cairo, Africa's most populous city, has a population that is larger than that of each of the 36 least populous countries on the continent (MIF 2018a).

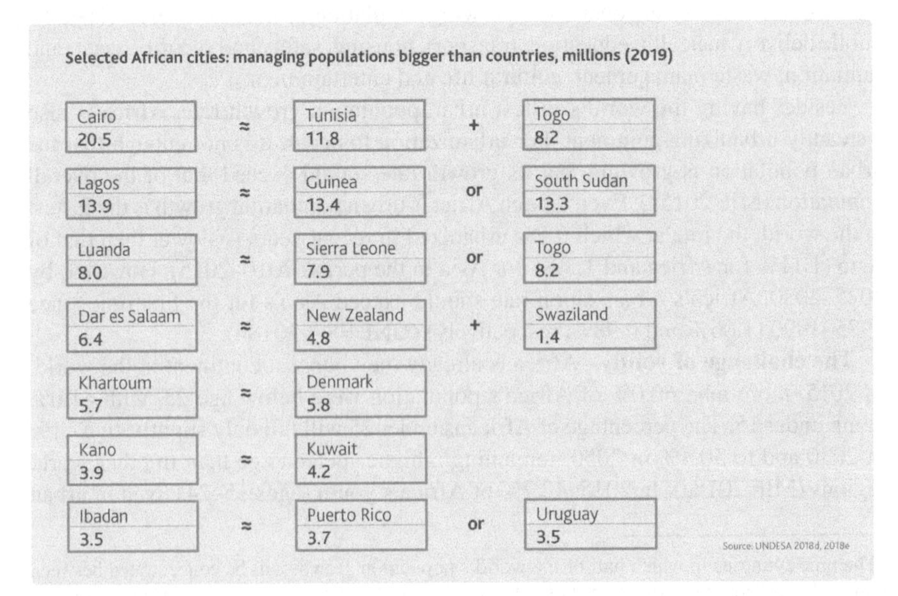

Selected African cities: managing populations bigger than countries, millions (2019)
Cairo 20.5 ≈ Tunisia 11.8 + Togo 8.2
Lagos 13.9 ≈ Guinea 13.4 or South Sudan 13.3
Luanda 8.0 ≈ Sierra Leone 7.9 or Togo 8.2
Dar es Salaam 6.4 ≈ New Zealand 4.8 + Swaziland 1.4
Khartoum 5.7 ≈ Denmark 5.8
Kano 3.9 ≈ Kuwait 4.2
Ibadan 3.5 ≈ Puerto Rico 3.7 or Uruguay 3.5
Source: UNDESA 2018d, 2018e

**The megacity**—A megacity is a metropolitan area with a total population of more than 10 million people. Only 3 of the world's 33 megacities are in Africa: Cairo, Lagos and Kinshasa. China and India each have more megacities than Africa. But, by 2030, it is expected that 5 of the world's 43 megacities will be in Africa: Cairo, Kinshasa, Lagos, Luanda and Dar es Salaam (UNDESA 2018e).

**African cities are mainly middle-sized**—Africa counts 64 cities of 1 million inhabitants or more. Of Africa's 221 urban areas with over 300,000 people,[2] more than 1/3 of them have a population of less than 500,000 people (UNDESA 2018e). Almost 1/2 of Africa's urban dwellers live in settlements of fewer than 300,000 people (UNDESA 2018c, e).

Current trends show that in Africa the fastest urban growth will be in intermediate-sized cities (ETTG 2018), as between now and 2030, 6 of the 10 African cities expected to grow the most have a population between 300,000 and 500,000 (UNDESA 2018e). Africa's largest urban agglomerations are not absorbing the bulk of current urban population growth. Moreover, they are not predicted to do so in the future. Intermediate-sized cities may face constraints in dealing with continued rapid growth. These cities tend to lag behind their larger counterparts in institutional and capacity development. Without adequate urban planning, it is therefore possible that urban slum proliferation may become a feature of these middle-sized cities (MIF 2015a).

## 3 Decentralization and Urban Structures for Service Delivery

Decentralization processes imply various degrees of devolution of power to local authorities, including urban centers, to bring it closer to the citizens. Through decentralization, urban centers take on increased responsibility in managing the lives of people, on political, fiscal and decision-making matters, but also becoming the key providers of services, infrastructure, housing and urban planning.

**Decentralization in Africa**—Decentralization is the transfer of (or part of) the central government's functions to sub-national units or levels of government. There are several degrees of decentralization: (a) deconcentration: opening a branch office in a region; (b) delegation: tasking a sub-national government to carry out functions; (c) devolution: allowing a sub-national government to take over functions autonomously. Likewise, there are several types of decentralization: (a) political: involves the transfer of political power and authority to sub-national units, to give citizens and elected representatives more power in public decision-making; (b) administrative: involves

[2]Unless stated otherwise, when talking about African cities, we are referring to those urban areas with more than 300,000 population (those included in the file "Annual Population of Urban Agglomerations with 300,000 Inhabitants or More" from World Urbanization Prospects: The 2018 Revision").

the transfer of the delivery of social services—namely education, health, social services—to sub-national units, to redistribute authority, responsibility and financial resources for providing public services among different levels of government; (c) fiscal: increases the revenues of sub-national governments through tax-raising powers and grants, and the expenditure autonomy of sub-national governments; (d) economic: through privatization and deregulation, governments shift responsibility from public functions to the private sector, or community groups, cooperatives, private voluntary associations, and other Non-Governmental Organizations (NGOs), also in areas such as service provision and administration (MIF 2018a).

Every African country has at least one sub-national level of government. There is no direct relation between the size of a country's population and the number of its administrative units. For example, both Equatorial Guinea and Tanzania had 30 sub-national administrative units in 2010, with very different sizes of population: over 0.9 million in Equatorial Guinea and over 46 million people in Tanzania. Since 1990, sub-national administrative units in 25 African countries have increased by at least +20%. Eight have more than doubled them between 1990 and 2010, among which are Guinea (from 14 to 341), Niger (from 35 to 256) and South Africa (from 53 to 284) (MIF 2018a).

**Complex layers of government**—Actual powers and responsibilities wielded by the different levels of government differ widely. Despite the wave of decentralization policies during the 1990s, and of constitutional reforms in the 2000s, the actual implementation and devolution programs and plans has been incomplete, inconsistent and sporadic, albeit with some exceptions (e.g. Morocco, South Africa) (MIF 2018a).

Local and urban authorities can be organised in many different ways and given that Africa is a complex myriad of 54 separate countries, there are multiple urban governance models that have emerged within the continent.

In Mali, decentralization involved a redistribution of the national territory. Benin simply transformed existing administrative divisions into territorial units. Furthermore, many countries that undertook territorial reorganization opted for the creation of two or three levels of decentralization (typically: regions, departments/provinces or 'communes'), but states such as Nigeria created only one level of decentralization (Diep et al. 2016).

Kisumu Municipal Council in Kenya is an example of the complexity of urban governance in Africa. Initially a relatively autonomous local government body, located within Nyanza Province, its role and function changed after the restructuring of subnational government by the 2010 Constitution that made counties the main subnational level of government. Kisumu still has a municipal administration (Kisumu City), but although it is meant to have a Municipal Board (according to the 2011 Urban Areas and Cities Act) the Governor of Kisumu County has not officially established this board (as is generally also the case in other counties in Kenya). Kisumu City essentially functions as an agency of Kisumu County, with all Kisumu City officials being employed by Kisumu County. In addition, many national ministries also play a direct role in Kisumu; for example, the National Ministry of Land and Housing allocates land within the city area (Smit 2018).

**The international and continental agenda**—The UN-Habitat's Governing Council adopted the International Guidelines on Decentralization and Strengthening of Local Authorities in 2017 and the International Guidelines on Decentralization and Access to Basic Services for All in 2009. The Habitat III Conference (Quito, Ecuador, 2016) signed the UN's New Urban Agenda setting a new global standard for sustainable urban development with three main operational enablers: local fiscal systems, urban planning, and basic services and infrastructure (MIF 2018a).

At the continental level, the African Union (AU) promotes comprehensive decentralization to achieve the goals from the global and continental development agendas (UN's Agenda 2030 and AU's Agenda 2063, respectively). The 2014 African Charter on Values and Principles of Decentralization, Local Governance and Local Development is the reference for decentralization policies. However, the Charter has to date been ratified by only three countries: Burundi, Madagascar and Namibia. Within the AU, the Technical Committee on Public Service, Local Government, Urban Development and Decentralization gathers Ministers of Housing and Urban Development; and the High Council of Local Authorities represents the voice of local governments in the deliberations of the AU (MIF 2018a).

**'Localizing the Sustainable Development Goals (SDGs)'**—For the SDGs to be relevant to the majority of African people they must be relevant to cities. Unlike the previous Millennium Development Goals (MDGs), the SDGs contained in the UN's Agenda 2030 include a dedicated and standalone urban goal. Goal 11 is to accomplish the following: "Make cities and human settlements inclusive, safe, resilient and sustainable". Furthermore, other goals, such as those on poverty, health, sustainable energy and inclusive economic growth, are intimately linked to urban areas. An integrated approach is crucial for progress across the multiple goals (MIF 2018a).

In 2016 and 2017, 12 African countries, accounting for 6483 Local and Regional Governments (LRGs), submitted national voluntary reviews on the 'localization of the SDGs'. The involvement of LRGs happened at different levels. Sierra Leone involved 19 local councils to integrate the SDGs into their district-level and municipal development plans. Egypt has adopted the City Prosperity Index to monitor the implementation of SDG 11 in 35 cities. In Nigeria the responsibility of data mapping and supply for SDGs indicators is shared with the regions. Nigeria's Kaduna State integrated the SDGs into its State Development Plan for 2016–2020 (MIF 2018a).

**The impact of decentralization: better public service delivery or increased inequality?**—In Ethiopia, decentralization has reportedly improved public service delivery. For example, net enrolments in education, access to basic health services such as antenatal care, contraception, vaccination rates and deliveries by skilled birth attendants have all improved. Child mortality rates have fallen from 123 per 1000 live births in 2005 to 88 in 2010, and primary net enrolment rates rose from 68% in 2004/2005 to 82% in 2009/2010 (MIF 2018a). In Sierra Leone, the creation in 2014 of decentralized District Ebola Response Centers (DERC) made it possible to contain the epidemic by relying on social structures and networks established in local communities. The provision of a focal point for partners to work through in the field was regarded as one of the DERC's most important contributions to the fight against Ebola (MIF 2018a).

Meanwhile, in Uganda, decentralization reforms implemented in the 1990s contributed to growing inequality and inefficiency in education provision. A study of two districts shows that, as the central government still controls more than 90% of their local budget, local governments are severely constrained by the lack of funds and have no say on development priorities. Moreover, higher levels of private and donor funding in certain districts led to variable education provision amongst districts, and thus higher inequality levels. In 2009, Botswana transferred the management of clinics and primary hospitals from local to central government (Ministry of Health). Centralization came with difficulties, such as delays in delivery of drugs and low maintenance of equipment and hospitals (MIF 2018a).

## 4 Access to Urban Services: The Main Mission

The already strained capacities of African local governments, which have resulted in unequal access to services in many areas, are being put to the test by Africa's rapid urban expansion. For instance, poor capacities in waste management as well as in the provision of affordable urban land and housing have been made even more noticeable by the sustained and fast urban growth seen on the African continent. Extending access to urban services should constitute the main mission for local authorities, in a context where African citizens are voicing their dissatisfaction with how governments are handling the provision of basic services such as health and education.

**Citizen perceptions**—The 2018 Ibrahim Index of African Governance (IIAG) shows a growing citizen dissatisfaction with how African national governments are performing in terms of provision of some of the most essential public services. In this sense, both *Satisfaction with Education Provision* and *Satisfaction with Basic Health Services*, two Afrobarometer-sourced perception-based indicators, have followed a negative trajectory over the past decade (−9.0 and −6.7, respectively). Of the 34 African countries with data, while only 11 have improved their score in *Satisfaction with Education Provision* during the years 2008–2017, 14 have experienced an improvement in *Satisfaction with Basic Health Services* in the same time period. Contrary to this, citizen satisfaction with how African governments are doing when it comes to providing electricity supply appears to be on the rise. This is shown by the trajectory of the Afrobarometer-sourced IIAG indicator *Satisfaction with Electricity Supply*, which has improved by +7.6 over the past decade (MIF 2018c).

**Access to electricity**—In 2016, 75.7% of sub-Saharan Africa's urban population had access to electricity. This is lower than the averages for the Middle East & North Africa (99.8%), Latin America & the Caribbean (99.5%), South Asia (98.1%) and the World (96.9%) (World Bank 2018a).

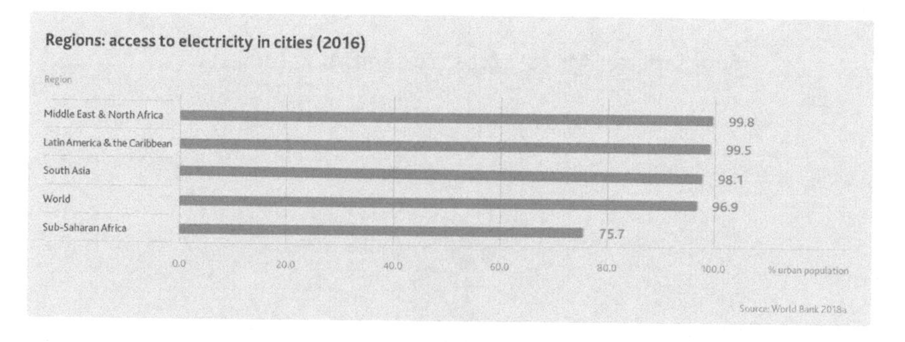


The ten African countries with the lowest urban access to electricity in 2016 were Somalia (57.2%), Burundi (49.7%), DRC (47.2%), Sierra Leone (46.9%), Malawi (42.0%), Central African Republic (34.1%), Liberia (34.0%), Chad (31.4%), Guinea-Bissau (29.8%) and South Sudan (22.0%). The case of DRC, which had the eight lowest rate of urban access to electricity on the continent in 2016, is particularly dramatic (World Bank 2018a). In that same year, DRC's urban population was the fourth largest in Africa (34.1 million) (UNDESA 2018c).

The ten African countries with the highest urban access to electricity in 2016 were Egypt (100.0%), Morocco (100.0%), Tunisia (100.0%), Algeria (99.6%), Seychelles (99.4%), Libya (99.1%), Gabon (96.7%), Cabo Verde (93.0%), South Africa (92.9%) and Comoros (92.1%) (World Bank 2018a).

In 2016, only Mauritius and Seychelles, had a higher percentage of population with access to electricity in rural areas than in urban areas (100% and 100% in rural areas, compared to 91.9% and 99.4% in urban areas, respectively). Only three North African countries reached a 100% access to electricity both among their urban and rural populations (Egypt, Morocco and Tunisia) (World Bank 2018a, b).

**Sanitation and water services**—In 2015, 7.4% of sub-Saharan Africa's urban population practiced open defecation (compared to 0.2% in the Middle East & North Africa region and 2.2% globally) (World Bank 2018c). In that same year, 41.7% of sub-Saharan Africa's urban dwellers used at least basic sanitation services (compared to 92.7% in the Middle East & North Africa region and 82.0% at the global level) (World Bank, 2018e). Moreover, in 2015, only 22.8% of the inhabitants of sub-Saharan African cities had handwashing facilities including soap and water (World Bank 2018f).

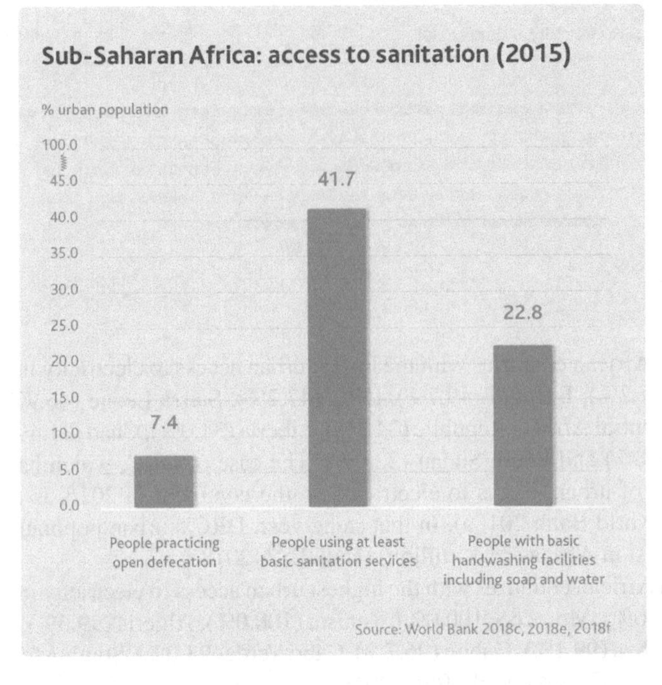


In 2015, the five African countries with the highest percentages of their urban population having at least basic sanitation services were either located in the North African region or islands: Tunisia (98.1%), Egypt (97.3%), Mauritius (93.9%), Algeria (89.7%) and Morocco (89.3%). Conversely, DRC (22.6%), Congo (20.0%), Ghana (18.8%), Ethiopia (18.5%) and Madagascar (16.2%) had the lowest urban populations with access to at least basic sanitation services (World Bank 2018e). This means that in the DRC and Ethiopia, whose capitals, Kinshasa and Addis Ababa, were the third and twelfth largest cities on the continent in 2015, only 7.4 million and 3.6 million urban dwellers had access to at least basic sanitation services in that same year (compared to total urban populations of 32.6 million and 19.4 million, respectively) (UNDESA 2018c, e; World Bank 2018e).

Only 81.4% of sub-Saharan Africa's urban population had access to at least basic drinking services in 2015. This is lower than the figures for the Middle East & North Africa and the World (95.9% and 95.2%, respectively) (World Bank 2018e). Accessing water through unsafe and untreated water sources poses a significant health risk. For instance, in Dar es Salaam, over 75% of the urban dwellers living in informal settlements only have access to informal and unsafe pit emptying services, where the risks of exposure to faecal sludge and contamination of drinking water sources are further exacerbated by flooding or heavy rains. However, faecal contamination can also affect the "served" urban population. In many African cities, formal water supplies to low-income areas are characterized by high degrees of discontinuity, which forces urban dwellers to resort to unsafe water sources, increasing the risk of diarrheal diseases (Dodman et al. 2017). In 2015, the five African countries with the highest urban access to at least basic drinking water services were Mauritius (99.9%), Tunisia (99.7%), Egypt (99.3%), South Africa (96.7%) and Namibia (96.7%). All five countries having the smallest urban populations with access to at least basic drinking water services managed to guarantee the basic water needs of more than half of their urban dwellers: Somalia (70.0%), DRC (69.7%), Eritrea (66.3%), Angola (63.3%) and South Sudan (59.6%) (World Bank 2018d).

In 2016, there was little urban-rural convergence in access to improved drinking water sources even in highly urbanized African countries. Ethiopia, one of the least urbanized countries, constitutes an outlier with a 57 percentage point difference. While in some of the more urbanized countries, such as Algeria, Cabo Verde, South Africa and Tunisia, the disparity in access was 5–20 percentage points, in Congo and Gabon, also highly urbanized, the percentage point difference was 50–60. This goes against the global trends where countries with high urbanization show almost no difference between urban and rural areas when it comes to access to basic services (UNECA 2017).

**Waste management**—Solid wastes constitute non-liquid or non-gaseous products (e.g. trash, junk, refuse) of human activities that are unwanted. Generation of municipal solid waste increases in line with the development rate of any country (Bello et al. 2016). An estimated 11.2 billion tons of solid waste is collected worldwide every year. In Africa, the main drivers of solid waste generation are urbanization and urban population growth (APHRC 2017). African cities generate between 0.3 kg and 0.8 kg of solid waste per capita per day compared to the global average of 1.4 kg per capita per day. By the time 50% of the population of sub-Saharan Africa lives in cities, the daily rate of production of waste is expected to rise to as much as 1 kg per capita. Waste generated in most urban areas in Africa could quadruple by 2025, fueling a potential waste management emergency (MIF 2015a). Poor waste management is linked to public health hazards and environmental damage, but also hinders broader economic growth (Adegoke 2018).

The solid waste management chain—which encompasses generation, collection, treatment, recycling and disposal—is complex and challenging for municipalities, whose financial capabilities are limited, paving the way for a variety of risks including the stagnation of economic development, proliferation of diseases, and degradation of the environment. In developing countries, existing empirical evidence points to disproportionate generation of waste compared to collection and disposal, possibly due to limited administrative capability and lack of public funding for municipalities. Less than 70% of generated waste is collected and more than 50% of the collected waste is disposed of through uncontrolled landfills, while 15% is processed through unsafe and informal recycling (APHRC 2017). In African cities, 57% of municipal solid waste is biodegradable organic waste, the majority of which is dumped. With an average waste collection rate of merely 55%, municipal solid waste collection services in most African countries have proved inadequate. 90% of the waste generated in Africa is disposed of to land, mostly to uncontrolled and controlled dumpsites. Only 4% of the waste generated in Africa is recycled, and mostly only by informal actors (Godfrey 2018).

The dimension of the problem is exemplified by the city of Dakar, the capital of Senegal. A survey conducted in 2016 in three sites in Dakar showed that only 27% of households were using safe means of waste storage and 11% had knowledge on composting. Moreover, only 3.3% of the surveyed households perceived themselves to be at high risk or very high risk of health and environmental-related hazards due to poor solid waste management (APHRC 2017).

**Housing**—Sustained and rapid urban growth on the African continent is placing enormous strain on the provision and affordability of urban land and housing. Across Africa, housing affordability in cities is very low due to the confluence of low urban household incomes, high mortgage interest rates, and short tenors. There are very few countries where the majority of the urban population can afford the cheapest and newest houses built by a formal developer (CAHF 2018). In 2018, for example, only in nine African countries were more than half of the urban population able to afford the cheapest (in $) newly built houses: Morocco (95%), Tunisia (95%), Egypt (89%), Algeria (79%), Gabon (70%), Sudan (68%), Equatorial Guinea (64%), Senegal (64%) and Kenya (51%) (CAHF 2018).

Population growth in African cities is occurring in an expansive rather than compact form, resulting in decreasing urban population densities. This is happening due to inadequate urban planning, rapid population growth and lack of financial or technical capacity to deliver large-scale infrastructure projects that are needed to support livable density. It is predicted that between 2000 and 2050 the area in urban use in sub-Saharan Africa will increase 12-fold. Urban sprawl comes with significant environmental and health costs. It is normally those who cannot afford cars or even formal public transport who are forced to find or build homes in hazardous areas (e.g. floodplains, swamps, coastal zones…) within cycling or walking distance of employment hubs. For instance, in Lagos 70% of the population lives in slums vulnerable to environmental hazards, such as regular flood events (Dodman et al. 2017).

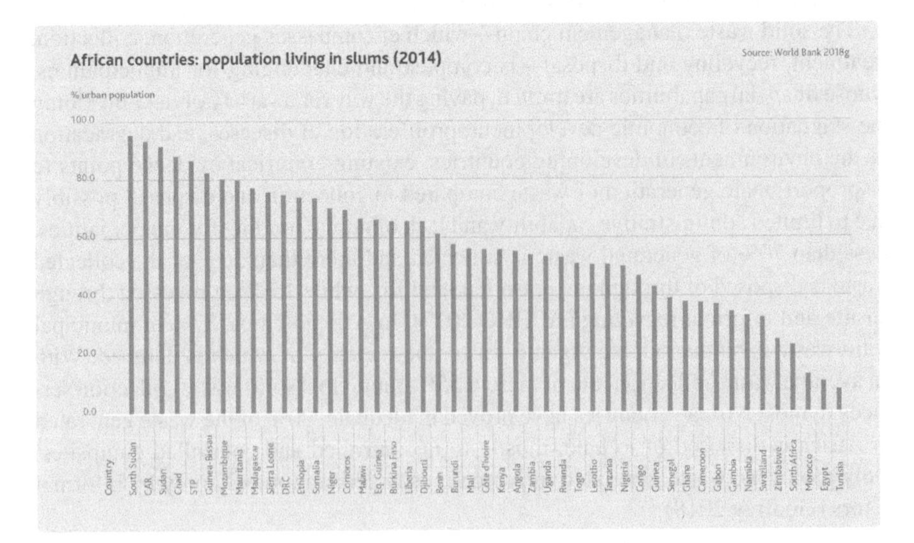

In 2014, more than half (55.3%) of sub-Saharan Africa's urban population lived in slums (compared to 20.3% in Latin America & the Caribbean and 30.4% in South Asia). In that same year, three African countries had more than 90% of their urban dwellers living in slums: Sudan (91.6%), Central African Republic (CAR) (93.3%) and South Sudan (95.6%). In opposition to this, only five countries had less than 30% of their urban population living in slums: Tunisia (8%), Egypt (10.6%), Morocco (13.1%), South Africa (23%) and Zimbabwe (25.1%) (World Bank 2018g).

**Transport**—Efficient and inclusive urban mobility is fundamental for economic and social development as it acts as an enabler of citizens' access to goods, services, jobs, markets, education opportunities and social contacts. Sustainable planning in the urban transport sector requires addressing mobility and climate issues altogether: the sector plays a critical role in reducing or stabilizing greenhouse gas (GHG) emissions globally. As a consequence of sustained urbanization and motorization, urban areas are expected to account for 76% of global greenhouse gases by 2030 (UN-HABITAT 2011).

Moreover, it also requires integrating transport and land use planning, which are frequently considered separately in African cities. Due to lack of coordination, construction of new transport infrastructure often forces urban residents, especially those from low-income neighborhoods, to relocate to the periphery, increasing their travel distances and expenditure on transport. The main constraints that African city managers face when attempting to design and implement effective sustainable transportation planning and policy include lack of reliable data on levels and trends in motorization and GHG emissions; limited financial resources and planning expertise; as well as inadequate institutional frameworks and limited experience at local government level. Other constraints are lack of political engagement in favor of sustainable urban transport; lack of coordination in policy development; inadequate learning and scaling up from existing projects; and weak capacity for monitoring and evaluating existing practices (UN-HABITAT 2011).

However, large infrastructure projects such as Addis Ababa's light rail transit system (LRT), the first of its kind in sub-Saharan Africa, are leading the way in tackling mobility challenges in African cities. At full capacity, Addis Ababa's light rail, which currently has two lines that run for around 11 miles each, should be able to carry 60,000 passengers an hour. Fares are about $0.30 per ride. The $475 million light rail project, 85% of which was financed by a loan from the Export-Import Bank of China, did not start to run until September 2015, six years after the contracts for its construction were signed (CPI 2016a). While there is empirical evidence showing that the Addis Ababa tram has improved residents' mobility, the lack of frequency, trains and integration into the existing transport network constitute obstacles preventing it from becoming a mass transit solution (Ifri 2018).

In Uganda, the Ministry of Works & Transport has initiated the process to build a light rail network in its capital Kampala. The light rail network will serve the Greater Kampala Metropolitan Area, which hosts approximately 3.5 million people. The population is growing at around 5% per annum and is expected to reach 15 million by 2040. Kampala accounts for more than 70% of Uganda's industrial production and more than 65% of its gross domestic product. The light rail will help to reduce pollution and crippling road traffic congestion. Commuters are estimated to lose about 24,000 man-hours each day due to congestion (Oirere 2018). Similar large public transportation projects that are currently being planned in other sub-Saharan African cities include a metro system in Abidjan and the Dakar TER (Train Express Regional) (Ifri 2018).

However, due to lack of affordable and accessible alternatives, most urban trips in Africa are still made by foot or bicycle (Bhattacharjee 2015). In the case of Kampala, Uganda's Ministry of Works & Transport estimates that 48% of people walk, 33% take taxis, 10% use minibuses, and only 9% have access to private cars (Oirere 2018). Recognizing the importance of non-motorized transport, the first pedestrian corridor was introduced in the busiest vehicular traffic road in the Central Business District of Kigali City in August 2015. The Corridor has 450 m of length and aims at becoming the main public hub in the city for people from different backgrounds to come together free from car traffic. In this sense, it places people at the center of the city's development model (Dalkmann 2018).

**Health and education**—Social indicators are a good measure of the extent of urban-rural inequalities. A well-functioning Civil Registration and Vital Statistics (CRVS) system registers all births and deaths, issues birth and death certificates, and compiles and disseminates vital statistics. The registration of births and deaths is fundamental for the realization of human rights, as well as for providing citizens with a legal identity and access to public services (MIF 2015b). The 2018 IIAG shows that Africa's progress in the indicator *Civil Registration* over the past decade (2008–2017) has been very marginal (+0.3). While 14 African countries have experienced an improvement over the past ten years, the scores of 31 countries have remained static and 8 have experienced a deterioration. Since 2008, São Tomé & Príncipe has experienced the largest improvement (+50.0) and Malawi has shown the largest decline (−62.5) (MIF 2018c). When comparing levels of birth registration in urban and rural areas in Africa, urban-rural parity holds in countries that are more than 60% urbanized. However, the difference is two times in countries that are less than 30% urbanized (UNECA 2017).

Urbanization appears to make little difference in the urban-rural variation in stunting, and the ratio moves within a very small range around 1.5 in all countries. Indeed, the largest differences are seen in some of the most urbanized countries. A potential explanation for this could be that basic non-food living expenses are much higher in more urbanized countries, which leaves a smaller share of poor households' budgets for food needs. Another is that the more urbanized countries have greater problems with congestion and inadequacy of public health and sanitation in poor areas, contributing to urban undernutrition and morbidity (UNECA 2017).

Highly urbanized countries are close to urban-rural parity in primary school net attendance ratios. Varying by only 0.5 or less based on extent of urbanization, they reflect success in meeting the Millennium Development Goal target of universal primary education in most countries. However, it is important to note that averages mask differences among countries (UNECA 2017).

**Poverty and urbanization**—When discussing Africa's progress in reducing poverty since 1990, two distinct phases should be mentioned. While the poverty headcount ratio increased from 1990 to 2002, from 54.3 to 55.6% of the continent's population, it declined by more than a quarter to 41% in 2013. Despite this, poverty in Africa fell much more slowly in 1990–2013 than in other world regions (UNECA 2017).

In absolute terms, the number of people living in poverty in Africa other than North Africa increased by +42% from 276 million in 1990 to 391 million in 2002. After 2002, thanks to the positive, albeit slow, impact of economic growth, the number of people in poverty remained almost constant at around 390 million. As a consequence, more than 50% of the world's poor in 2013 were in Africa, compared to 15% in 1990. The main reasons explaining the small impact of economic growth on reducing poverty in Africa are: depth of poverty; high initial inequality; mismatch between sectors of growth and of employment; as well as rapid population growth and delayed demographic transition. In 2013, Africa's poverty gap, a measure of how far below the poverty line the poor on the continent are, was nearly twice the global gap (15.2% compared to 8.8%) (UNECA 2017).

In that same year, nine African countries had a depth of poverty that is more than twice the African average: Madagascar, DRC, Malawi, CAR, Burundi, Lesotho, Zambia, Mozambique and Guinea-Bissau (UNECA 2017).

For the 22 African countries for which there is data in the period 2010–2015, the average (using latest data year) level of urban poor (at national poverty lines, as a % of their urban population) was 24.4%. The five countries with the highest levels of urban poor were Guinea-Bissau (51.0%), Lesotho (39.6%), Côte d'Ivoire (35.9%), Togo (35.9%) and Guinea (35.4%). Only three African countries had a level of urban poor lower than 10% of their urban populations: Uganda (9.6%), Cameroon (8.9%) and Algeria (5.8%) (World Bank 2018h).

Linked to the urban-rural inequalities in social indicators highlighted in the previous section, between 1996 and 2012 poverty declined in all subregions and faster in urban than rural areas; except in Southern Africa, where rural poverty declined marginally faster (UNECA 2017).

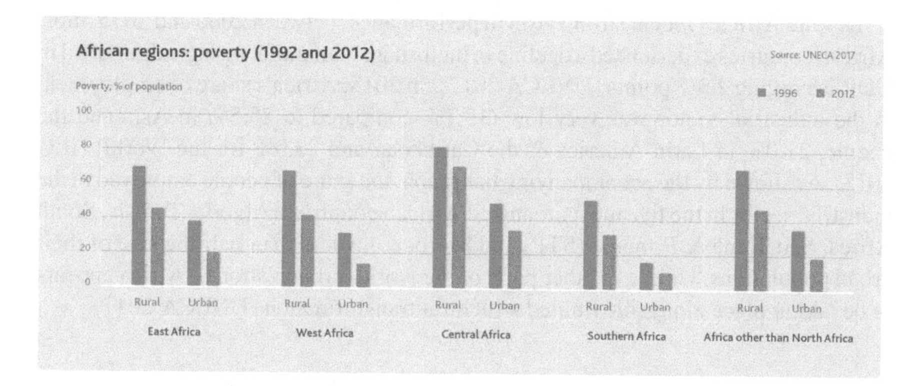


The multidimensional nature of urban poverty is exemplified by the role that access to food plays in it. Empirical evidence from three secondary African cities sheds light into the high levels of households experiencing either moderate or severe food insecurity in urban areas: 90% in Kitwe (Zambia), 88% in Epworth (Zimbabwe) and 71% in Kisumu (Kenya) (Battersby and Watson 2019).

Urban food insecurity heavily impacts low-income residents who often have to purchase food from informal street vendors at higher costs and more variable quality. Household surveys suggest that the average urban household in Africa spends 39–59% of its budget on food; however, for households in the poorest quintile, the share of food expenditure reaches 44–68%. Similarly, they frequently have to purchase water at high cost: studies in four cities show that buying sufficient municipal water can cost a five-person household without piped supplies more than 13% of their income if they are to meet even minimal water needs (20 l per person per day). While poverty raises the cost of meeting basic needs, the cost of living is also higher in real terms in urban Africa than in cities in other low- and middle-income countries. A conservative estimate suggests that, controlling for per capita GDP and other factors, residents in sub-Saharan African cities pay 11–18% more for goods and services than comparable cities worldwide (Dodman et al. 2017).

Urbanization is generally expected to take place alongside an "urban dividend" and structural transformation. "Urban dividend" refers to the acceleration of economic growth in cities due to the existence of scale and agglomeration economies. Structural transformation is the process characterized by the movement of labor and other productive resources from low-productivity (e.g. agricultural) to high-productivity economic activities (e.g. manufacturing). At the global level, the share of manufacturing activities in total output tends to rise with per capita income until countries reach upper-middle-income status, and then starts to decline as countries become service economies at higher incomes (UNECA 2017).

The relationship between urbanization and income appears generally weaker in Africa than in other parts of the world. However, due to commodity price increases, economic reforms and improved governance, growth and income have rebounded in many African countries since 2000, reviving the association between urbanization and income (UNECA 2017).

Despite Africa's recent strong growth performance, between 2000 and 2015, most African countries experienced a decline in their share of manufacturing value added in GDP, averaging 2.3% points (UNECA 2017). In 2018, Africa's share of employment in the industrial sector was very low (13.1% compared to 25.5% in Asia and the Pacific, 21.0% in Latin America & the Caribbean and 23.0% for the World) (ILO 2018). As shown by the accompanying bar graph, the share of people employed in the industrial sector in the five most urbanized African countries (Algeria, Tunisia, South Africa, São Tomé & Príncipe (STP) and Morocco) is less than half the size of their urban populations. Unlike in other parts of the world, urbanization in Africa appears to be taking place alongside limited structural transformation (UNECA 2017).

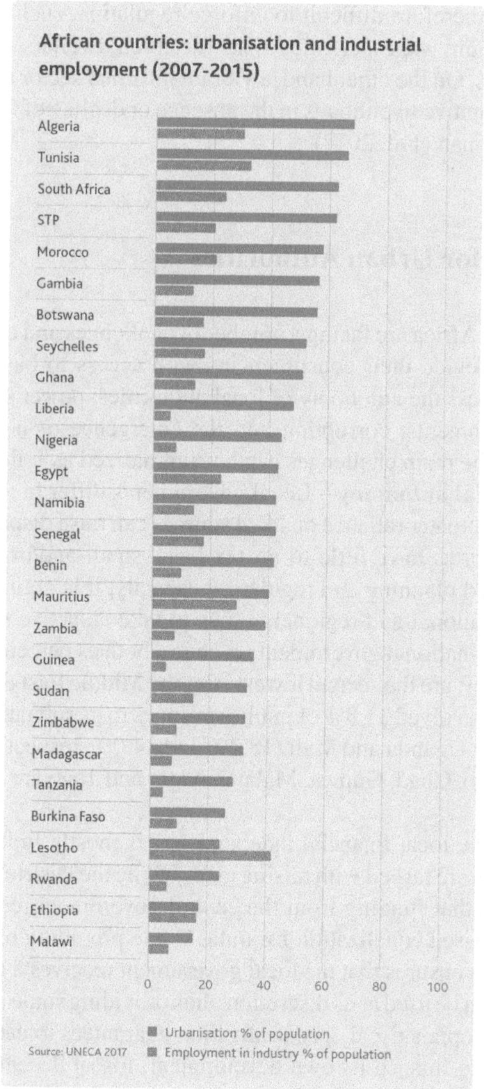

**African countries: urbanisation and industrial employment (2007-2015)**

Algeria
Tunisia
South Africa
STP
Morocco
Gambia
Botswana
Seychelles
Ghana
Liberia
Nigeria
Egypt
Namibia
Senegal
Benin
Mauritius
Zambia
Guinea
Sudan
Zimbabwe
Madagascar
Tanzania
Burkina Faso
Lesotho
Rwanda
Ethiopia
Malawi

0 20 40 60 80 100

■ Urbanisation % of population
■ Employment in industry % of population

Source: UNECA 2017

On top of this, informality is widespread: 76% of sub-Saharan Africa's workforce is in informal employment, financially or politically excluded from formal property and labor markets. This generates a vicious cycle. The large share of informality makes it hard for governments to collect taxes, severely limiting their capacity to provide affordable public transport, water, sanitation or power. At the same time, without access to these services, people are trapped in poverty and informality (Lazer 2018). While the informal economy has the potential to exacerbate local environmental degradation, it can also allow it to respond flexibly and contribute solutions to a variety of challenges. On the one hand, informal providers lack formal state oversight and it is therefore difficult to enforce regulation, such as water treatment standards or minimum wages, creating risks for urban dwellers as prospective consumers and workers. On the other hand, a vibrant informal sector also provides urban residents with alternative livelihoods in the absence or decline of formal employment opportunities (Dodman et al. 2017).

## 5 Challenges for Urban Authorities

Cities and towns in Africa are facing a number of challenges and obstacles impacting their mission to provide their constituencies with access to basic services. These challenges encompass the autonomy of local authorities, power sharing agreements with central governments, corruption and the emergence of non-public actors in service delivery. The main challenges can be summarized as follows:

**Varying financial autonomy**—Local governments differ in their ability to raise revenue, hence the greater reliance on local sources can raise disparities. In instances where local authorities have little to no tax base, small and/or weak institutional capacity and limited planning and regulatory capacity, it is arguably more efficient and effective for national and regional bodies to take an active role in their governance. Africa's sub-national government revenues, both as percentage of total public revenues and of GDP, are the second lowest after the Middle East & West Asia region. In 2017, Tanzania devolved 21.8% of public revenues to its subnational governments, followed closely by Uganda and Mali (18.2% and 14.0%, respectively). Meanwhile, Benin, Burkina Faso, Chad, Guinea, Malawi, Niger and Togo are all below 6% (MIF 2018a).

Overall in Africa, local financial independence is mostly limited. In Ghana, the District Assemblies are tasked with raising taxes, while the District Assemblies Common Fund ensures that funding from the central government reaches each district, based on a needs-based equalization formula. While providing only 37% of district income, this system ensures that the local government receives a guaranteed amount of income which can be used at its discretion, thus providing some amount of financial independence. Ethiopia's fiscal decentralization guarantees to each level of government the capacity to finance its own development. Fiscal decentralization remains limited however as the central government controls 80% of income resources, such as taxes on international trade, leaving only 20% for the regions (MIF 2018a).

**Politicization**—The method by which mayors come to power differs between urban areas. In Cape Town each party nominates a candidate for mayor. The winning party of the local government elections then positions its chosen candidate as mayor. In Dakar the process is by indirect election: the mayor is elected by the municipal council which is itself elected. The Mayor of Accra is appointed by the President and approved by the Accra Metropolitan Assembly. The Accra Metropolitan Assembly is made up of elected and appointed Assembly members. The Mayor can be dismissed by the President of Ghana, or by the Metropolitan Assembly with 2/3 of votes of the members (MIF 2015a).

*Vertically divided authority:* For large cities and city-regions especially, models with a directly elected mayor appear to have greater potential to provide a coherent city vision, mobilize coalitions of stakeholders and provide profile and accountability for citizens. Providing adequate services in urban areas, where the responsibility has been transferred to sub-national authorities, can be more complex when the local authorities are controlled by the opposition. Vertically divided authority can be a problem if the initiatives of a sub-national authority controlled by the opposition do not receive the adequate funding or political backing by a central government led by the main party in the country. Such vertically divided authority appears to be a growing trend in Africa, with a number of important cities in the region in the hands of the opposition, including Nairobi, Dakar, Cape Town and Kampala (MIF 2015a).

*Potential power struggles:* In some cases, both national and local governments are addressing shared issues such as security, health, transport and employment. Without a clear partition of both decision-making power and resources, overlapping agendas and priorities are bound to result in some form of power struggle.

**Corruption**—Decentralization has theoretical advantages and disadvantages related to governance and development, including when it comes to corruption. On the one hand, local elites may "capture" the benefits of decentralization and are not necessarily more pro-poor than national elites. More people have political influence, therefore the risk of corruption is higher. Decentralization can pose a threat to coordination, generating overlapping extractive incentives leading to "overgrazing", and can overburden administrations with low capacity. Indeed, decentralization may be related to more, or at least more decentralized forms of corruption, due to the fact that institutional hybridity and weak accountability mechanisms tend to be prevalent at the local level. On the other hand, by bringing government "closer to the people" it serves, decentralization has the potential to reduce bureaucracy and increase competition, increase citizen voice and participation, as well as strengthen accountability. All of these positive effects, when in place, are poised to reduce corruption (ODI 2018).

The empirical evidence on the interlinkages between decentralization and corruption is very mixed. Case studies and comparative work have also pointed at the fact that this interaction varies in different conditions. For instance, a recent study of political decentralization and corruption in various countries, including Côte d'Ivoire and Ghana, has found that in the short run, decentralization makes a dent on grand theft but increases petty corruption, while in the long run, both may be reduced (ODI 2018).

**Crowding out**—Partly to answer an exponential demand, partly to substitute failing public supply, a growing range of non-state actors have become key providers of public goods and services. Foreign bilateral and multilateral donors have for a long time played a key role in delivering security, health and education, to an extent that may have sometimes prevented public actors from sufficiently owning these key policies. Private sector, as well as a complex galaxy of NGOs, are equally extending their involvement in these sectors, sometimes themselves also crowding out national public services (MIF 2018a).

*Case study: health provision in Africa:* In the case of health provision on the continent, multiple actors overcrowd the role of public services. Donors play a key role as one fifth of total ODA to Africa goes to health. According to the OECD Development Assistance Committee (DAC), in 2016, 21.2% of the ODA to Africa was allocated to the health sector. The top ten donors were, in order, US (38.2%), Global Fund, the UK, GAVI (Vaccine Alliance), International Development Association (IDA), EU, Germany, Canada, France and United Nations Children's Fund (UNICEF) (1.2%). In 2016, the share of total World Health Organization (WHO) disbursement allocated to Africa was the second largest (21.0%) after Asia's (26.0%) (MIF 2018a). The private sector is also increasingly engaged in health provision in Africa, a sector offering growing business opportunities. The sub-Saharan African average for 2014 for total private health expenditure was 57.4%. As lifestyles progressively change on the continent, non-communicable diseases such as diabetes, cardio-vascular pathologies and cancer are spreading. Those who can afford it often travel to places such as India, Turkey, Gulf countries and Europe. The lack of health systems on the continent is seen as a growing opportunity for private healthcare investments. Lastly, civil society organizations (CSOs) also play a key role in health provision across the continent, from small scale NGOs to larger actors such as the Gates Foundation. In the health sector, which is a key focus of the Gates Foundation, the Foundation works in ten countries, Burkina Faso, DRC, Ethiopia, Ghana, Kenya, Nigeria, Senegal, South Africa, Tanzania, and Zambia, and mainly in the fight against infectious diseases, malaria, HIV and tuberculosis. The Gates Foundation has recently committed to invest $5 billion between 2017 and 2022 in Africa in support of health and anti-poverty initiatives (MIF 2018a).

## 6 Factors and Good Practices Affecting Urban Governance and Access to Services

A number of factors and good practices at the country and city levels influence urban local governance and access to services in Africa. Starting from the availability of disaggregated data to recognizing the importance of a comprehensive approach to urbanization that includes all dimensions of governance, the main avenues to explore are the following:

**The tyranny of averages and the need for disaggregated data to "leave no one behind"**—The absence of sufficiently granular information makes it difficult for governments to address spatial inequality. The strong inclination towards national-level averages risks masking important sub-national variations, such as hotspots of deprivation in rich countries and pockets of affluence in poor countries. Similarly, disaggregated data (by area, gender, age, income, employment status, etc.) at the city level are fundamental to design and monitor the impact of authentically pro-poor interventions and policies that really reach those who are the most in need. If local governments are to succeed in "leaving no one behind", they must avoid the "tyranny of averages" (MIF 2018a). For instance, despite being among the top ten performers globally in average yearly GDP growth per capita between 2000 and 2015, in Ethiopia, nighttime lights data suggest that a small number of administrative regions that are centrally located and close to the capital appear to account for the vast majority of economic activity gains. Likewise, in Kenya, even though national-level averages show that under-5 mortality is following a downward trajectory, georeferenced data show that there are subnational hotspots where child mortality levels have actually gone up between the 1980s and the 2000s (AidData 2017). Another example is Data-Driven Lab's Urban Environment and Social Inclusion Index (UESI), which uses neighbourhood-level income data to show that Johannesburg has poor environmental performance and the inequality burdens poorer parts of the city (Data-Driven Lab 2018).

**BOX: Data at the city-level**

*Data at the city-level:* IESE Business School's Cities in Motion Index (CIMI) 2019 is composed of 96 indicators grouped into nine dimensions: human capital, social cohesion, economic indicators, governance indicators, environmental indicators, mobility and transportation indicators, urban planning indicators, international outreach indicators and technology indicators (IESE Business School 2019). Of the 174 cities covered in the CIMI 2019, only 5% (a total of 9 cities) are African: Douala (Cameroon), Cairo (Egypt), Nairobi (Kenya), Casablanca (Morocco), Rabat (Morocco), Lagos (Nigeria), Cape Town (South Africa), Johannesburg (South Africa) and Tunis (Tunisia) (IESE Business School 2019). Casablanca heads the African ranking, followed by Tunis. Cape Town, Nairobi and Cairo complete the list of the top five in the region in 2018. As shown in the table, all of the African cities included in the ranking are among the lowest positions in the overall ranking and they have been consistently so since the first iteration of the CIMI in 2016 (IESE Business School 2019). In terms of the overall score, African cities tend to perform best in the economy dimension and the social cohesion dimension. For all of the five best ranked African cities, the economy dimension is either the first or second highest scoring, while the social cohesion dimension is between first and third (IESE Business School 2019). Within a global context, some African cities stand out when it comes to the environment dimension, with Nairobi (40th), Douala (50th) and Tunis (76th) among the top 50% of performers globally. Additionally, Casablanca features in the top third of performers when it comes to the technology dimension, ranking 58th out of 174 (IESE Business School 2019).

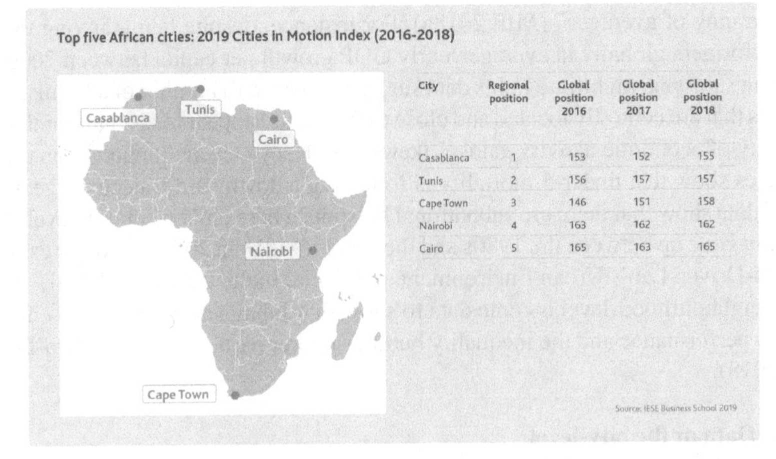


| City | Regional position | Global position 2016 | Global position 2017 | Global position 2018 |
|---|---|---|---|---|
| Casablanca | 1 | 153 | 152 | 155 |
| Tunis | 2 | 156 | 157 | 157 |
| Cape Town | 3 | 146 | 151 | 158 |
| Nairobi | 4 | 163 | 162 | 162 |
| Cairo | 5 | 165 | 163 | 165 |

**A balanced approach to all governance dimensions and a stronger focus on accountability, citizens' rights and welfare**—The IIAG measures governance performance both holistically and through specific indicators that span a variety of broad topics that measure outcomes of governance, with the same indicators applied to all countries. The indicators showing the strongest relationships with high governance scores span across all four categories of the IIAG, indicating that a comprehensive and balanced approach is needed in bringing about stronger governance. The factors most associated with high governance scores are citizen centered factors involving strong property rights, civil rights and liberties, an accountable government and effective public service, and policies focused on social safety nets and environment. High scores in indicators measuring these issues are the most common factors among the best scoring countries in the IIAG. For instance, Mauritius, which ranks 1st at the *Overall Governance* level in the 2017 data year in the 2018 IIAG, is one of the highest scoring countries in seven of the ten indicators that are the most correlated with high governance scores, spanning all four dimensions of the Index. The Southern African island receives the best score on the continent in *Sanctions for Abuse of Office* (85.7 out of 100.0), *Social Safety Nets* (96.2) and *Environmental Policies* (85.7). On top of this, while it ranks 2nd in *Property Rights* and *Civil Rights & Liberties* (with 2017 scores of 79.6 and 99.0, respectively), it obtains the third highest score in *Transport Infrastructure* and *Independence of the Judiciary* (with 2017 scores of 67.6 and 91.5, respectively). Similarly, Botswana, which ranks 5th at the *Overall Governance* level in the 2017 data year in the 2018 IIAG, is one of the best performing countries in five of the indicators showing the strongest correlations with high governance scores. While it obtains the best score on the continent in *Property Rights* (88.3), *Sanctions for Abuse of Office* (85.7) and *Environmental Policies* (85.7), it ranks 3rd in *Social Safety Nets* and 5th in *Civil Rights & Liberties* (with 2017 scores of 79.5 and 91.3, respectively) (MIF 2018c).

**Enhanced local capacity and autonomy**—The challenges of delivering public services at local level reveal that in order to meet demand, public services must be equipped properly, both in terms of resources and in terms of competence and capacity. This is particularly important in the case of Africa's middle-sized cities, which are where the fastest urban growth is expected to happen and already lag behind their larger counterparts in institutional and capacity development. Urban political and financial autonomy are key for effective public service delivery at the local level and can only result from better power sharing agreements and collaboration with central governments. In South Africa, the Inter-Governmental Relations Framework Act (Act 13 of 2005) provides for collaboration among different levels of government and includes mechanisms for settling intergovernmental disputes (Cartwright et al. 2018).

In terms of strengthening capacity, programs in Liberia and Ghana for instance aim at attracting youth talent into public service. Rwanda has been at the forefront of ensuring performance in public service through the use of performance contracts (Imihigo) signed between the president of Rwanda and local government institutions and line ministries, rooted in the country's cultural practice and beneficial for results as well as for accountability practices (MIF 2018a). Inspired by Sir Michael Barber's Delivery Unit in Downing Street, Cape Town's Strategic Policy Unit (SPU) was set up shortly after Patricia de Lille was elected mayor in 2011. Aiming at refocusing the city's governance around strategic policy planning, performance monitoring and evaluation, its tasks are mainly to identify problems in policy implementation and drive strong delivery of the administration's objectives across the city (CPI 2016b).

**Urban policies and planning**—Urban planning is one of the most important tasks for African governments, faced with the multiple and intertwined challenges of urban expansion. If a city develops without planning, the introduction of planning processes afterwards will be much more difficult and will be likely to cause social conflicts. Without inclusive planning the numbers of inhabitants without access to shelter and basic services such as water and sanitation, energy or formal employment opportunities will explode. Informal housing and development of land, already a feature of urban Africa, will continue unrestrained. This will compound existing hazards such as waste, air pollution and the effects of climate change. Unequal planning exacerbates social disparities by marginalizing the urban poor from the inner most spaces of cities, where there are paved roads with street lamps, a regular power supply, an adequate water supply, infrastructure and amenities (MIF 2015a).

In Koudougou, Burkina Faso, as in many intermediate African cities, the urban planning process is exogenous, not really aligned with the requests of its inhabitants, nor with the human, material and financial capabilities available at the city level, and therefore rarely applied. Despite the many plans drawn up for this city, the fact that their design is the result from a collaborative framework between the central government and foreign donors has limited their execution significantly. In this case, the local plans have mostly only served to reassure external donors during financial negotiations (Bolay 2015).

As in Koudougou, the participation of civil society in planning decisions in most African cities is insufficient. However, there are also some examples of more inclusive planning processes. For instance, in Mali, "communes" have been established at a grassroots level and have been given the exclusive competency to implement local development authority. In this case, the community, to a certain extent, constitutes both the user and the delegated contracting authority for infrastructure and equipment (UN-HABITAT 2013).

While in most cases planning policies and frameworks formulated are not relevant to the actual development problems of African cities, and the link between planning and monitoring of implementation of planning projects is not always existent, there are exceptions to this. For example, South Africa has achieved some success with its Integrated Development Plans for municipalities (UN-HABITAT 2013).

**Harnessing innovation**—Africa has demonstrated a unique capacity to leapfrog and applying innovation and technology to public service delivery. In Rwanda, drones are used to deliver blood and medical supplies and in Côte d'Ivoire they ensure the maintenance of the country's electricity network. In Kenya, through Huduma, public services are available electronically and offline communities are reached by mobile offices (MIF 2018a). In Nigeria, user-friendly digital applications have been developed for smartphones to allow citizens to monitor the implementation of government's projects (MIF 2018b). Innovative solutions are also applied to urban self-financing, for instance green bonds help fill gaps in development finance for climate-friendly projects: Johannesburg issued Africa's first municipal green bond in 2014 to help finance emissions-reducing projects including biogas energy, solar power and sustainable transportation (MIF 2015a). Another example of Africa leapfrogging is smart cities, putting modern technology and infrastructure at the center of public service delivery. Cape Town gathers data from sensors around the city for a more effective policy response in areas from traffic monitoring to waste management, crime detection and fire response. Some countries such as Kenya, Rwanda and Nigeria have projects for satellite cities, built around existing cities, providing modern infrastructure and luxury amenities to attract tech-savvy entrepreneurs (The Borgen Project 2018).

**Committed leadership**—Cities such as Johannesburg and Kampala are emerging as new African models for urban service delivery, having modelled their institutions on corporate work, with structures and strategies for delivering results. In both cases, mayors built a team of young and bright people and pride themselves on being favorable to innovation. This shows that, if the right leadership is in place, public services can become attractive work places, where capable young Africans are motivated to serve their constituencies; as well as hubs for innovation, where new technologies are used to improve work processes and better reach citizens, including through the use of social media platforms. Commitment can also be shared, as shown by the initiative of the mayors of 9 African cities—Accra, Addis Ababa, Cape Town, Dakar, Dar es Salaam, Durban, Johannesburg, Lagos and Tshwane—towards climate action, with the pledge to deliver on their share of the Paris Agreement. Nairobi and Abidjan have also joined the common efforts and planned to submit their climate action commitments (C40 Cities 2018).

**Bottom up checks and balances**—Participatory budgeting allows the population to define the destination of public resources and help oversee the implementation of the agreed budget, to tackle corruption, boost citizen participation in the decision-making process, as well as help the municipal authorities to engage with the population. In 2006, Yaoundé VI was the first municipality in Yaoundé, Cameroon, to implement participatory budgeting. In terms of impact, this process led to the construction of much needed infrastructure projects in poor neighborhoods, such as a water tap which serves a community of 50,000 people in Yaoundé IV. Apart from improving basic services, participatory budgeting in Yaoundé also had a positive impact on municipal-citizen relations, citizen participation, fiscal transparency and local tax revenues. The introduction of participatory budgeting in Cameroon, although in its infancy, paves the way for a new social contract between the municipality and the population, aiming to place the aspirations of citizens at the forefront of local development (MIF 2018a).

**Networks and south-south cooperation**—For African cities to better respond to their current and future challenges, including on the path towards achieving the SDGs and the AU's Agenda 2063, networks for sharing best practices and for cross-fertilization both within and outside the continent can prove fundamental. Among these, United Cities and Local Governments (UCLG) supports international cooperation between cities and their associations, and facilitates programs, networks and partnerships to build the capacities of local governments. UCLG Africa gathers 40 national associations of local governments as well as 2000 cities of more than 100,000 inhabitants. UCLG Africa represents nearly 350 million African citizens (MIF 2018a). Moreover, 12 African cities have joined C40, a network of the world's global cities committed to addressing climate change. C40 supports cities to collaborate effectively, share knowledge and drive meaningful, measurable and sustainable action on climate change (C40 Cities 2019).

## 7 Conclusion

Growing urban populations put cities at the heart of governance, with their main mission being service provision to their constituencies. Decentralization in some African countries such as Kenya and South Africa has provided local and urban authorities with increased power and responsibilities. However the picture across the continent remains fragmented, as different levels of transfer of power exist. It is thus difficult to formulate a one size fits all judgement on the impacts of decentralization for better serving Africa's citizens. That said, public satisfaction with basic services that are key for prosperous societies is a good indicator of success, and in Africa this is declining.

Local and urban authorities face several obstacles in their mission, related to their relations with central governments, their autonomy and capacity, as well as the wider challenge of ensuring that the best potential of decentralization is realized, which is bringing government closer to citizens for improved transparency and accountability, rather than government further away from central institutions, free to appropriate resources.

A number of avenues can be explored on the way forward, that range from clearer power sharing agreements to harnessing new technologies and innovation for better public services. However, the core of strong and effective local and urban authorities relies on a committed leadership, and in the acknowledgement of the often underestimated role of these authorities, in many cases managing populations bigger than those of countries. Local and urban authorities cannot achieve their mission without enabling and strengthening processes for citizen participation and bottom up monitoring and oversight; nor without continental and international networks for exchanges of best practices among these key actors, whose role is not only increasingly essential in the daily life of African citizens but also in achieving goals such as the SDGs and Agenda 2030.

## 8 Notes

The opinions expressed in this publication are those of the authors. They do not necessarily reflect the opinions or views of the Mo Ibrahim Foundation or its members.

This chapter makes extensive use of research publications previously published by the Mo Ibrahim Foundation (MIF), especially the reports 2015 Facts & Figures: African Urban Dynamics and 2018 Ibrahim Forum Report: Public Service in Africa.

A reference list containing all the sources used is provided at the end of this chapter. Sources used are not always the primary data sources. For instance, the primary data source of variables from the World Development Indicators is not always the World Bank.

When necessary, figures from our reports have been updated. In those cases, the original data source has been referenced. Data were correct at the time of research (the last access date for each variable is provided in the references).

Some figures provided here are the result of our calculations using different variables available from source.

Comparisons of regional averages are provided. The composition of regions may vary according to source. When data in the report is presented disaggregated for North African and sub-Saharan African countries, this is done reflecting the choices made at source.

All population statistics are taken from the 2018 revision of the World Urbanization Prospects from the United Nations Department of Economic and Social Affairs (UNDESA). There is one main exception to this: UNDESA's Urban and Rural Population by Age and Sex, 1980-2015 dataset.

The Ibrahim Index of African Governance (IIAG) is an annual statistical assessment of the quality of governance in every African country, produced by MIF. The IIAG focuses on outputs and outcomes of policy and is used throughout this chapter as a measure of public service delivery across the continent. All IIAG data included here come from the 2018 IIAG, our latest released dataset. To distinguish the IIAG, all measures taken from it are italicised, as opposed to measures obtained from other sources. You can explore the full Index dataset here: http://mo.ibrahim.foundation/iiag/

To get in touch with MIF about this chapter, please contact: rocca.c@moibrahimfoundation.org or fernandez.d@moibrahimfoundation.org.

## References

Adegoke Y (2018) When will Africa's fast-growing cities get on top of their garbage problems? Quartz. https://qz.com/africa/1237012/lagos-nairobi-and-accra-are-among-africas-big-cities-to-struggle-with-poor-waste-management/. Accessed 15 Nov 2019

AidData (2017) Beyond the tyranny of averages: development progress from the bottom up. AidData, Williamsburg. https://www.aiddata.org/publications/beyond-the-tyranny-of-averages-development-progress-from-the-bottom-up. Accessed 15 Nov 2019

APHRC (2017) Solid waste management and risks to health in urban Africa. A study of Dakar City, Senegal. APHRC, Nairobi. https://aphrc.org/wp-content/uploads/2019/07/Urban-ARK-Dakar-Report.pdf. Accessed 15 Nov 2019

Battersby J, Watson V (2019) Introduction. In: Battersby J, Watson V (eds) Urban food systems governance and poverty in African cities. Routledge, London and New York, pp 1–26

Bello IA, Ismail MNB, Kabbashi NA (2016) Solid waste management in Africa: a review. Int J Waste Resour 6(2):1–4

Bhattacharjee D (2015) Urbanization and mobility in Africa. In: Africa climate resilient infrastructure summit, 27–29 Apr 2015. UN-HABITAT, Addis Ababa

Bolay JC (2015) Urban planning in Africa: which alternative for poor cities? The case of Koudougou in Burkina Faso. Curr Urban Stud 3:413–431. https://doi.org/10.4236/cus.2015.34033

Cartwright A, Palmer I, Taylor A, Pieterse E, Parnell S, Colenbrander S (2018) Developing prosperous and inclusive cities in Africa—national urban policies to the rescue? Coalition for Urban Transitions/African Centre for Cities Working Paper. https://newclimateeconomy.report/workingpapers/wp-content/uploads/sites/5/2018/09/CUT18_Africa_NatUrbanPolicies_final.pdf. Accessed 15 Nov 2019

C40 Cities (2018) Mayors of 9 African megacities commit to delivering their share of the Paris agreement. C40 Cities. https://www.c40.org/press_releases/mayors-of-9-african-megacities-commit-to-delivering-their-share-of-the-paris-agreement. Accessed 15 Nov 2019

C40 Cities (2019) History of the C40. C40 Cities. https://www.c40.org/history. Accessed 15 Nov 2019

CAHF (2018) Housing finance in Africa. A review of Africa's housing finance markets. Johannesburg, South Africa. http://housingfinanceafrica.org/documents/housing-finance-in-africa-yearbook-2018/. Accessed 15 Nov 2019

CPI (2016a) Light rail transit in Addis Ababa. CPI. https://www.centreforpublicimpact.org/case-study/light-rail-transit-in-addis-ababa/. Accessed 15 Nov 2019

CPI (2016b) Cape Town's crusaders: planning to deliver. CPI. https://www.centreforpublicimpact.org/cape-towns-delivery-crusaders/. Accessed 15 Nov 2019

Dalkmann H (2018) Sustainable urban mobility principles and planning. GIZ: Transformative Urban Mobility Initiative (TUMI) and Mobilize Your City (MYC). SSATP, Abuja. https://www.ssatp.org/sites/ssatp/files/2018_Annual_Meeting_Abuja/Day1/UTM/Sustainable%20Urban%20Mobility%20Principles%20and%20Planning%20-%20Holger%20Dalkmann.pdf. Accessed 15 Nov 2019

Diep L, Archer D, Cheikh G (2016) Decentralisation in West Africa: the implications for urban climate change governance. The cases of Saint-Louis (Senegal) and Bobo-Dioulasso (Burkina Faso). IIED Working Paper, IIED, London. http://pubs.iied.org/10769IIED/. Accessed 15 Nov 2019

Dodman D, Leck H, Rusca M, Colenbrander S (2017) African urbanisation and urbanism: implications for risk accumulation and reduction. Int J Disaster Risk Reduct 26:7–15

ETTG (2018) Africa's future is urban. Implications for EU development policy and cooperation. ETTG, Brussels. https://ettg.eu/wp-content/uploads/2018/07/Africas-future-is-urban.pdf. Accessed 15 Nov 2019
Godfrey L (2018) Overview of the Africa Waste Management Outlook. In: Launch of the UN Africa Waste Management Outlook. CSIR, Pretoria. http://sustainabilityweek.co.za/assets/files/Day%203%20-%20AWMO%20-%20Waste.pdf. Accessed 15 Nov 2019
IESE Business School (2019) IESE cities in motion index 2019. IESE Business School, Barcelona. https://media.iese.edu/research/pdfs/ST-0509-E.pdf. Accessed 15 Nov 2019
Ifri (2018) The challenge of urban mobility: a case study of Addis Ababa light rail, Ethiopia. Ifri, Paris. https://www.ifri.org/sites/default/files/atoms/files/nallet_urban_mobility_addis_ababa_2018.pdf. Accessed 15 Nov 2019
ILO (2018) ILO modelled estimates, Nov. 2018. Employment by sector. ILO, Geneva. https://www.ilo.org/ilostat/faces/oracle/webcenter/portalapp/pagehierarchy/Page33.jspx?locale=EN&MBI_ID=565&_afrLoop=2438245628731506&_afrWindowMode=0&_afrWindowId=null#!%40%40%3F_afrWindowId%3Dnull%26locale%3DEN%26_afrLoop%3D2438245628731506%26MBI_ID%3D565%26_afrWindowMode%3D0%26_adf.ctrl-state%3Dailpj8jjm_295. Accessed 6 Mar 2019
Lazer L (2018) Africa's urban future: the policy agenda for national governments. TheCityFix. http://thecityfix.com/blog/africas-urban-future-policy-agenda-national-governments-leah-lazer/. Accessed 15 Nov 2019
MIF (2015a) 2015 facts & figures: African urban dynamics. MIF, London. http://static.moibrahimfoundation.org/u/2015/11/19115202/2015-Facts-Figures-African-Urban-Dynamics.pdf. Accessed 15 Nov 2019
MIF (2015b) Strength in numbers. MIF, London. http://s.mo.ibrahim.foundation/u/2016/05/16162558/Strength-in-Numbers.pdf. Accessed 15 Nov 2019
MIF (2018a) 2018 Ibrahim forum report: public service in Africa. MIF, London. http://s.mo.ibrahim.foundation/u/2018/06/21170815/2018-Forum-Report.pdf?_ga=2.176124386.1712197560.1551885727-2031218662.1525792781. Accessed 15 Nov 2019
MIF (2018b) 2018 Ibrahim forum summary: public service in Africa. MIF, London. http://s.mo.ibrahim.foundation/u/2018/10/05154012/2018-Forum-Summary.pdf?_ga=2.115793094.1712197560.1551885727-2031218662.1525792781. Accessed 15 Nov 2019
MIF (2018c) 2018 Ibrahim index of African governance. MIF, London. http://mo.ibrahim.foundation/iiag/downloads/. Accessed 6 Dec 2018
ODI (2018) Local governance, decentralisation and corruption in Bangladesh and Nigeria. ODI, London. https://www.odi.org/sites/odi.org.uk/files/resource-documents/12177.pdf. Accessed 15 Nov 2019
Oirere S (2018) Ugandan capital embarks on light rail project. Int Railw J. https://www.railjournal.com/in_depth/ugandan-capital-embarks-on-light-rail-project. Accessed 15 Nov 2019
Smit W (2018) Urban governance in Africa: an overview. Int Dev Policy 10:55–77
The Borgen Project (2018) Benefits of smart cities in Africa. The Borgen Project, Seattle. https://borgenproject.org/benefits-of-smart-cities-in-africa/. Accessed 15 Nov 2019
UNDESA (2014) Urban and rural population by age and sex, 1980–2015 (version 3, August 2014). UNDESA, New York. http://www.un.org/en/development/desa/population/publications/dataset/urban/urbanAndRuralPopulationByAgeAndSex.shtml. Accessed 6 Dec 2018
UNDESA (2018a) World urbanization prospects: the 2018 revision. Average annual rate of change of the urban population by region, subregion, country and area, 1950–2050 (per cent). UNDESA, New York. https://population.un.org/wup/Download/. Accessed 5 Dec 2018
UNDESA (2018b) World urbanization prospects: the 2018 revision. Average annual rate of change of the percentage urban by region, subregion, country and area, 1950–2050 (per cent). UNDESA, New York. https://population.un.org/wup/Download/. Accessed 5 Dec 2018
UNDESA (2018c) World urbanization prospects: the 2018 revision. Annual urban population at mid-year by region, subregion, country and area, 1950–2050 (thousands). UNDESA, New York. https://population.un.org/wup/Download/. Accessed 5 Dec 2018

UNDESA (2018d) World urbanization prospects: the 2018 revision. Annual total population at mid-year by region, subregion and country, 1950–2050 (thousands). UNDESA, New York. https://population.un.org/wup/Download/. Accessed 5 Dec 2018

UNDESA (2018e) World urbanization prospects: the 2018 revision. Annual population of urban agglomerations with 300,000 inhabitants or more in 2018, by country, 1950–2035 (thousands). UNDESA, New York. https://population.un.org/wup/Download/. Accessed 6 Dec 2018

UNDESA (2018f) World urbanization prospects: the 2018 revision. Annual percentage of population at mid-year residing in urban areas by region, subregion, country and area, 1950–2050. UNDESA, New York. https://population.un.org/wup/Download/. Accessed 5 Dec 2018

UNECA (2017) Economic report on Africa 2017. Urbanization and industrialization for Africa's transformation. UNECA, Addis Ababa. https://www.uneca.org/sites/default/files/uploaded-documents/ERA/ERA2017/era-2017_en_fin_jun2017.pdf. Accessed 15 Nov 2019

UN-HABITAT (2011) Sustainable mobility in African cities. UN-HABITAT, Nairobi

UN-HABITAT (2013) The state of planning in Africa, an overview. UN-HABITAT, Nairobi

World Bank (2018a) World development indicators. Access to electricity, urban (% of urban population). The World Bank, Washington, DC. https://data.worldbank.org/indicator/EG.ELC.ACCS.UR.ZS. Accessed 5 Dec 2018

World Bank (2018b) World development indicators. Access to electricity, rural (% of rural population). The World Bank, Washington, DC. https://data.worldbank.org/indicator/EG.ELC.ACCS.RU.ZS. Accessed 5 Dec 2018

World Bank (2018c) World development indicators. People practicing open defecation, urban (% of urban population). The World Bank, Washington, DC. https://data.worldbank.org/indicator/SH.STA.ODFC.UR.ZS. Accessed 5 Dec 2018

World Bank (2018d) World development indicators. People using at least basic drinking water services, urban (% of urban population). The World Bank, Washington, DC. https://data.worldbank.org/indicator/SH.H2O.BASW.UR.ZS. Accessed 5 Dec 2018

World Bank (2018e) World development indicators. People using at least basic sanitation services, urban (% of urban population). The World Bank, Washington, DC. https://data.worldbank.org/indicator/SH.STA.BASS.UR.ZS. Accessed 5 Dec 2018

World Bank (2018f) World development indicators. People with basic handwashing facilities including soap and water, urban (% of urban population). The World Bank, Washington, DC. https://data.worldbank.org/indicator/SH.STA.HYGN.UR.ZS. Accessed 5 Dec 2018

World Bank (2018g) World development indicators. Population living in slums (% of urban population). The World Bank, Washington, DC. https://data.worldbank.org/indicator/EN.POP.SLUM.UR.ZS. Accessed 5 Dec 2018

World Bank (2018h) World development indicators. Urban poverty headcount ratio at national poverty lines (% of urban population). The World Bank, Washington, DC. https://data.worldbank.org/indicator/SI.POV.URHC. Accessed 5 Dec 2018

**Camilla Rocca** is Head of Research at the Mo Ibrahim Foundation, where she leads on the Foundation's research agenda and publications. Her main interest is the African political, economic and social landscape, with a focus on the nexus between fragility, development and state-society relations. She has worked extensively on these issues as independent analyst as well as political officer for the European Union based in Central Africa. With a background in International Relations, she holds two MAs from the College of Europe and the University of Padua, and professional certifications including in political analysis and mediation.

**Diego Fernández Fernández** is Senior Analyst at the Mo Ibrahim Foundation, where he leads on the methodological/analytical aspects of the Foundation's publications. His main areas of interest encompass quantitative research methods, political economy of development topics such as taxation and inequality, as well as urban governance. Before joining the Foundation's Research Team, he worked as a Research Associate to The Governance Report and as an Intern at the Costa Rica office of the German Development Agency (GIZ). He holds a Master of Public Policy from the Hertie School of Governance in Berlin, as well as a Master's in International Relations from the Barcelona Institute of International Studies (IBEI).

# Local Governance and Access to Urban Services: Conclusions and Policy Implications

**Shabbir Cheema**

**Abstract** Studies in this volume show that inclusive, participatory and sustainable urban service delivery and access require a set of policy and program responses: It is essential to distribute resources equitably to urban local governments and strengthen their planning and management capacity in order to enable them to perform their responsibilities. Institutional arrangements should be restructured to promote collaborative governance and stock-taking of functional gaps and overlaps among multiple agencies and departments located within a city. Participatory mechanisms should be provided for the engagement of civil society, local governments, citizen groups and other stakeholders in local decision-making processes. The need is for greater use of widely recognized instruments of accountability and transparency including participatory budgeting and right to information. One of the challenges of urban policy implementation is political and social inclusion and engagement of marginalized communities including women, youth, migrants, ethnic minorities and the urban poor in the structures and processes of local governance including access to urban land and housing through revised land use regulations, effective land density and mixed use, and housing finance and land titles reforms. Cities have been laboratories of experimentation, innovations and good practices to improve service delivery and access. Recent surveys have highlighted a number of innovations and good practices in cities in terms of their content, rationale and impact on urban residents. These need to be replicated. Information and communication technology (ICT) should be used to enhance quality, performance and interactivity of urban services; to reduce costs and resource consumption; and to improve contact between citizens and government. Finally, peri-urbanization is a critical issue in access to services for the urban poor. It requires an integrated planning and coordination of urban areas.

**Keywords** Urban local government · Urban services · Planning and management capacity · Accountability and transparency · Innovation · Gender · Migrants · ICT · Participatory budgeting · Peri-urbanization

S. Cheema (✉)
Harvard Kennedy School, Ash Center for Democratic Governance and Innovation, Harvard University, 79 John F. Kennedy Street, Cambridge, MA 02138, USA
e-mail: shabbir_cheema@hks.harvard.edu



S. Cheema (ed.), *Governance for Urban Services*,
Advances in 21st Century Human Settlements,
https://doi.org/10.1007/978-981-15-2973-3_13

The world urban population has been growing rapidly over the past fifty years. Over the next decade, two-thirds of the demographic expansion in the world's cities will take place in Asia. By 2020, 2.2 billion of the world's 4.2 billion city dwellers will live in Asia. Nine out of the ten largest megacities and fourteen out of the top twenty megacities of the world are already in Asia. Population growth in intermediate and small-sized cities is even faster. Africa too is urbanizing rapidly. By 2050, Africa's population will more than double, from 1.2 billion to more than 2.5 billion.

Urbanization has contributed to economic development though expanded economic opportunities. However, it has also led to increasing urban poverty and inequality, deteriorating quality of urban environment and unequal access to urban services such as water supply and sanitation, shelter, waste management, energy, transport, and health care. Specifically, slum household suffer from insecure land tenure, unreliable power supply, intermittent water availability, insufficient treatment of wastewater, flooding, and uncollected garbage. In view of the above, the need for equal access to economic and political opportunities in cities and to urban services was recognized in the 2030 Agenda for Sustainable Development endorsed by the World's Heads of State and Government. The Agenda consists of Sustainable Development Goals (SDG), including SDG11 to promote inclusive, participatory and resilient cities, SDG 16 on accountable institutions and justice and SDG 5 on gender equality (UN 2016). The New Urban Agenda adopted by the United Nations Conference on Housing and Sustainable Urban Development (HABITAT III) endorsed the central role of cities and urban governance in achieving inclusive and sustainable development.

This chapter presents main conclusions and policy implications of collaborative studies on how processes of participation, accountability and transparency in local governance affect access to urban services in cities, especially for marginalized groups. It is based on research conducted by a group of national research and training institutions in nine cities in five Asian countries—India (New Delhi and Bangalore), Indonesia (Bandung and Solo), China (Chengdu), Vietnam (Hanoi and Ho Chi Min City), and Pakistan (Lahore and Peshawar)—as well as concept papers by leading scholars on access, participation, accountability and gender equality and a regional review of access to urban services in Africa. These conclusions and policy implications are also informed by regional dialogues organized with the support of the Swedish International Center for Local Democracy (ICLD). This chapter discusses key issues and implications that were applicable across countries and cities.

## 1 Deficits in Urban Services and Citizens' Perceptions

There are alarming deficits in urban services in Africa and Asia (Chaps. 7–12 in this volume). National reviews, household surveys of the residents of slums and squatter settlements in Asian cities and regional survey of urban areas in Africa also show an increasing dissatisfaction of urban residents about performance of national and local governments in terms of access to urban services.

Deficits in access to urban services are the highest in Africa (Rocca and Fernandez 2020).

*Electricity*: In 2016, 75.7% of sub-Saharan Africa's urban population had access to electricity which is much lower than 98.1% in South Asia. Nine African countries with lower than 50% access were Burundi (49.7%), DRC (47.2%), Sierra Leone (46.9%), Malawi (42.0%), Central African Republic (34.1%), Liberia (34.0%), Chad (31.4%), Guinea-Bissau (29.8%) and South Sudan (22.0%).

*Sanitation*: In 2015, 41.7% of sub-Saharan Africa's urban residents used at least basic sanitation services compared to 82.0% at the global level; and 7.4% of sub-Saharan Africa's urban population practiced open defecation compared to 0.2% in the Middle East & North Africa region and 2.2% globally.

*Waste Management*: About 50% of the collected waste is disposed of through uncontrolled landfills; 15% is processed through unsafe and informal recycling.

*Housing*: In 2014, 55.3% of sub-Saharan Africa's urban population lived in slums (compared 30.4% in South Asia), with three African countries with more than 90% of their urban dwellers living in slums: Sudan (91.6%), Central African Republic (93.3%) and South Sudan (95.6%).

*Urban transport*: Because of lack of affordable and accessible alternatives, most urban trips in Africa are still made by foot or bicycle.

Citizen perceptions indicate high level of dissatisfaction of urban dwellers about overall performance of central and national governments in terms of access to urban services. The 2018 Ibrahim Index of African Governance (IIAG), for examples, shows a growing citizen dissatisfaction in terms of provision of some of the most essential public services. *Satisfaction with Education Provision* and *Satisfaction with Basic Health Services*, two Afrobarometer-sourced perception-based indicators, have shown a negative trajectory over the past ten years (−9.0 and −6.7, respectively). Of the 34 African countries for which data was collected, only 11 have improved their score in *Satisfaction with Education Provision*, and 14 have experienced an improvement in *Satisfaction with Basic Health Services* (MIF 2018).

The UN HABITAT Report on The State of Asian and Pacific Cities 2015 shows exclusion of a large percentage of the urban population from access to urban services (UN HABITAT 2015). For example, in 2012, the number of urban residents living in slum conditions in Asia was 35% in Southern Asia. In the Least Developed countries of the region, less than 15% of the urban population relies upon unimproved sources of drinking water (with the exception of Lao (UN HABITAT 2015). In nine countries of the region, the share of urban population relying on unimproved sanitation facilities (e.g. septic tanks, pit latrines, buckets and open defecation) ranges from 53 to 34%. In ten countries, 0% of wastewater is treated. In other ten countries, it ranges from 5 to 19%. The urban household use of solid fuels for cooking in ten countries in the region ranges from 91 to 18%.

There are wide variations in the extent of deficits in urban services among the Asian countries examined in this study. In China, improving public services has been the focus of governments in during transition to state-led market economy and mechanisms for service delivery have undergone significant changes. The Municipality of Chengdu, for example, invested heavily in improving such services as public safety, transportation, food and drug safety, ecology and the environment, health care, compulsory education, recreational activities, legal services, and agricultural security. This is reflected in satisfaction of urban dwellers as shown in surveys undertaken in Chengdu as well as selected low-income settlements (Qin and Yang 2020). Yet, low-income floating migrants lacking local urban household registration (urban hukou) do not have adequate access to services as a result of their household registration (*hukou*) status. Vietnam too has achieved high levels of access to urban services including almost 100% to electricity, 67% for clean water and sanitation, and 90% in primary education. Household survey results show high level of satisfaction with the government programs (Chap. 11). As in the case of China, Vietnam's residence-based social policy (*Ho khau*) has also created many barriers for migrants and other marginalized groups to access urban services including water and sanitation, health, education, and other socio-political rights enjoyed by permanent residents.

In India, 70.6% of the urban households are covered with tap water; 19% either have no toilet within their premises or defecate in the open; and 13% households have no bathing facilities within the home. These conditions are worse in the slums where 17.4% of the urban population lives. Household survey of selected slums in Delhi and Bangalore showed very low level of satisfaction with the overall performance of different levels of government, including quality of services, transparency, and gender sensitivity (Kundu 2020).

In Pakistan, the household survey in selected slums and squatter settlements in Lahore and Peshawar showed high levels of deficits in access to services as well as low level of the satisfaction of the residents about over-all performance of government related to different aspects of service delivery (Javed and Farhan 2020). The survey shows that only 52% of respondents have access to safe drinking water, 42% to community level sanitation and waste disposal, 58% to primary health care facilities, 76% to education, 47% to food, 65% to electricity and 42% in terms of restrooms. In terms of overall assessment of respondents about institutional features of local governance, 57% of the respondents think that politicization of service delivery is high; 41% respondents said that corruption is high; nearly half thought that work efficiency of local government departments and equity in resource distribution is low; Other aspects of overall performance that were rated low by respondents were transparency of activities (81%), responsiveness to special needs of marginalized groups (83%), quality of service (40%), and resource mobilization capacity (43%).

The case studies in Asia show that access to urban services can be improved in both multi-party democratic systems as well as one party systems as long as, among others, citizens are engaged at the local level, decentralization policies and programs are clearly designed and implemented, public officials at the local level are accountable and sufficient investments are made to improve living conditions of residents of low-income urban settlements. Local political context, however, determines mechanisms and processes of citizen engagement, local accountability and fiscal and political decentralization.

## 2 Factors Influencing Access to Urban Services

Chapters 2–12 in this book show that there are a number of factors that have influenced successes and failures in effectively coping with access to urban services for residents of slums and squatter settlements: local government resources and capacity; agency overlaps and coordination; local participatory mechanisms; accountability and transparency; access of migrants, women and minorities; replicating innovations and good practices; information and communication technologies; and peri-urbanization. The chapters have attempted to provide a road-map to influence urban policies and programs to ensure equitable access to urban services for inclusive development.

### *2.1 Local Government Resources and Capacity*

Distributing resources equitably to urban local governments and strengthening their planning and management capacity are essential for ensuring access to urban services.

Cities in Asia are financially dependent on higher tiers of government that control the bulk of tax revenues and are often reluctant to share with urban authorities—despite the strained budgets and unmanageable service loads that come with increasing urban density. In a contemporary governance context, the need for problem solving and interaction across actors, agencies, levels of government, and sectors means there must be mechanisms for resources to flow to the urban local governments that are best situated to identify and respond to deficits in services. Equally important is the need for more investments to strengthen planning and management capacities of local governments.

In Asia, four different approaches to urban decentralization have been adopted to expand powers and capacity of local governments.

In India, for example, political powers have been decentralized to local governments through constitutional amendments. In practice, however, the ability of urban local governments to raise resources continues to be weak. This requires capacity development programs to make local governments as catalyst for urban development in cities and towns (Kundu 2020).

The "big bang" decentralization in Indonesia has transformed the central-local relations. The policy is aimed at improving public services, increasing community participation and ensuring the accountability of local governments. It has led to greater weight to devolution than deconcentration, shift from vertical to horizontal responsibility and the provision of increased allocation of funds from central to local governments (Salim and Drenth 2020).

Through the 1980s and the early 1990s, China implemented a series of reform to decentralize its fiscal system so as to provide more incentives for local governments to promote economic growth. As Qin and Yang (2020) point out in Chap. 9, it led to local governments accounting for 51.4% of the national expenditure in 2006.

China's remarkable economic performance since 1978 has been attributed in part to the country's fiscal system.

In Pakistan, the 18th Amendment has decentralized numerous important Ministries previously held by the Federal Government. The Amendment provides a legal framework for the structural reshaping of the state into a decentralized federation, with the Provinces, the second tier of the federation, taking over legislative and policy making power in key areas such as health, education, and social welfare. However, local governments, the third tier of the Federation, still suffers from inadequate resources and capacities due to reluctance of Provinces to decentralize powers to local governments (Javed and Farhan 2020).

Urban local governments in African countries also face obstacles related to their powers and resources, their capacity to provide urban services, and effectively implementing decentralization policies and programs to bring government closer to citizens. Africa's sub-national government revenues, both as percentage of total public revenues and of GDP, are the second lowest after the Middle East & West Asia region (Rocca and Fernandez 2020). However, as Rocca and Fernandez (2020) point out in Chap. 12, there are some good examples such as the Inter-Governmental Relations Framework Act (Act 13 of 2005) in South Africa which provides for collaboration among different levels of government through specific mechanisms to settle intergovernmental disputes; capacity development programs in Liberia and Ghana; and ensuring performance in public service through the use of performance contracts between the central government and local governments in Rwanda. In 2017, Tanzania devolved 21.8% of public revenues to its subnational governments, followed closely by Uganda and Mali (18.2% and 14.0%, respectively). The African experience suggests that disaggregated data (e.g. by area, gender, age, income, employment status) at the city level are essential to design, monitor and implement pro-poor policies and programs that can reach the residents of low-income urban settlements. For local governments are to succeed in "leaving no one behind", policy makers and planners avoid the "tyranny of averages" (MIF 2018).

At macro-level, resources and capacities of urban local governments have been influenced by decentralization policies adopted by government around the world. With the evolution in thinking about development and governance and the rapid pace of democratization, the concepts and practices of decentralization too have changed over the past few decades (Cheema and Rondinelli 2007).

In developing countries, debates over the structure, roles, and functions of government focused on the effectiveness of central power and authority in promoting economic and social progress. Decentralization efforts, thus, focused on deconcentration of government functions from central to local levels and delegation of some of the functions to semi-autonomous development authorities and enterprises. The second wave of decentralization beginning in the mid-1980s broadened the concept to include political power sharing, democratization, and market liberalization, expanding the scope for private sector decision-making. During the 1990s, decentralization was seen as a way of opening governance to wider public participation through the organizations of civil society. With democratization of the political systems, governments were pressured to decentralize by political, ethnic, religious, and cultural groups seeking greater autonomy in decision-making and stronger control over national resources. In Africa, for example, the tribal minorities and economically peripheral ethnic groups sought decentralization of decision-making (Mawhood 1993). Pressures for decentralization increased partly due to the inability of central government bureaucracies to effectively deliver services (Smoke 1994).

To position cities better to provide services efficiently and equitably, reform agendas should prioritize the devolution of financial resources and authority to cities, investments in urban social economies and local enterprises, and implementation of participatory budgeting processes. Reform should also focus on securing tenure for slums and squatter settlements, and working with other progressive cities and non-governmental organizations to scale up service delivery and access programs.

## *2.2 Agency Overlaps and Coordination*

To ensure access to urban services, institutional arrangements in cities should be restructured to promote collaborative governance and inter-agency coordination.

A large number of entities and agencies are responsible for providing urban services in Asian cities. These include urban local government, offices of national ministries and departments, offices of state (sub-national) governments in federal systems, semi-autonomous government organizations providing infrastructure, civil society organizations, and the private sector. The management of urban services in Asia often suffers from lack of coordination, as sectoral departments (e.g. health) of central government based in cities compete with urban local governments. This is one of the factors for "bureaucratic dysfunction" that impedes access of citizens to urban services (De Jong and Monge 2020). As Kundu (2020) argues in Chap. 8, one of the factors negatively affecting access to services is a lack of coordination among multiple agencies from the national, state and local levels to perform their tasks, leading to roads and pavements dug up and remade several times and delayed agency responses to service delivery in slums.

There is a critical need for institutional alignment, particularly with regard to land-use allocation and regulation and developing a risk-reduction orientation in planning around disaster management and climate change adaptation. Challenges to coordination at the policy level include the absence of legal, regulatory, and institutional systems; fragmented mandates; and haphazard and sprawling urbanization complicated by ambiguous urban boundaries. Promoting better coordination is the major task to ensure marginalized groups' access to services. Furthermore, urban planning and organizational coordination can help integrate a broad array of interests within and beyond the city scale for policymaking and implementation.

Multi-stakeholder partnerships are also essential to promote a coherent approach to provide urban services and infrastructure. As Dahiya and Gentry (2020) argue in Chap. 4, government, business and civil society have their respective overlaps, weaknesses and assets in terms of service delivery and access. Therefore, Public-Private Partnerships (PPPs) with active engagement of civil society can promote collaboration and coherence in providing urban services and infrastructure. Important features of successful PPPs are: individual champions who can serve as drivers to bring different individuals from various organizations to forge partnerships to achieve shared objectives; creating "partnership space" in different contexts based on shared rewards, and shared investments; and the establishment of different structures and processes ranging from fully public to fully private with varying degrees of civil society engagement in accordance with the tasks to be undertaken. Another mechanism is for government to work with private sector service providers in a manner that allows two or more private firms to compete against each other in different parts of a city, thereby providing incentives to perform better than the competition.

To enable innovative institutional arrangements and reorientation of policy and practice necessary to promote access to city services, countries need to formulate coherent national urbanization frameworks; streamline institutional roles, responsibilities, and coordination both horizontally and vertically; and strengthen collaborative governance in urban local governments with the engagement of civil society. Equally important is the need to pursue public-private partnerships for providing services; organize local communities; and establish flexible models for post-disaster resilience.

## *2.3 Local Participatory Mechanisms*

Local participatory mechanisms, including elected local governments and engagement of civil society, are essential to get local stakeholders fully engaged in service delivery and access.

As cities grow in population and wealth, the burden on service delivery increases and ensuring adequate access becomes increasingly important. Local governments and municipal service providers can be ill equipped to work with residents and civil society organizations to meet this growing demand in an inclusive fashion. In low-income and most middle-income nations, reducing most of the deprivations associated with urban poverty depends on changes in approach by city and municipal governments. Most of the measures needed to address such deprivations in urban areas fall within the responsibilities of local governments, even if these governments so often lack the capacities to meet them. One key aspect of the needed change in approach is shift in local government relations with their citizens living in informal settlements and working in the informal economy.

If we review the examples of where local governments have changed approaches and become more successful at reducing poverty and providing better service delivery and access, four paths can be identified—although in many cities, there is evidence of more than one path.

The first is through democratization and decentralization within national government, so urban governments get more power and resources and structures that are more accountable and transparent—for instance as mayors and city councils are elected. This is most evident in political and financial devolution in Indonesia and

India's 73rd and 74th amendment that specified roles to be played by community based organizations and women in local organizations. These changes were certainly driven by citizen pressures and demands and urban poor organizations and movements had considerable importance in this.

The second path is from changes in local governments (and governance) driven by the organizations formed by urban poor groups. These include specific local examples such as a group of waste pickers and recyclers negotiating a contract with the local government so they become part of the formal waste management system, a savings group formed by homeless women who negotiate a plot of land on which they design and build homes, and partnerships formed between the police and resident committees in informal settlements to provide policing there.

The third path is the government-led provision of basic urban services within a highly centralized political system led by one party. In Vietnam, for example, public services are provided by "public service companies" and "state non-business organizations", which are established under state agencies such as the government Ministries, Departments, and People's Committees. Similar pattern is followed in China.

The fourth path is proactive roles of national and urban local governments in engaging local communities in the process of planning, monitoring and implementing service delivery and related local urban development programs and projects. In Indonesia, for example, "Musrenbang" is a tool for participatory development. It refers to the process of community discussion about local development needs. This bottom-up process is participatory in nature as it attempts to give communities a voice and a chance to influence the development planning that will be implemented. The process was introduced to replace Indonesia's former centralized and top-down government system. The Regional Development and Empowerment Program in Bandung is aimed to increase the community participation in the development process. It is implemented by the entire local government organization and through community institutions.

To sustain partnerships with local governments, urban poor's organizations need to be established and policymakers should be prepared to deal with what are often slow processes that do not produce perfect outcomes. Even when senior government officials are supportive of partnerships, promised support for initiatives can take a long time to come—or face unexpected blockages due to vested interests at the local level.

## 2.4 Accountability and Transparency Mechanisms

Local accountability and transparency mechanisms are needed to promote effective service delivery and access.

Effective accountability in local governance is the single most important vehicles for establishing a country's economic and social priorities within the scarce resources to ensure that benefits of government's initiative and local development programs reach urban residents, especially of slums and squatter settlements. There is no simple formula for the proper sequencing of accountability enhancement activities. Sequencing should be developed in response to the particular constraints identified. Sequencing of reforms should be designed to enhance the credibility of the leadership, to ensure early tangible results and to strengthen the constituency for accountability reform.

Local government accountability should not be viewed in isolation, but as part of the broader issue of governance and public management. The international community's recognition in the late 1990s of the corrosive effect of inadequate accountability at the all government levels is a logical extension of the link between governance and development created earlier in the decade. While progress has been made in a number of areas toward local government accountability with the establishment of public sector budget frameworks, much remains to be done to enhance and sustain these reforms.

What can be done to better achieve these outcomes? One mode of reform is targeting more equitable distribution of resources in cities through developing collaborative approaches between citizens and municipal governments. Such collaborations should be premised on building new capacity and political will to reform outdated practices. The poor and marginalized citizens should be directly engaged in planning processes to better understand their needs and the most appropriate delivery mechanisms for providing essential services.

Many aspects of poverty reflect exclusion from government processes, whether within a democratic or authoritarian state. In some cases, such exclusion is related to the fact that the residents of informal urban settlements may not be entitled to be on the electoral roll, as they might lack a legal address or the required documentation. But the core problems are more substantive. As an outcome of the deficiencies in provision for essential urban investments and services, clientelist relations between politicians and local communities are commonplace. Such relations may deliver some public investments or services that partly address needs—for instance communal water taps may be installed, and concrete pathways provided, but such investments do little to address the scale of need. The outcomes of these relations do not provide long-term comprehensive investment because clientelism is based on managing resource scarcity in the interests of the political elite and, in some cases, government officials.

One of the prominent approaches to combine citizen involvement and state accountability in delivering public services is participatory budgeting (PB) which started in the city of Porto Alegre, Brazil and spread throughout the world. PB can facilitate citizens' access to information and voice their needs and demands, ensure that citizen needs and public services can better match each other, and enhance well-being of people. As Blair (2020) in Chap. 2 argues, participatory budgeting can serve as an engine for accountability. He uses as a lens the World Bank's principal-agent model of state accountability for public service delivery. He discusses relationship between participatory budgeting and social accountability, and its implementation in Kerala and some of the other Indian cities. He argues that PB has been successful in Kerala but its implementation has not been sustainable in other Indian cities and that essential ingredients of success of are strong state support, CSO engagement, competitive politics and educated populace.

Other Asian countries have also promoted transparency in local governance. In Indonesia, for example, the introduction of elected local governments and mayors after decentralization of government and access to information legislation provided an institutional framework for transparency. Specifically, several mechanisms and tools have been used to ensure that local leaders and public officials are transparent in providing services. They include website of government agencies such as Provincial Development Planning Agency, presence of government agencies on social media, Bandung Planning Gallery and Information Management and Documentation Office.

There are a number of instruments of accountability and transparency that can facilitate access to urban services: local leadership commitment to accountability and transparency, effective anti-corruption bodies, and transparent and accountable systems of public procurement. Other instruments are participatory budgeting and auditing, engagement of civil society in local decision-making, right-to-information legislation, and the promotion of ethics and integrity among local public officials at all levels across public agencies.

## *2.5 Access of Migrants, Women and Minorities*

One of the core issues in access to services is addressing challenges faced by marginalized groups including migrants, women, and minorities.

Urbanization in Asia has led to an unprecedented diversity and social change including different forms of social exclusion in cities and towns of the region. For every international migrant, there are many domestic migrants (UNDP 2016). Yet, greater humanitarian assistance tends to be offered to those crossing international borders. Even if one is solely concerned with international migration, it should be remembered that a large proportion of international migrants begin as domestic ones, as crossing borders is sometimes a consequence of a failure to integrate into new domestic host communities. Urbanization represents the most widespread form of voluntary internal migration. Most commonly, rural youths are drawn to the excitement and perceived opportunities of urban life, relocating to cities and sometimes sending remittances to rural families.

Migrant workers constituting the majority of China's floating migrant population, for example, are low-income residents. Because they do not enjoy access to urban minimum living allowances as a result of their unique legal status, their situation is more dire than is the case for a city's local low-income residents. Migrant workers in China totaled 280 million in 2016, of which 110 million were specifically migrant workers from within a given province, while 170 million of them were migrant workers from outside a given province. These urban poor are at the bottom of the society, left behind in the accelerating achievements since the Chinese opening reforms of the late 1970s (Qin and Yang 2020). They are to this day facing the risk of being further marginalization due to the institutional and structural mechanisms that leave them at a disadvantage.

The household registration system in Vietnam, which was used as a tool for social control in the pre-1986 period, proves to be one of the major barriers that discriminates migrants from non-migrant population (Thanh et al. 2020). The general social exclusion and isolation of migrants from rural areas is evident in several ways: difficulties in finding employment, low and unstable income, poor living arrangements, home sickness, poor healthcare and labor exploitation.

The volume of migration is significantly higher in Delhi and Karnataka than in many other cities in India because of availability of better employment opportunities, educational institutes, health and other facilities (Kundu 2020). In case of tenure status, the percentage share of house ownership among the non-migrant households is expectedly higher (62% for urban India in 2002) than migrants (14.6% for urban India in 2012). Similarly, the percentage share of the non-migrant households who owned toilet facility was higher in comparison to migrant households in 2002. In 2012, 11.2 and 3.9% of migrant households in Karnataka and Delhi had no drainage facility at their living place.

Gender equality is focus of the Sustainable Development Goal #5 in the 2030 Agenda for Development. As Björkdahl and Somun-Krupalija (2020) argue in Chap. 5, the need is to translate global ideas into local practices. However, this continue be a challenge in different local and national contexts. Based on their review of the global framework to implement SDG5, and case study of Bosnia and Herzegovina, they analyse institutional mechanisms and tools to change perceptions and behaviour, mainstream gender equality processes, and cooperation for gender equality. They argue that obstacles to the implementation of SDG #5 include inadequate political will, lack of adequate funding to implement relevant activities, inadequate awareness of SDGs, patriarchal structures and instruments, and ineffective strategies for gender mainstreaming.

Asian countries have adopted various policy instruments to promote engagement of women in political and economic activities—including electoral quotas for women in Pakistan, gender mainstreaming through administrative and legislative reforms in Cambodia, and mobilization of political support to cope with gender discrimination in Indonesia. Family Hope Program in Indonesia, for example, is a social protection program that involves conditional cash transfers to very poor households that have a pregnant or breastfeeding mother, and/or children aged 0–18. The recipients of the program receive help on the condition that the children go to school, go to a clinic when necessary and the recipients ensure that the children and pregnant mother have adequate nutrition and a healthy lifestyle. The goal of the program is to break the cycle of poverty that many poor families experience because they cannot afford health services and education for their children.

After the devolution of women's development to the provinces under the 18th Amendment to the Constitution of Pakistan, the Punjab Commission for the Status of Women was conceived as an oversight body to ensure policies and program of the government to promote gender equality in Punjab. The Punjab government has launched various programs for the systematic inclusion of women in all tiers of governance and economic life. The Commission's objectives are stated to be the elimination of discrimination against women in all forms and the empowerment of women.

Yet, access to urban services continues to be an issue of serious concern in low-income urban settlements. Widowed, separated and unmarried single household women in urban India, for example, are economically poorer and live in precarious conditions. Level of asset ownership among women is either absent or negligible. Most women in these categories who participate in economic activities are reported to draw income from informal sector work, characterized by job insecurity, low and irregular wages and poor working conditions. Women tend to engage as casual wage labors with extremely low payments or are self-employed in petty business.

The way forward must include strengthening mechanisms to enable the genuine participation of all segments of the population including migrants, women and minorities in the co-production of public services and urban planning, the evaluation of public policies and decision-making, and in ensuring the accountability of governments at all levels. Desirable results in relation to equity include development and integration of methods for citizen dialogue, handling complaints and securing the participation of women and vulnerable groups. Community mapping and participatory budgeting result in more informed and appropriate budget allocations and have been a vital tool in addressing critical needs of all groups in cities.

## 2.6 *Replicating Innovations and Good Practices*

Cities have been laboratories of experimentation, innovations and good practices to improve service delivery and access. Recent surveys have highlighted a number of innovations and good practices in terms of their content, rationale and impact on urban residents.

In Chap. 2, De Jong and Monge (2020) discuss innovations that have dealt with "bureaucratic disfunction" and improved access to urban benefits in cities. To empower people to access elementary education in Mumbai, for example, the *Balsakhi* program for remedial elementary education worked to deal with the entrenched inequalities in the Indian education system. An NGO, Pratham, in collaboration with municipal governments, hired women from local urban communities to teach basic skills to children in upper elementary school who still needed help with their reading and math. Due to this, the costs were kept low and community was engaged. The *Balsakhi* method was easily replicated and spread to 20 cities in India, leading to improvement in children's learning.

In Quezon City, the Philippines, in 2016 the city hall passed a law establishing a fine of up to 200 US dollars and jail sentences of up to a year for sexual harassment in public spaces. To generate support in the process of program design and implementation, the initiative was launched in partnership with UN Women and women's groups, community groups, and training of frontline workers such as policemen.

Also, a media and awareness campaigns were launched about the impact of fear on women's mobility.

To bring services closer to citizens, the Bahia State Government in Brazil established Citizen Assistance Service (SAC) Centers, which were designed as "one stop shops" in partnership with federal, state, and municipal agencies as well as private companies. These centers were located in shopping malls, public transportation centers and low-income neighborhoods to make it easier for citizens to access services. The Centers improved relations between state authorities and citizens and led to a new level of professionalism in access to public services. This innovation was adopted in 22 out of 26 states in Brazil as well as in Portugal and Colombia.

Another approach to examine innovations and their replication is to focus on different aspects of sustainable and inclusive development: (1) leadership, new competencies and changing mindsets of public servants at all levels; (2) institutional and organizational arrangements; (3) partnership building and people's engagement; (4) knowledge sharing and management; and (5) digital transformation (Alberti and Senese 2020).

For example, in Colombia "Cambia Tu Mente…Construye Paz (change your mind… build peace)" promotes dialogue among members of rival gangs to change the mindset of young people in neighborhoods of Manizales to address armed conflict. The initiative provides support from public and private entities such as employment and university placements. In Singapore, the Urban Redevelopment Authority (URA) emphasizes consultations with stakeholders in public and private sectors, and in-house capacity building programs to create a culture of data driven and evidence-based decision-making. With new skills and techniques of data analytics, the URA staff have been able to collaborate with more agencies, enabling them to strengthen the culture of digital planning and partnerships. The provincial government of Chungcheongnam-do Province, Korea, created an online fiscal information system on its website to strengthen the disclosure of its revenues, budget, expenditure and settlement information to the public. Fifteen local governments in the province signed an agreement to disclose their current revenues and expenditures including information on all contract methods, and parties. This initiative promoted transparency and accountability in urban development. Melbourne, Australia is promoting innovations across the city by engaging the community residents, workers, businesses, students and visitors to design, develop and test the best ways to live, work and play in Melbourne. In the Netherlands, the City of Amsterdam has invested in public-private partnerships, which has allowed the city to be transformed into an open source urban lab, where new solutions aimed at improving the quality of life for citizens and tourists are developed.

A number of innovations in service delivery and access have emerged in Pakistan over the past few years. These include: Lahore Waste Management Company which signed an agreement with the Lahore City District Government to plan, implement and manage different public-private partnership programs to provide solid waste management services to Lahore City; Government of Punjab has developed an ICT based system of "Citizens Feedback Monitoring Program CFMP" which works on a simple mechanism of reaching proactively to all the users of public services; and

Orangi Pilot Project in Karachi which is community designed and managed urban sanitation project in one of the largest low-income settlements in the world.

Replicating any of these good practices and innovations in service delivery and access entails major shifts—from small pilot projects to widespread implementation or from one aspect of the governance process to the systemic level. This poses many challenges, including the opposition of various groups with vested interests in the status quo, lack of political support at national and subnational levels, and local power structures that often impede the implementation of equity-oriented service delivery initiatives. There are, however, several ways to promote the replication of innovations. The first is to ensure that the content, process, and results of the innovation are regularly documented and disseminated among stakeholders—especially the decision-makers at local and national levels. Other approaches include training and capacity development programs to educate stakeholders about the content and process of an innovation, identification of constraints and opportunities to promote replication, and mobilizing the support of champions of an innovation to build consensus about the need for replication at systemic level.

## 2.7 *Information and Communication Technology*

Information and Communications Technology (ICT) can help provide effective solutions to challenges of service delivery and access.

Information and communication technology (ICT) is used to enhance quality, performance and interactivity of urban services, to reduce costs and resource consumption and to improve contact between citizens and government (Goldsmith and Crawford 2014). Smart city applications are developed to manage urban flows and allow for real-time responses. For example, the Integrity Pact of Seoul Metropolitan Government was designed to ensure transparency in procurement. The Public Record of Operations and Finance (PROOF), for example, was launched in Bangalore, India in 2002 to monitor the financial performance of the City Corporation in Bangalore. The Bandung Command Center was launched by Bandung City Government in Indonesia, to monitor the conditions of the city including traffic congestion and street vendors and reviewing bureaucratic performance in making decisions. The Center is connected with the city surveillance (CCTV) cameras installed in 80 strategic locations. In 2015, Municipality of Surakarta, Indonesia, launched a public complaint service with a website-based electronic system called the *Unit Layanan Aduan Surakarta* (ULAS) to improve government performance through a web-based complaint system, increasing the effectiveness of services to the community, which enabled the community to monitor the response of the government unit to a complaint (Salim and Drent in Chap. 7).

A number of "smart city" initiatives have emerged in both the developed and the developing countries. The use of ICT and smart city methods and approaches have gone through three phases: "smart cities 1.0" which is technology driven such as Songdo in South Korea and the on-line service delivery system in Singapore; smart cities 2.0 which is led by city leadership but is enabled by technology solutions such as Rio's initiative to create a sensor network to reduce the role of landslides in hillside favelas (slums), the Integrity System in Seoul to promote on-line procurement and combat corruption and the Dengue Activity Monitoring System in Lahore, Pakistan to combat the deadly infectious disease; and smart cities 3.0 in which citizens are partners and are fully engaged such as in Medellin to promote growth with equity and social inclusion.

Smart city initiatives are taking place at two levels. The first level is specific project or service such Beijing's Monitoring Devices and Equipment for City's Drainage System with Central Control Panel for analysis and decisions; and Shanghai's Smart Education Data Center. The second level aims to transform the city to use ICT solutions to cope with all urban challenges including the environment, service delivery and access, public safety, sustainable livelihoods and transportation system. Beijing is leading the efforts to apply technology in operationalizing the smart city principles in such areas as transportation, electronic medical records, and the environment and security systems.

In China, the concept of Smart Cities and a Smarter Planet has been developed since around 2008 when underlying technologies including RFID sensors, wireless connectivity, electronic payments, and cloud-based software services enabled new approaches to collaborative solutions for urban challenges based on extensive data collection. Urban infrastructure projects, including significant Smart City elements in their construction, have been implemented since 2010. In January 2013, the Ministry of Housing and Urban-Rural Development (MOHURD) formally announced the first list of national pilot Smart Cities. By April 2015, there were over 285 pilot Smart Cities in China, as well as 41 special pilot projects.

The Government of Gujrat in India launched the School Mapping—Innovative use of GIS Technology in Access to Education in order to ensure universal enrolment and retention in schools. The State also introduced the Migration Card initiative to track students who were migrating along with their parents for seasonal employment. The main objective was to avoid dropout and ensure the continued education of children during the period of migration. Participatory budgeting was initiated in Pune in 2006 for citizens to directly make suggestions to the urban local bodies for projects, developmental work or any other civic services enhancement. Public Grievance and Redressal System was launched with the aim to strengthen service delivery mechanisms in urban local bodies in Karnataka through enhanced community participation in governance. Through this mechanism, citizens can register their grievances and as well as track the progress of redressal (over the internet or through a phone call) using a complaint number generated by the "Helpline" at the time of registration of grievance. The Provincial Government of Punjab in Pakistan has developed an ICT based system of "Citizens Feedback Monitoring Program CFMP" which works on a simple mechanism of reaching proactively to all the users of public services.

Initiatives like these give cities tools to cope with urban challenges including environmental management, service delivery and access, public safety, and ensuring sustainable livelihoods and safe and efficient transportation.

## 2.8 Peri-urbanization

Peri-urbanization is a burgeoning issue in access to services for the urban poor. It can be defined as the process of transition and change from rural to urban. It is often attributed to urban population growth, price of rural vs urban land, mixed land uses, availability of labor, and to various forms of public policy interventions related to economic and employment structures, dispersal of manufacturing and spatial development. There are many definitions of peri-urbanization, reflecting the degree of emphasis on one or more of the above attributes.

Peri-urbanization, and planning practices to respond to this phenomenon, can be examined from three broad perspectives, as Johan Waltjer argues—spatial, functional and drivers of change. The spatial dimension emphasizes the transition or interface between cities and agricultural areas. The functional dimension focuses on uses and activities of space that trigger economic and social change. The third perspective emphasizes such drivers of change as investments and land use (Woltjer 2014).

Peri-urban areas occupy large portions of the national landscape in Asia and Africa and are home to hundreds of millions of people. Rural to urban migration without the residence registration system in China and Vietnam and proliferating peri-urban settlements in India and Indonesia have led to rapidly increasing population in these areas. One of the key challenges for urban planners and development practitioners is to effectively manage this physical, economic and social transformation to ensure inclusive and sustainable urban development. African experience, for example, shows that urban planning that does not reflect existing realities within a city results in social disparities and marginalizes the urban poor from spaces with roads, regular power supply and adequate infrastructure and amenities (Rocca and Fernandez 2020).

The growth of population in peri-urban areas can be attributed to low cost of agricultural land for shelter, interests and priorities of industry as sources of materials for urban life, and local governments to use these areas for infrastructural and industrial development programs. In Southeast Asia, for example, the expansion of peri-urban areas is taking place mostly beyond the core of the cities. The main determinants of peri-urbanization in the region are auto-centered transport systems that have transformed urban mobility, and hinterland acting as a resource frontier providing such inputs as water, food, building materials, labor, and land. In China, one of the reasons for the expansion of peri-urbanization and urban sprawl has been the tendency of local governments to convert agricultural land to urban uses to earn revenue from land sales and leases which has also benefitted developers and villagers.

Peri-urban areas are often characterized by marginalization, inequality, and exclusion. Specifically, they face enormous deficits in access to urban services and are often dumping grounds for various kinds of urban waste from city centers, leading to health risks. Depending upon the country context, a number of factors constrain their access to urban services. For example, their jurisdictions are sometimes undefined, resulting in institutional fragmentation and lack of coordination among sectoral agencies often organized for rural or urban functions. Furthermore, lack of comprehensive urbanization including fringe zones policy, low capacity of local government and local offices of government agencies to cope with social service delivery, and weak mechanisms for citizen engagement severely constrain access of residents of peri-urban areas to services.

To bring about change in peri-urban areas that leads to inclusive development, urban planners and development practitioners need to focus on establishing stakeholder partnerships with the private sector and other organization and formulate development strategies from a holistic perspective. To ensure access to services, the need is to change administrative boundaries and jurisdiction to formalize peri-urban areas as units of government and administration; strengthen the financial, administrative, and technical capacity of local governments in peri-urban areas; identify mechanisms for inter-regional coordination and inter-sectoral integration; and promote the process of citizen engagement in local-level planning and management. Another way forward is to support positive economic, social, and environmental links between urban, peri-urban, and rural areas by strengthening national and regional development planning.

## 3 Conclusion

Developing countries have many challenges and opportunities in establishing urban local governance systems that are participatory, accountable and transparent but are also effective in improving access to services for all segments of the urban society. Despite alarming deficits in urban services in Asia and Africa, cities provide opportunities to promote positive economic and political change through variety of ways they have been built and managed within the national context. This chapter has identified eight sets of factors and institutional reforms at national and local levels to fully utilize the potential role of cities in promoting political and social inclusion and access of residents of low income urban settlements to urban services: local government resources and capacity; inter-agency coordination; participatory mechanisms; accountability and transparency; engagement of women, minorities and migrants; innovations; information and communication technologies, and peri urbanization. At the global level, Sustainable Development Goal 11 and the New Urban Agenda have provided a framework and a road map for actions by national and local governments to promote inclusive and sustainable cities. The future of cities is in urban local governance reforms and experimentation with what works, realizing no one approach or blueprint will work everywhere, and that ongoing social learning will always be necessary as urban centers evolve.

## References

Alberti A, Senese M (2020) Developing capacities for inclusive and innovative urban governance. In: Cheema S (ed) Governance for urban services: access, participation, accountability and transparency. Springer, Singapore

Björkdahl A, Somun-Krupalija L (2020) Gender equality for sustainable urban development: translating global ideas into local practices. In: Cheema S (ed) Governance for urban services: access, participation, accountability and transparency. Springer, Singapore

Blair H (2020) Accountability through participatory budgeting in India: only in Kerala? In: Cheema S (ed) Governance for urban services: access, participation, accountability and transparency. Springer, Singapore

Cheema GS, Rondinelli DA (2007) Decentralizing governance: emerging concepts and practice. Brookings Institution Press, Washington, DC

Dahiya B, Gentry B (2020) Public-private partnerships to improve urban environmental services. In: Cheema S (ed) Governance for urban services: access, participation, accountability and transparency. Springer, Singapore

De Jong J, Monge F (2020) The state of access in cities: theory and practice in: Cheema S (ed) Governance for urban services: access, participation, accountability and transparency. Springer, Singapore

Goldsmith S, Crawford S (2014) The responsive city: engaging communities through data-smart governance. Wiley, New York

Javed N, Farhan K (2020) Access to urban services for political and social inclusion: the case of Pakistan. In: Cheema S (ed) Governance for urban services: access, participation, accountability and transparency. Springer, Singapore

Kundu D (2020) Political and social inclusion and local democracy in India: a tale of two cities. In: Cheema S (ed) Governance for urban services: access, participation, accountability and transparency. Springer, Singapore

Mo Ibrahim Foundation (2018) Ibrahim forum report: public service in Africa. MIF, London. http://s.mo.ibrahim.foundation/u/2018/06/21170815/2018-Forum-Report.pdf?_ga=2.176124386.1712197560.1551885727-2031218662.1525792781

Mawhood IP (ed) (1993) Local government in the third world: experience with decentralization in tropical Africa. Africa Institute of South Africa, Johannesburg

Qin B, Yang J (2020) Access of low-income residents to urban services for inclusive development: the case of Chengdu, China. In: Cheema S (ed) Governance for urban services: access, participation, accountability and transparency. Springer, Singapore

Rocca C, Fernandez D (2020) Serving Africa's citizens: governance and urban service delivery. In: Cheema S (ed) Governance for urban services: access, participation, accountability and transparency. Springer, Singapore

Salim W, Drenth M (2020) Local governance and access to urban services: political and social inclusion in Indonesia. In: Cheema S (ed) Governance for urban services: access, participation, accountability and transparency. Springer, Singapore

Smoke PJ (1994) Local government finance in developing countries: the case of Kenya. Oxford University Press, Nairobi

Thanh ND, Long PV, Giang NK (2020) Governance for urban services in Vietnam. In: Cheema S (ed) Governance for urban services: access, participation, accountability and transparency. Springer, Singapore

United Nations (2016) The sustainable development goals report. United Nations, New York

United Nations Development Programme (2016) UNDP human development report. UNDP, New York

UN-HABITAT (2015) The state of Asian and Pacific cities 2015. UNESCAP, Bangkok

Woltjer J (2014) A global review on peri-urban development and planning. Jurnal Perencanaan Wilayah dan Kota 25(1)

**Shabbir Cheema** is Senior Fellow at Harvard Kennedy School's Ash Center for Democratic Governance and Innovation. Previously, he was Director of Democratic Governance Division of United Nations Development Programme (UNDP) in New York and Director of Asia-Pacific Governance and Democracy Initiative of East-West Center in Hawaii. Cheema prepared the UNDP policy papers on democratic governance, human rights, urbanization and anti-corruption and provided leadership in crafting UN-assisted governance training and advisory services programs in over 25 countries in Asia, Africa, Latin America and the Arab region. He was Program Director of the Global Forum on Reinventing Government and the Convener of the Harvard Kennedy School's Study Team of Eminent Scholars on Decentralization. He has taught at Universiti Sains Malaysia, University of Hawaii and New York University. As the UN team leader, he supported the International Conference on New and Restored Democracies, the Community of Democracies and UN HABITAT II. He has undertaken consultancy assignments for Asian Development Bank, the World Bank, U.S. Agency for International Development, Swedish International Development Agency, Dubai School of Government and United Nations. He holds a Ph.D. in political science from the University of Hawaii. He is the co-author of *The Evolution of Development Thinking: Governance, Economics, Assistance and Security* (Palgrave Macmillan 2016) and the author of *Building Democratic Institutions: Governance Reform in Developing Countries* (Kumarian Press, 2005) and *Urban Shelter and Services* (Praeger 1987). He is the contributor and co-editor of the four-volume Series on *Trends and Innovations in Governance* (United Nations University Press, 2010); *Decentralizing Governance: Emerging Concepts and Practices* (Brookings Institution Press in cooperation with Harvard University, 2007); *Reinventing Government for the Twenty First Century: State Capacity in a Globalizing Society* (Kumarian Press, 2003) and *Decentralization and Development* (Sage Publications, 1984). Cheema has been a member of the advisory committees of the Swedish International Center for Local Democracy, UNHABITAT III, and the Pacific Basin Research Center and editorial boards of Urbanization and Environment and Third World Planning Review. A featured speaker at global and regional forums, he served as an advisor to the Dubai School of Government, Pakistan Institute for Economic Development, the Malaysian Academy for Leadership in Higher Education, and the UN Governance Center in Seoul, Korea.